T0140361

Studies in Computational Intelligence

Volume 495

Series Editor

J. Kacprzyk, Warsaw, Poland

For further volumes:
http://www.springer.com/series/7092

Fatos Xhafa · Nik Bessis

Editors

Inter-cooperative Collective Intelligence: Techniques and Applications

 Springer

Editors
Fatos Xhafa
Departament de Llenguatges I Sistemes
    Informàtics
Universitat Politècnica de Catalunya
Barcelona
Spain

Nik Bessis
School of Computer Science and
    Mathematics
University of Derby
Derby
UK

ISSN 1860-949X   ISSN 1860-9503 (electronic)
ISBN 978-3-642-42945-3   ISBN 978-3-642-35016-0 (eBook)
DOI 10.1007/978-3-642-35016-0
Springer Heidelberg New York Dordrecht London

Printed on acid-free paper

Springer is part of Springer Science+Business Media (www.springer.com)

Preface

The emergence of various computational paradigms, from Sensor Networks and Internet of Things (IoT) to Social Networking, Cloud and Data Centers, is manifesting the need for their integration into larger computational systems to support end users with collective intelligence and decision-supporting systems in an inter-cooperative mode. Such integration and collective building of intelligence is raising many research and development issues and challenges due to the heterogeneous nature of computational resources (sensors, devices, servers, data centers, etc., of the interconnecting networks) of data sources as well as different computational intelligence layers on top of them. The inter-cooperative collective intelligence is emerging as a new cross-cutting feature in many multi domain applications, in which knowledge can be built in a collective manner, being it from information and actions of millions of Internet users worldwide in social networks, the intelligent networks of smart devices and intelligent things, networks of producers and consumers, or combinations of them. Among most salient features of collective intelligence is the decentralized co-creation and building of knowledge on certain activity, problem, phenomena, or complex system of interest driven by creativity, innovation, and collaborative problem solving. As an example, in the business domain, the inter-cooperative collective intelligence promotes the collaborative value creation as part of the business model.

This book investigates and provides insights on research issues in the field of collective intelligence and focuses especially on the new paradigm shift of collective intelligence built inter-cooperatively in large-scale Internet-based systems. This shift is being driven by the profound change of the Internet from the vision of a network of networks of computers to a global platform of people, computers, networks, and devices offering services and enabling interactions, and collaborations globally. Indeed, unlike the recent past when collective intelligence used to be built from single platforms, e.g., separate social networks nowadays, users can provide information from many interconnected platforms as well as from their mobile and wireless devices, which enrich the generated information with the context, and thus, enables to extract and build valuable knowledge in an inter-cooperative way. Additionally, such large amount of generated information can be nowadays accommodated in Cloud computing, inter-Clouds, and data centers, which make possible its massive, intelligent processing, and extraction of useful information and knowledge.

Main Contributions of This Book

This book covers the latest advances in the rapid growing field of inter-cooperative collective intelligence aiming at the integration and cooperation of various computational resources, networks, and intelligent processing paradigms to collectively build intelligence and advanced decision support and interfaces for end users. The book brings a comprehensive view of the state of the art in the field of integration of Sensor Networks, IoT, Social Networks, Cloud Computing, Massive and Intelligent Querying, and Processing of Data. As a result, along its 15 chapters, the book presents insights and lessons learned so far and identifies new research issues, challenges, and opportunities for further research and development agenda. Emerging areas of applications are also identified and usefulness of inter-cooperative collective intelligence in multi-domain applications is envisaged.

Specifically, the contributions of this book focus on the following research topics and development areas of inter-cooperative collective intelligence:

The Internet of Intelligent Things: IoT is rapidly expanding to different application domains due to ever-increasing number of mobile and wireless devices connected to Internet and their increasing capability not only to observe but also to actuate on the real word. Therefore, the new paradigm of Internet of Intelligent Things (IoIT) has emerged. Through the cooperative sensing and coordination and action, the IoIT enables the development of collective intelligence applications.

Cloud Computing and Data Centers: The amount of information generated through Internet-based systems, users, devices, etc., as well as their interaction and collaboration require large computing and storing capacities due to the unprecedented amount of data. On the one hand, due to the advanced services delivered through the Cloud, the interactive collaboration and information generation can be at global large-scale systems. On the other hand, the Cloud computing and data Centers make possible to store and process very large amounts of data, known as "big data" and deliver *knowledge as a service* to end users.

Integration of IoT and Cloud Computing Platforms: The shift from collective intelligence to inter-cooperative collective intelligence is being pushed by the integration of various paradigms and technologies into large-scale systems. Indeed, the systems of today are able to embrace from low level layers of smart devices and intelligent things to mobile computing and the Cloud in a transparent and seamless way.

Inter-cooperative Collective Intelligence: The global nature of Internet-based systems and their conception in terms of various heterogeneous systems, data sources, social networks, smart networks, clouds, and inter-clouds add to the collective intelligence, the dimension of inter-cooperativity by which knowledge can be extracted from and delivered at global level.

Social Networks and the Internet of Intelligent Things: The combination of social networks and intelligent things enables not only sharing and problem-solving collaboratively, but also providing more accurate solutions to complex problems.

Innovative Business Models and Services: Businesses of today are driven by "business intelligence," which among others promotes the need to create value in a decentralized, inter-cooperative, and collective way. Thus, the collective intelligence is seen as the way to obtain new innovative business models in which consumers and producers are intergrated into one business model.

Emerging Applications: The inter-cooperative collective intelligence is fast expanding to many application areas such as *World Sensing, Smart Cities, Intelligent Buildings, Ambient Intelligence, eHealth,* etc.

Targeted Audience and Last Words

The contributions in the book are from researchers from both academia and industry, who bring interesting views and experiences in the field of inter-cooperative collective intelligence. Thus, the audiences of this book—academics, instructors, and senior students as well as researchers and senior graduates from academia, networking and software engineers, data analysts, business analysts from industries and businesses—can benefit from such views as well as from the interdisciplinary nature of the contributions in the book.

We hope that the readers find this book useful in their academic and professional activities and that it can foster their collaboration and inspire them for innovative approaches in this rising field of inter-cooperative collective intelligence!

April 2013

Fatos Xhafa
Nik Bessis

Acknowledgments

The editors of this book wish to thank all the authors of the chapters for their interesting contributions and for their timely efforts to provide high quality chapter manuscripts. We are grateful to the reviewers of the chapters for their generous time and for giving useful suggestions and feedback to the authors. We would like to acknowledge the support and encouragement received from Prof. Janusz Kacprzyk, the editor in chief of the Springer series "Studies in Computational Intelligence", Dr. Thomas Ditzinger, Ms. Heather King and the whole Springer's editorial team for their support in the preparation of this book.

Finally, we wish to express our gratitude to our families for their understanding and support during this book project.

Contents

Contributors

Muath Alrammal Laboratoire d'Informatique Fondamentale d'Orléans, Universié d'Orléans, Orléans, France

João Andrade Instituto Superior Técnico/UTL, Lisbon, Portugal

Artur Arsénio YDreams Robotics, and Instituto Superior Técnico/UTL, Lisbon, Portugal

Rihards Balodis Institute of Mathematics and Computer Science, University of Latvia, Riga, Latvia

Guntis Barzdins Institute of Mathematics and Computer Science, University of Latvia, Riga, Latvia

Rabia Bilal Department of Electronic, School of Engineering and Informatics, University of Sussex, Brighton, UK

Radu-Ioan Ciobanu University Politehnica of Bucharest, Bucharest, Romania

Ciprian Dobre University Politehnica of Bucharest, Bucharest, Romania

Paul Dohmen Department of eBusiness, School of Business, Economics and Statistics, University of Vienna, Vienna, Austria

Beniamino Di Martino Department of Industrial and Information Engineering, Genoa, Italy

Rui Francisco YDreams Robotics, and Instituto Superior Técnico/UTL, Lisbon, Portugal

Marco Gaudina Circle Garage, Genoa, Italy

Antonio Gentile Dipartimento di Ingegneria Chimica Gestionale Informatica Meccanica, Università degli Studi Palermo, Palermo, Italy

Spyridon V. Gogouvitis Electrical and Computer Engineering, Athens Technical College, Elberton, Greece

Gaétan Hains Laboratoire d'Algorithmique, Complexité et Logique, Faculté des Sciences et Technologie, Université Paris-Est, Paris, France

Adrians Heidens Institute of Mathematics and Computer Science, University of Latvia, Riga, Latvia

Antonio J. Jara Department of Information and Communications Engineering, Facultad de Informática, University of Murcia, Murcia, Spain

C. Karagiannidis Special Education Department, Argonauton and Filellinon, University of Thessaly, Vó-los, Greece

C. Kouroupetroglou Research Programmes Division, Thessaloniki, Greece

Adamantios Koumpis Research Programmes Division, Thessaloniki, Greece

Natalia Kryvinska Secure Business Austria, Department of eBusiness, School of Business, Economics and Statistics, University of Vienna, Vienna, Austria

Dimosthenis Kyriazis National Technical University of Athens, Athens, Greece

Radu-Corneliu Marin University Politehnica of Bucharest, Bucharest, Romania

Pavel Merzlyakov Institute of Solid State Physics, University of Latvia, Riga, Latvia

Mauro Migliardi Centro Ingegneria Piattaforme Informatiche, University of Genoa and University of Padua, Genoa, Italy

Philip Moore Faculty of Technology, Engineering and the Environment, School of Computing, Telecommunications and Networks, Birmingham City University, Birmingham, UK

María V. Moreno-Cano Facultad de Informática, Department of Information and Communications Engineering, University of Murcia, Murcia, Spain

Bilal Muhammad Khan Department of Electrical and Power Engineering, National University of Science and Technology, Karachi, Pakistan

Fernando Nabais YDreams Robotics, Fundão, Portugal

Inara Opmane Institute of Mathematics and Computer Science, University of Latvia, Riga, Latvia

Dario Pirrone Dipartimento di Ingegneria Chimica Gestionale Informatica Meccanica, Università degli Studi Palermo, Palermo, Italy

Roberto Pirrone Dipartimento di Ingegneria Chimica Gestionale Informatica Meccanica, Università degli Studi Palermo, Palermo, Italy

Desanka Polajnar University of Northern British Columbia, Prince George, BC, Canada

Jernej Polajnar University of Northern British Columbia, Prince George, BC, Canada

Giuseppe Russo Dipartimento di Ingegneria Chimica Gestionale Informatica Meccanica, Università degli Studi Palermo, Palermo, Italy

Maria Salama Faculty of Informatics and Computer Science, Department of Computer Science, British University in Egypt, Cairo, Egypt

José Santa Facultad de Informática, Department of Information and Communications Engineering, University of Murcia, Murcia, Spain

Hugo Serra YDreams Robotics, and Instituto Superior Técnico/UTL Av. Prof. Dr. Cavaco Silva, Taguspark, Portugal

Eduardo Serrano Instituto Superior Técnico/UTL Av. Prof. Dr. Cavaco Silva, Taguspark, Portugal

Ahmed Shawish Faculty of Computer and Information Science, Department of Scientific Computing, Ain Shams University, Cairo, Egypt

Antonio F. Skarmeta Facultad de Informática, Department of Information and Communications Egineering, University of Murcia, Murcia, Spain

Christine Strauss Department of eBusiness, School of Business, Economics and Statistics, University of Vienna, Vienna, Austria

Dimitris Tektonidis Research Programmes Division, M.Kalou 6, 54629, GR, Thessaloniki, Greece

Leo Truksans Institute of Mathematics and Computer Science, University of Latvia, Riga, Latvia

Salvatore Venticinque Department of Industrial and Information Engineering, Caserta, Italy

Martin Watmough Sheffield Hallam University, Sheffield, South Yorkshire

Fatos Xhafa Department of Languages and Information Systems, Technical University of Catalonia, Barcelona, Spain

Miguel A. Zamora-Izquierdo Facultad de Informática, Department of Information and Communications Engineering, University of Murcia, Murcia, Spain

Mohammad Zubayer University of Northern British Columbia, Prince George, BC, Canada

Internet of Intelligent Things: Bringing Artificial Intelligence into Things and Communication Networks

Artur Arsénio, Hugo Serra, Rui Francisco, Fernando Nabais, João Andrade and Eduardo Serrano

Abstract This chapter introduces the Internet of Intelligent Things (IoIT), the future Internet of Things (IoT) with significant intelligence added to "things". We discuss the importance of Artificial Intelligence approaches to enable such Intelligent Communication Networks. Nowadays, sensor networks are becoming a reality, especially for remote monitoring of events in fields such as healthcare, military, forest integrity or prediction of seismic activity in volcanoes. Intelligent devices and sensors are also appearing, besides electronic home appliances and utilities, as gadgets to mobile phones or tablets. And some of these devices have capability to actuate on the world. This chapter is focused on surveing current approaches for the Internet of all these intelligent things connected and communicating. It addresses artificial intelligence techniques employed to create such intelligence, and network solutions to exploit the benefits brought by this capability.

A. Arsénio (✉) · H. Serra · R. Francisco
YDreamsRobotics, and Instituto Superior Técnico / UTL, Taguspark, Oeiras, Portugal
e-mail: Artur.arsenio@ist.utl.pt

H. Serra
e-mail: hugo.serra@ist.utl.pt

R. Francisco
e-mail: rui.francisco@ist.utl.pt

F. Nabais
YDreamsRobotics, Edifício A Moagem—Cidade do Engenho e das Artes, Largo da Estação
6230-311 Fundão, Portugal
e-mail: fernando.nabais@ydreamsrobotics.com

J. Andrade · E. Serrano
Instituto Superior Técnico / UTL, Taguspark, Oeiras, Portugal
e-mail: joao.andrade@ist.utl.pt

E. Serrano
e-mail: eduardo.serrano@ist.utl.pt

F. Xhafa and N. Bessis (eds.), *Inter-cooperative Collective Intelligence:*
Techniques and Applications, Studies in Computational Intelligence 495,
DOI: 10.1007/978-3-642-35016-0_1, © Springer-Verlag Berlin Heidelberg 2014

1

1 Introduction

Internet changed the way we communicate. And as time passed by a large number of objects were connected to the Internet, giving rise to the Internet of Things (IoT), which refers to a multitude of uniquely identifiable objects (things) that are connected through the Internet [1]. For instance, fridges might be connected to the Internet and present people with information about current fresh food stock, clothes might read relevant biomedical information and alert people if any action is needed, or umbrellas might receive information from the weather service and glow when it is about to rain so people do not forget to take them.

The concept behind IoT is *the pervasive presence around us of a variety of things or objects—such as Radio-Frequency Identification (RFID) tags, sensors, actuators, mobile phones, etc.—which through unique addressing schemes are able to interact with each other and cooperate with their neighbours* [2]. The IoT paradigm is therefore changing, and will continue to change, the lives of people worldwide, whether its effects are obvious to the user or not. Application areas are various (see Fig. 1), spanning from home automation, assisted living, e-health, smart energy management, logistics, automation, etc.

Especially due to cost and energy issues, deployed hardware, such as sensors, are usually simple, with low computational processing capabilities. However, new application requirements, such as energy savings and operation autonomy, is

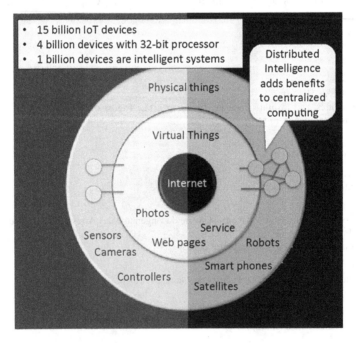

Fig. 1 Internet of things and devices [5]

pushing for the deployment of more intelligent sensors, or even mechanisms with actuators, such as forest sensors capable to move and act on the environment for energy harvesting, in order to become autonomous with respect to energy [3]. On the other hand, world population is aging, and there is increasingly more demand for intelligent robots at peoples' home to support elderly living.

Furthermore, current smartphones are equipped with a large plethora of sensors, being as well an aggregator of network traffic from personnel sensors. These devices have now significant processing capabilities, acting as a gateway to platforms offering various services such as education and healthcare. Increasingly, smartphones will play a crucial role on the Internet of things, as aggregators of information from multiple sensors (either internal to the smartphone or else available on the surrounding environment), or as computing platforms for artificial intelligence algorithms to learn from the acquired data.

We will present a survey of different platforms and applications, employing varying elements such as RFID technology, healthcare sensors and robots, discussing the main challenges brought by these systems.

1.1 Internet of Things Challenges

According to a joint study by Global System for Mobile Communications Association (GSMA) and Machine Research, there are 9 billion connected devices in the world today. By 2020, there will be 24 billion and over half of them will be non-mobile devices such as household appliances (see Table 1). The importance of the IoT paradigm [4] has already been recognized worldwide. For instance, the U.S. National Intelligence Council lists the IoT among the six technologies that may impact U.S. national power by 2025.

IoT enables a state of ubiquitous networking in which every object in our daily lives could be a potential connected thing on the internet, with its own set of sensors, actuators and, possibly, these could even have a sense of purpose and intelligence to react to their surroundings, which leads us to the IoIT. But before we dive into the IoIT, we will describe some of the challenges that the IoT still has yet to overcome [5], such as:

- Connectivity: the still on going deployment of IPv6, which has not reached most end users, and the fact that the range of available IPv4 addresses is clearly insufficient. In addition, the possibility of having billions of things uniquely addressable over the Internet, presents a challenge to the goal of ubiquitous computing, which IPv6 should solve with the advantages of allowing easier network management and improved security features.
- Energy management of sensors and actuators, although there have been large improvements in this aspect with the appearance of technologies that allow to recharge batteries without the need of an external power source—such as producing energy through bodily movements and through the usage of

Table 1 Top ten 2020 connected applications (study by GSMA and machine research)

Application	Description
Connected car	Vehicular platforms behind today's satellite navigation or stolen vehicle recovery devices
Clinical remote monitoring	Reducing healthcare costs by remote monitoring and immediate, local corrective actions, for instance
Security remote monitoring	Using wide-area connected security systems for providing higher quality and cheaper solutions
Assisted living	Enabling better quality of life and independence for old and infirm people, as well as improved health tracker systems
Pay-as-you-drive car insurance	Insurance policy that charges users according to driving behaviour
New business models for car usage	Exploiting car connectivity for providing value-added services such as vehicle tracking and remote access
Smart meters	Number of smart meters deployed globally is expected to increase by an order of magnitude. Benefits include remote meter reading and a reduction in fraud and theft
Traffic management	Cities traffic management includes connected road signs and tolling for improving the efficiency of traffic flow
Electric vehicle charging (EVC)	EVC as a *connected* activity, for locating, booking and paying for vehicle charging
Building automation	Energy savings, better quality of life by intelligent usage of domestic appliances to ensure optimum use of energy

environmental energies. Improvements in this area are even more evident in the new paradigm since these objects need to have computing capabilities to enable them to take decisions and act accordingly.

• Creation and usage of standards for the various systems that are connected by this paradigm. This will enable IoT to evolve so that networks and various sensors become integrated and standardized.

Turning connected objects into robots requires small, low-weight and cheap sensors, actuators and processing hardware as components for robots that exhibit complex behaviours, integrating mechanisms for social interactions, autonomous navigation and object analysis, among others.

An Artificial Intelligence (AI) approach suitable for IoT in order to create such robots consists on the development of robots whose embodiment and situatedness in the world evoke behaviours that obviate constant human supervision [6]. Robots become therefore capable of acting autonomously, demonstrating some degree of artificial intelligence. Pervasive robotics—robots present everywhere to perform a variety of tasks—are thus becoming currently a reality, enabling the next stage of evolution, the Internet of Intelligent Things (IoIT).

New paradigms for the Internet of Things are therefore crucial for migrating from nowadays sensor networks into networks of intelligent sensors enabled with actuation mechanisms. Such future networks will consist of the "Internet of Intelligent Things". This paradigm is the next step in the evolution of networking, for creating the experience of an ubiquitous [7], and intelligent, living, internet.

It appears from the need to give commonplace objects the ability to comprehend their surroundings and to make decisions autonomously. Hence, decisions do not need to be sent to core decision making nodes, by giving intelligence to sensors and giving them the ability to act according to the stimulus perceived by sensors, enabling the IoIT to respond better to time-critical situations because the decisions are made in a non-centralized way [8].

The introduction started by presenting the current drive for IoT, and the motivation behind IoIT. Section 2 will discuss Artificial Intelligence (AI) approaches for intelligent sensing, discussing how state of the art knowledge on AI and intelligent robots can be brought into sensor networks in order to increase its "intelligence" and to enable new applications.

Section 3 discusses the role of social networks in the Internet of Intelligent Things. The number of things connected, with some sort of intelligence, in the future will outnumber significantly the number of people currently connected by social networks. This section will therefore discuss the main challenges, and current proposals, in order to interconnect socially the IoIT.

Pervasive intelligent things, as elements of the IoIT (such as smart devices or pervasive robots) will be discussed on Sect. 4, focusing on applications and platforms for building's technologies, home automation and ambient intelligence.

Section 5 discusses intelligent sensing with a large number of sensors. It presents as well developments on people-centric networking and cooperative sensing. Section 6 presents the role of cloud computing for enabling communications among machines on the Internet of Intelligent Things, analyzing several cloud platforms for integration of wireless sensor networks.

Finally, we present the main conclusions, and discuss future research directions for IoIT.

2 Artificial Intelligence for Intelligent Sensing

This section reviews the main artificial intelligence techniques that have been proposed for the Internet of Things. It discusses the usage of machine learning techniques for the identification of interesting patterns or prediction, from information gathered from sensors.

In general, learning algorithms can be classified among three categories: Supervised Learning, Semi-supervised Learning and Unsupervised Learning.

- Supervised Learning assumes a fixed closed set of labels [9]. Only labelled training data is allowed. It is feasible for small-scale sensing applications, but unlikely to scale to handle the wide range of behaviours and contexts exhibited by a large community of users [10]. In multi-instance learning, a variation of supervised learning, labels are associated with a set of data points that includes at least one positive data point but may also include points that are not from the labelled class.

- In semi-supervised approaches, only a portion of the data is labelled, reducing the need for labelled examples [10], while class labels are considered to be fixed. It has been described to face issues that come up with a limited or inconsistent set of labels.
- Unsupervised Learning techniques need no labels on data to be provided, implying a much larger spectrum of application, but a more demanding learning process. Unsupervised learning techniques, such as clustering, latent semantic analysis and matrix factorization, can be applied to mine context based on behavioural similarities [11]. However, they can potentially lead to classes that do not correspond to activities that are useful to the application. Furthermore, they may require for unlabelled data to come only from already labelled class categories.

Activity Classification has also been proposed in the literature, in which sensor data is fed into learning algorithms in the form of extracted features, which are calculations on raw data that emphasize characteristics that more clearly differentiate classes [10].

Features for classification are extracted during pre-processing either from an individual sample or from a n-sample long window. Not all samples have to be considered for processing. An admission control phase may be applied, where filters exclude samples that are outside a given threshold or whose classification cannot be done accurately due to context. Filtering may consider a set of samples, instead of just one, to be able to cope with more complex patterns [12].

Mislabelled data is usually impossible to avoid, due to simple errors or malicious behaviour, resulting in boundary errors between classification groups [9]. Labels may be specified a-priori by application users to improve classification. However, labelling a large dataset is extremely time-consuming and obtaining accurate results is difficult. It is highly desirable to have systems that can infer models from a relatively small amount of labelled data [11].

Features selection is critical in building a robust classifier. Classification may be either done on individual samples or with respect to the complete n-sample window. A classifier may behave less accurately, when classifying periodic patterns with sample windows, if the considered samples contain transitions between other patterns. One possible solution is to smooth the sample window during the classification process. Category classifiers may either provide a coarse or hierarchical classification, where multiple levels of classification are applied [12].

An activity classifier fetches sensed data and classifies it in order to match an activity. It encompasses the pre-processing, the training and the classification algorithms. The former extracts features from raw data and should have a low computational impact. The training and classification algorithms effectively learn and assign an activity to the data, respectively. Due to its computational costs, the classifier's training process is often run offline [13].

Classifier Training consists on the process of building general inference models. Challenges include the lack of appropriate sensor inputs and the time and effort required to setup training models that yield enough classification accuracy [14]. The authors in [14] propose two possible solutions for model creation:

- *Opportunistic Feature Vector Merging.* Method argues for the performance of classification possible with devices that are poor in sensing capabilities is boosted towards the ones that are rich. Features available from more capable device are lent and merged with the features available from a weaker device. This process can occur in two opportunistic scenarios. In a direct scenario, a sensing device that lacks a certain sensor borrows data features from another sensor that possesses the lacked functionality, and inputs it to its classifier. In an indirect scenario, devices sample data within their own capabilities. Subsequently, a centralized matching between the collected features provides a relation between the feature vectors, importing missing features via this connection.

- *Social-network-driven Model and Data Sharing.* Even when devices provide an appropriate set of features to build accurate models, a large set of training data is needed. This solution shares training data among devices, reducing training time and labelling efforts, as the training cost is distributed over the system. However, this is likely to reduce model accuracy, as different people may execute a given activity in different ways and may describe it with different labels. To address this challenge, training data is shared only within social groups, where commonalities lead to more consistent labelling.

Classifier Location in the system must be carefully considered. Classifying data from a large number of devices is computationally expensive and can compromise the scalability of the system [15].

One solution [13] is to divide the classification between the mobile sensors and backend servers. A primitive is the result of the classification process on a mobile device. When primitives arrive at a backend they are persisted, becoming ready to be considered in more complex classifications. Facts represent higher-level classification, including social context and significant places or data statistics (e.g. is a person sick more often than others). Facts are returned by the classification operation on the backend, stored locally and retrieved when necessary. In this approach, classification can be done on the mobile sensing device with backend support. Local classification implies bigger demands on power and computational capabilities on the mobile sensing device. However, by classifying primitives locally, primitives are obtained and buffered even without connectivity. When connectivity is present, primitives are uploaded in batches, minimizing data sent to the backend and improving system efficiency by avoiding sending raw sensor data, which is more bandwidth demanding. Energy costs are also reduced, as fewer consecutive upload operations are needed. As raw sensor data is not sent to the backend, privacy and data are enhanced. The backend classifier generates facts based on primitives or facts produced by other classifiers. Classification can either be invoked periodically or event triggered.

Active Sensor Learning dynamically increases the classes recognized by a model [10]. As data can be acquired in real-time, a developmental strategy for knowledge acquisition has to be followed. One way is to selectively querying users for labels. Nonetheless, sourcing labels from users is a complex process [12] and there are computational costs associated with periodic re-training, compromising the scalability of this approach. For models that need to be adapted in real-time, learning may be forced to occur on the sensing device, leading to a significant increase in computational requirements [10].

To achieve active learning, unsupervised learning approaches may be used. It aims to categorize events there is no prior knowledge of, such as the vast variety of different ambient sounds. One possible approach is for learning events based on encounter frequency and duration [16], excluding less frequent ones, as it is impossible to store all possible patterns, while adapting the classifier to identify these events. This solution achieves scalability for modelling a personal context, without relying on user triggered re-trainings, thus simplifying the learning problem [16],

In community-based models (namely Wikipedia and SETI@home) individual faults are compensated by contributions from the community. To facilitate learning and overcome issues that arise from incorrectly labelling data in unconstrained scenarios, [9] proposes Community-Guided Learning (CGL). CGL learns activities based on the actual behaviour of the community and adjusts transparently to changes in how the community performs them, making it suitable for large scale sensing applications [10]. Intelligent classification of inconsistently labelled crowd-sourced data is made possible by exploring the inherent data structure. Crowd-sourced data is incorporated without neither focusing on labels nor discarding them completely [9]. Labels are treated as soft hints to class boundaries in combination with the observed data similarity, improving the classification accuracy of the system [10]. By employing supervised and unsupervised classifiers to group data, robustness and classifier performance is gained, i.e. both distinguishable and undistinguishable classes are properly grouped. This solution was shown to overcome the limitations associated with existing techniques in coping with inconsistent labelling. By dynamically regrouping modelled classes, it can adapt to a wide range of classes in a robust fashion. However, repeated re-trainings may not be scalable in all mobile inference systems [9].

Social and community pattern mining relies on the gathering of individual activity contexts to identify characteristics belonging to a certain community [11]. It aims to bridge the gap between individual activities and high-level social events, through the use of machine learning techniques. Once data is delivered to these systems, intelligence can be extracted with the appropriate learning techniques and classifiers.

3 Social Networks and the Internet of Intelligent Things

Millions of people regularly participate within online social networks. In [13] the usage of phone sensors to automatically classify events in people lives was investigated. These classifications can be selectively shared using online social networks, replacing manual actions that are now performed daily [9]. On the other hand, the IoT is an enabling paradigm for other forms of networking and computing, such as IoIT and Robotics as a Service (RaaS) paradigms. These new paradigms propose to add intelligence to the things that are connected to the Internet, or consider things as robots that are available as a service to the users, respectively.

But one can take solutions like social networking to the IoT and have connected intelligent things to solve collaboratively complex problems autonomously. By connecting and sharing ideas, large numbers of people and/or machines can provide more accurate answers to complex problems than single individuals [17]. Applying principles that have been thoroughly studied in Social Networking to the IoT may therefore bring several advantages. The structure can be shaped as required to guarantee network navigability, so as to permit an effective discovery of things and services through the network and to make it scalable. Different levels of trustworthiness can be established between the things in the IoT by determining their relationships with one another [18, 19].

Things in Social Internet Of Things (SIoT) are therefore intelligent objects that publish themselves as services and can be discovered through the network by searching through friends or in the neighbourhood, adapting the content to users. Humans and robots, or a mixture of both forms web communities. But such communities will also be formed by intelligent avatars that reside in the virtual world of the Internet of Things. Some long term research is also looking for connecting other biological organisms (like plants), programmed to be capable of intelligent processing, into social networks. The relationships between things in the SIoT can be summarized as follows:

- "Parental object relationship"—established among objects belonging to the same manufacturer and produced in the same batch;
- "Co-location object relationship"—established among objects that are always used in the same shared space;
- "Co-work object relationship"—established among objects that work to accomplish a common objective;
- "Ownership object relationship"—established among objects of the same owner;
- "Social object relationship"—established among object that communicate with each other, with some kind of frequency.

SIoT is therefore expected to have a great impact in the way people do things, for instance, (1) people will not need to worry about leaving appliances turn on when they leave home (because these would understand that and would turn themselves off); (2) if people have a misconfigured appliance, that equipment could communicate with other machines in the house or office to know the preferences of that particular user or company, in order to correctly configure itself; (3) these things detect malfunctions and call for assistance without human intervention.

4 Pervasive Intelligent Robots

This section describes the challenge of home connected intelligent objects, or robots, addressing first IoT platforms for Home Automation, and afterwards IoT applications for data acquired from a large number of sensors.

4.1 IoT and Urban Technologies

Architects have long relied on knowledge of existing buildings both to establish the foundations of current building practices, and to assess the extents and possibilities for innovation. As a discipline that is grounded in the hard reality of large-scale built artefacts, architecture needs to be in touch with technological and building materials innovations, and with the current practices of the industry and peers. However, it is common to visit a building and not be able to know what are the materials that are applied there, either because they are not easily recognizable (type of window frames, glass or finishing materials like ceramics, wood, stone or paints) or because they are not visible at all (structural elements, insulation materials, and other non-finishing, not visible materials) (Fig. 2).

On the other hand, the global economy is currently highly interconnected. The transfer of products from suppliers to consumers depends on a variety of supply chains. In the context of Intelligent Buildings, the materials used on the construction are diverse, ranging from a multitude of specific types of wood,

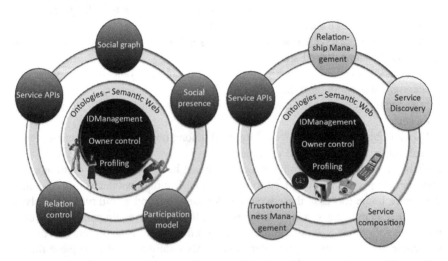

Fig. 2 Basic components of social network platforms [19] for humans (*on the left*) and for things (*on the right*)

aluminium, paints, ceramics, minerals, etc. There is a currently the need for creating building and urban information systems with the identification and description of the components inside these environments [20]. Example use-cases include automatic information systems for expositors, as well as for the common house owner in order to replace broken or out-dated elements.

Urban Sensing

There has been currently a surge of urban related research work, such as energy and environment sustainability [20], or city and buildings design and maintenance, among other areas. In the field of computer networks, Intel has developed an intense research program denominated Urban Atmospheres by Intel, aiming at using sensor networks and community information to automatically build city environmental information available to all users.

Other alternative works, such as PEIR [21], the Personal Environmental Impact Report, or BIKETASTIC [22], have recently proposed to exploit personalized estimates of environmental exposure and impact using people' mobile phones to gather environmental related data, sharing such data for determining how people impact the environment.

Radio frequency identification (RFID) is an automatic identification technology that extends the reach of supply chain information systems in such a way that it becomes now possible (and economically feasible) to tag many physical objects, so that we can locate and track them overtime [23, 24]. These technologies promise significant improvements in several sectors, namely industrial and logistics (such as Warehousing, Retail, Maintenance, Pharmaceuticals, Medical Devices, Agriculture, Food, Retailing, Defense). They are currently employed to monitor physical goods and/or keeping a tracking history of goods, and they are also increasingly appearing on everyday objects that people use.

Virtualization

Building's models have also been used on virtual environments, such as SecondLife, in order to make these structures available to users through their virtual personal avatars. In the sensor networks field, current work sponsored by Nokia and Intel [10, 25] is proposing the integration of people real-world experiences into virtual worlds using Mobile Phones as an edge gateway [26], that collects information from people sensors and forward such information to a network platform. Bridging real building environments and materials and these virtual worlds together is currently challenging, but enables completely new application scenarios with respect to architecture, urbanism, and building environment sustainability. In order to have a seamless integration between the physical and virtual worlds, and to maximize ease of integration and scale of adoption, the system should rely on devices that people already use every day.

4.2 IoT Platforms for Home Automation/Ambient Intelligence

The usage of IoT technology in Home automation and Ambient Intelligence environments is a powerful means to interconnect objects and materials, supported by wireless network solutions enabling the automatic acquisition of such data. We will first give a brief overview of IoT Platforms and equipment that are already available that can be employed as a starting point in IoIT implementation, presenting their comparative evaluation.

There are platforms that enable the creation of IoIT projects and extensions of these—such as Social Internet of Things and Robot as a Service projects—like Cosm [27] (formerly Pachube) that lets sensors and other equipment post and read data to feeds—much like twitter works for people—which allows them to trade messages and take action. Other Home Automation platforms include openRemote [28] and openHAB [29]. These solutions allow users to connect virtually any device and have the platform take action when a command is given or when a condition is met. Ninja Blocks [30] provides a set of open source hardware parts for people to create their own custom devices that connect to the platform, whereas openHAB and openRemote use more standard hardware platforms like Arduino.

Platforms like IFTTT [31] and SmartThings [32] are platforms that are more oriented towards defining intelligent behaviour from devices, like informing the thermostat that the users are close to their home, through the GPS in their smartphone, and thus turning on the Air Conditioning.

Funf [33], developed at MIT Media Lab, is another open source sensing and data processing framework, now a commercial product. Funf consists on software modules (Probes) that act as controllers and data collectors for each sensor's data. It also allows intelligent processing of the personal captured data (e.g. monitoring a person's physical activity by an activity monitor probe that already incorporates motion logic over the accelerometer sensor, and sharing such data on the network). With respect to backend storage, it allows saving data on a remote backend (e.g. on a cloud) or via Wireless Sensor Networks (WSN).

SenseWeb [34] is a large-scale ubiquitous sensing platform, aimed at indexing globally sensor readings. Its open-source layered modular architecture allows to register sensors or sensor data repositories, using web-based sensing middleware.

Table 2 presents a comparative overview for these platforms.

4.3 Smart Devices

Devices like LIFX [35] and Phillips Hue [36] are network controlled light bulbs, and are good examples of objects that can be integrated in a smart environment. They can communicate with the aforementioned platforms, thus enabling a seamless experience and effectively making them intelligent. We can have them

Table 2 Home automation platforms comparison

Platforms	Supported OS	Programming languages	Communication protocols	Hardware	License
Cosm	Multiplatform	javascript,ruby	HTTP (RESTfull API)	Any hardware	Proprietary
OpenHAB	Multiplatform (JVM)	Java-OSGI	Any protocol (depends on bindings)	Any hardware	Open Source (GPLv3)
NinjaBlocks	Multiplatform (the blocks run Ubuntu)	Ruby, Node.js, PHP, Python	HTTP (RESTfull API), Any protocol (depends on plugins)	NinjaBlocks (OpenSource)	Proprietary, Open Source hardware
openRemote	Multiplatform	Java	HTTP (RESTfull API)	Any Hardware	Open Source
SmartThings	iOS, Android	SMART	HTTP (RESTfull API)	Only supported hardware	Proprietary
IFTTT	Web App	Non-programable	HTTP	No hardware required	Proprietary
Funf	Android	Java	–	Any Hardware	Open Source
SenseWeb	Web App	–	HTTP, web service interface	Any Hardware	Open Source

connected to a network and know their status as well as changes in their status—for instance a change in the colour or intensity of light. This enables these bulbs to be connected, through the IoT, to a GPS sensor or a daylight sensor to make them react to the user's proximity to home and to ambient light, enabling as well large energy savings from this intelligent control.

Ubi [37] is a personal assistant, commanded by voice, for the home environment which performs tasks requested by the user, tasks such as searching the web, finding recipes, or commanding things in the home, and can turn the common home into a smart home by making it listen to the user's commands and responding to them.

5 Intelligent Large Scale Sensing

This section addresses the usage of artificial intelligence techniques on data acquired from a large number of sensors, such as mobile devices.

5.1 People-Centric Sensing

The sensing of people constitutes a new application domain that broadens the traditional sensor network scope of environmental and infrastructure monitoring. People become the carriers of sensing devices and both producers and consumers of events [13]. As a consequence, the recent interest by the industry in open programming platforms and software distribution channels is accelerating the development of people-centric sensing applications and systems [10, 13].

To take advantage of these emerging networks of mobile people-centric sensing devices, Mobiscopes, i.e. taskable mobile sensing systems that are capable of high coverage, were proposed. They represent a new type of infrastructure, where mobile sensors have the potential to logical belong to more than one network, while being physically attached to their carriers [15]. By taking advantage of these systems, it will be possible to mine and run computations on enormous amounts of data from a very large number of users [10].

A people-centric sensing system imbues the individuals it serves in a symbiotic relationship with itself [14]. People-centric sensing enables a different approach to sensing, learning, visualizing and data sharing, not only about ourselves, but about the world we live in. The traditional view on mesh sensor networks is combined with one where people, carrying sensors turn opportunistic coverage into a reality [14]. These sensors can reach into regions, static sensors cannot, proving to be especially useful for applications that occasionally require sensing [15]. By employing these systems, one can aim to revolutionize the field of context-aware computing [11].

The worldwide coverage of static sensors to develop people-centric systems is unfeasible in terms of monetary costs, management and permissions [14]. In addition, it is extremely challenging in static sensing models, due to band limits and issues that arise from covering a vast area, to satisfy the density coverage requirements. Thanks to their mobility, mobile sensors are a better fit to overcome spatial coverage limitations [15, 38].

The modelling of behaviour requires large amounts of accurately labelled training data [9]. These systems constitute an opportunity for machine learning systems, as relevant information can be obtained from large-scale sensory data and employed in statistical models [9, 10]. Great benefits can be taken from this unconstrained human data, in opposition to the traditional carefully setup experiments. With these developments it is now possible to distribute and run experiments in a worldwide population rather than in a small laboratory controlled study [10].

By leveraging the behavioural patterns related to individuals, groups and society, a new multidisciplinary field is created: Social and Community Intelligence (SCI) [11]. Real-time user contributed data is invaluable to address community-level problems and provide an universal access to information, contributing to the emergence of innovative services [10, 11].

5.2 Ubiquitous Sensing

Distributed systems have been used as a platform to allow the interaction between groups of individuals and a set of devices. As technology advances in sensing, computation, storage and communications become widespread, ubiquitous sensing devices will become a part of global distributed sensing systems.

Pervasive computing has been mostly used to build systems encompassing a small number of devices that interact with single users or small groups, ranging from wearable computing platforms to ubiquitous computing infrastructures. As technology becomes truly pervasive, low-cost sensing systems may be built and easily deployed based on diverse sensing devices, such as mobile phones, which are carried by a large number of people. Indeed, mobile devices are being increasingly used as the personal gateway to a large plethora of sensor and actuator networks , collecting user and community data. Such large amounts of data can be processed either centrally (on dedicated servers or else on the Cloud) or else distributed over many mobile and fixed computing devices. Large-scale sensing systems may support a diverse set of applications, mostly related to urban scenarios, from predicting traffic jams and modelling human activities, social interactions and mobility patterns, to community health tracking and city environmental monitoring.

As the usage of the mobile phone starts to be pervasive, the exploitation of this personal platform as the centre for sensorial systems is expected to have a direct impact in our daily life. Large-scale sensing applications will enable not only

assisting people on their daily tasks, but also assisting communities and govern-
ments in dynamically monitoring resources.

However, there is little or no consensus on the sensing architecture, which may
include personal device and the cloud. It is not clear what architectural components should run on the phone and what should run in the cloud. Lane et al. [10]
proposes a simple architectural for the mobile phone and the computing cloud. The
proposed architecture comprises building blocks for sensing, learning, sharing and
persuading.

The advent of ubiquitous networks of mobile sensing devices constitute
therefore a paradigm shift, offering researchers challenges in network architecture,
protocol design and data abstractions. Results from mobile sensor networks, pervasive computing, machine learning and data mining can be exploited. However,
new unaddressed challenges arise. These challenges range from growing volumes
of multi-modal sensor data, dynamic operating conditions and the increasing
mobility of the sensing devices [10].

As most people possess sensing-enabled phones, the main obstacle in this area
is not the lack of an infrastructure. Rather, the technical barriers are related to
performing privacy and resource respecting inference, while supplying users and
communities with useful feedback [10].

The main challenges in this area are therefore as follows:

– Managing user participation in the system (participatory vs opportunistic
  sensing);
– Managing trust in users as not to compromise the whole system;
– Adapting to a device's changing resource availability and sampling context
  [14];
– Dealing with mobile sensing devices resource restrictions;
– Coping with sensing device mobility (e.g. when there aren't enough mobile
  sensors or the sensor is moving and jeopardizing sampling for a given context)
  [14, 38];
– Enabling devices to share sensory data while protecting user privacy [38];
– Relating and optimizing diverse resources and application-relevant metrics to
  define data collection and dissemination methods [15];
– Managing large amounts of generated data;
– Defining an approach to collect data capable of assessing the accuracy of
  algorithms that interpret sensor data;
– Performing robust and accurate activity classification in a dynamic real-time
  environment [15];
– Providing the user's with useful information feedback;
– Sensing system scaling from a personal to a population scale;

5.3 Location-Aware Sensing

Mobile sensing devices are user-centric, and as such, they are appropriate to sense location-aware user activities and interactions [11]. Applications might require collected data to be associated to a specific area and time window. Location can be obtained through WiFi or to be based on the triangulation of signal strength measurements from cellular base stations. However, in the former method internet connectivity is required and in the latter, accuracy varies drastically depending on the base stations within range [38]. Consequently, location-based activity recognition benefits from the wide deployment of GPS sensors [11, 38], as they allow the production of location estimates. GPS based location information is accurate to several meters if there is good GPS satellite visibility, i.e. under open sky. Inaccuracy can be compensated by applying filters based on quality measures and rejection metrics relative to the estimates on previous and subsequent samples [13]. Other sensors, namely the compass and gyroscope, provide increased position awareness enhancing location-based applications.

According to [13], location classification can be based on bindings maintained between a physical location and a generic class of location type. Bindings can be sourced from Geographical Information System (GIS) databases. Social Networks can improve this process by providing other metrics, such as significant places [10].

Ensuring privacy in this process is a challenge that involves routing and resource of management, as a mobile sensor is not restricted to a private secure location, but may pass through multiple locations. Moreover, a mobile sensor might be used by different applications over time [15].

5.4 Opportunistic and Participatory Sensing

Recent research efforts aim to make large-scale sensing a reality by leveraging the increasing sensing capabilities found in personal devices such as cell phones. Architectural challenges include methods for data collection, analysis and sharing data, as well as protecting the privacy of people involved. However, one major requirement should be considered: owners of sensorial devices will like to be agnostic of any sensing activities, which may actually be quite frequent. Systems that require a constant participation of humans to support sensing activities may not reach the desirable adoption factor.

There is therefore the need to define an individual's role in the sensing system. Two modalities are considered: participatory and opportunistic sensing [11].

In participatory sensing [21, 39], individuals are incorporated in the decision making process over the sensed data. They can decide which data to share, enjoying control over data privacy issues. In this approach the target is restricted to a group of users willing to participate in the system [11]. As a consequence, a participatory sensing application should have community appeal.

On the other hand, with opportunistic sensing the owner of the sensorial device (e.g. smart-phone) can remain agnostic of any sensorial application. In opportunistic sensing [14, 39, 40], a system automatically takes advantage a device's resources whenever its state (e.g. location or user activity) matches the context requirements of the application. Nevertheless, any sensing activity needs to respect the privacy (e.g. owner's location) and transparency (e.g. avoid impact on user experience) of the owner of the device. Thus, one major feature of any opportunistic sensing system is to sample when device state and privacy and transparency requirements are met.

Opportunistic sensing becomes possible by the system's ability to modify its state in response to a dynamic environment. Sampling only occurs if requirements are met and it is fully automated, with individuals having no involvement in the data collection process. A result of this is that the decision burden is shifted away from users and moved into the system, resulting in more resources being demanded in this decision-making process [11, 14]. This heavier resource demand should not noticeably impact the normal usage experience of the sensing devices. This issue can be tackled if opportunistic sensing is considered a secondary, low priority operation on the sensing devices. Nonetheless, as devices might only be able to meet sensing requirements for short and intermittent periods, a trade-off between availability and resource management should be considered.

5.5 Contextual Sensing

Classification models can exploit sensed data to infer for instance people actions and environment. Significant challenges exist in the real-world usage of human activity sensing. For example, a major challenge is the differentiation of contextual conditions and personal characteristics (e.g. age and lifestyle) encountered in large-scale mobile sensing systems. This challenges increases with the scale of the sensed systems.

These challenges can be addressed by training personalized classification models, which are tuned to the particular sensor data encountered by each user. However, training a personalized model for each user requires a large amount of data labelled by each user, which increases the participatory level of the systems, reducing its transparency to the user. The burden of manually labelling sensor data makes personalization difficult to achieve in large-scale mobile sensing systems.

Context is the metadata that describes the conditions to which sensors are exposed, affecting both data and sensors' ability to perform sensing operations. In opportunistic sensing, context contributes to the evaluation of potential sensor candidates, indicating when sampling should be started and stopped [14]. Context affects data sensing, while sensing devices with mobility can be used in unpredictable ways [10, 12].

Context is important for analyzing sampled data, especially when samples might be taken under suboptimal conditions [14]. In these environments, statistical models may fail to generalize. Also, sensors may be exposed to events for a too short time period, i.e. if the user is traveling too quickly or the sensor's sampling rate is too low. Possible solutions are sharing sensors of available neighboring devices temporarily if they are best suited to sense events [14]. Devices exchange context information and data is selected from the device whose context most closely matches application requirements. A mobile sensor detects the event target using its sensors and forwards the task to better suited neighbors. To recover a lost event source, the area to which the source is predicted to be in is estimated and the task is forwarded to sensors in the predicted area. Another approach is to use super-sampling, where data from nearby sensors is collectively used to lower the noise in an individual reading [10]. One challenge is determining a metric for context matching that provides samples with enough fidelity to respect application requirements.

The reliability of machine learning algorithms may decrease under the dynamic and unexpected conditions presented by mobile sensor usage (e.g. different individuals execute the same activity differently). These problems can be overcome by gathering sufficient samples of the different usage scenarios, i.e. training data. However, acquiring training data is costly and anticipating the different scenarios that might be encountered is not possible for all applications, compromising the scalability of large-scale learning models [12]. Existing solutions are based on borrowing model inputs, i.e. features, from nearby sensors and performing collaborative inference between the associated models. These models might have evolved based on different scenarios, so it is possible to discover new events that were not considered during application design [10]. Other approaches consider a combination of supervised and unsupervised learning techniques, where the learning method to apply depends on data classification stage [12].

5.6 Machine Learning for Large-Scale Sensing

The recent technological developments on mobile technologies allied with the growing computational capabilities of sensing enabled devices have given rise to mobile sensing systems that can target community level problems. These systems are capable of inferring intelligence from acquired raw sensed data, through the use of data mining and machine learning techniques. However, due to their recent advent, associated issues remain to be solved in a systematized way.

Recently, the predominance of mobile phones equipped with sensors, the explosion in social networks and the deployment of sensor networks have created an enormous digital footprint that can be harnessed by employing machine learning techniques [11]. Furthermore, developments in sensor technology, communications and semantic processing, allow the coordination of a large network of devices and large dataset processing with intelligent data analysis [15].

There is a tendency to augment devices with sensing, computing and communication functionalities, connecting them together to form a network, and make use of their collective capabilities. This sensing network is made possible across various ranges, including: single individuals, groups with common interests and the entire population of a city.

Users become a key system component, enabling a variety of new application area such as personal, social and community sensing. Each of these scenarios has its own challenges on how to understand, visualize and share data with others. In personal sensing, the focus is on monitoring the individual. In these applications, information is generated for the sole consumption of the user and is generally not shared with others. In social sensing, information is shared within the group. Individuals who participate in these applications have a commonality of interests, forming a group.

In community sensing (SCI) [11], data is shared for the greater good of the community. Considering data source origin, SCI has its source in three fast growing research fields: mobile sensor based activity recognition, context inference in smart spaces and social network analysis. The key idea behind mobile sensor based activity recognition is to acquire the mathematical model behind human activities after a series of observations. It takes advantage of the prevalence of sensors that accompany users in their mobility patterns. Context inference in smart spaces relies on already deployed infrastructures of static sensors. Static sensors allow the detection of activities, enabling space context. Social network analysis has been studied by physicists and social scientists for a couple of decades and is a major source of information and relationships among a group of individuals. Aggregation of data from these sources constitutes an opportunity for the extraction of intelligence in a community. Applications only become useful once they have a large enough number of individuals participating. The growing diversity of communications technologies enables large amount of data to become available. An infrastructure capable of integrating heterogeneous data sources is required, combining the resulting multimodal data and extracting behavioral patterns from it, through data mining and machine learning methods.

Authors in [10] present the prevalence and development of SCI, which still faces challenges ranging from multi-modal data gathering, to complex intelligence inference and privacy issues. Intelligent inference may be achieved by using machine learning and data mining methodologies, which are fundamental to extract useful information from sensorial data. The authors in [41] applied methodologies such as neural networks, weighted cluster modelling, genetic algorithms, locally receptive fields, piece-wise linear adaptive models to a broad spectrum of machine learning problems along a diverse categorical scope: people actions, objects, scenes and people, using real-time data acquired from human-robot interactions. Such machine learning strategies have also been applied to data acquired from a large number of sensors in order to extract information concerning different categories, such as environment, people identification and activities, expected people and objects' behaviours.

Conventional ways of evaluating environmental impact rely on aggregated statistical data that applies to a community. In [42] a personalized environmental impact approach is described, allowing the tracking of human actions and their impact towards urban problem exposure and contribution. In [43] continuous physical activity data is captured and related to personal health goals in the form of user feedback. These applications have been proven to be effective in impacting the way health is assessed, helping people improve behavioral patterns.

In [9] a lightweight and scalable hierarchical audio classification system, designed with resource limited mobile phones in mind, while remaining capable of recognizing a broad set of events was provided. In opposition with off-line audio context recognition systems, classification was performed online at a lower computational cost, while yielding comparable results.

Despite machine learning strategies, significant challenges exist in the real-world usage of human activity modelling, such as: (1) capacity to process large amount of data; (2) capacity to detect differences in contextual conditions and user characteristics (e.g. age, and lifestyle) encountered in large-scale mobile sensing systems.

The former can be mitigated by a cooperative system of personal devices, or by applying cloud computing. For instance, Nokia Siemens Networks proposed a Cloud middleware to enable easy connection of diverse sensors [44] to gather the acquired data and the simple set-up of a cloud necessary to process the bio-signals. The use of Cloud Computing means that data collection, processing, interface and control can be separated and distributed to the most appropriate resource and device, removing current processing limitations [42, 43]: for instance, applications are not restricted by tiny processors, low-power consumption, and special purpose programming languages.

Detection of differences in contextual conditions and user characteristics is important to avoid discriminative features in sensor data, used by classifiers to recognize different human activities, varying from user to user. Training personalized classification models [9, 45] can mitigate such problems. However, training a personalized model for each user requires a large amount of labelled data from each person. The burden of manually labelling sensor data makes personalization a problem for the deployment of large-scale mobile sensing systems.

5.7 Sensing Architectures

Technologies such as *Mobiscopes* [15] are still recent, leading to a lack of normalized architectures. As these systems have no control over human mobility patterns, the coverage of spaces, events and human interactions becomes opportunistic. In order to face mobility, decisions must be taken in real-time.

Sensing devices enjoy a high degree of heterogeneity. Typically, sensed data has varying time-space resolutions and may become biased depending on the sensing context. Nonetheless, the heterogeneity in sensing data can be harnessed to increase system robustness by exploiting distinct views that may complement each other.

Sensing devices have typically resource limitations that require careful consideration as to where data processing takes place [14]. One approach is to persist data by employing local buffering capabilities. However, for analyzing large amounts of data, local storage limitations may require to have persistency on remote servers. Privacy issues also need to taken into account, as it may be inappropriate to store sensitive data in a remote untrusted system.

Connectivity issues in the system affect sensing performance. In sensing networks, at a given time, a greater amount of data is gathered when compared to data that can be delivered. To circumvent these issues and avoid resource waste, data prioritization schemes to be used when multiple nodes cover the same area, have been suggested [14]. Opportunistic data diffusion schemes between sensing devices, with possible data aggregation, aim to improve connectivity and data quality despite data incongruences. Since information needed by an application may only be available by integrating data from multiple sensing modalities, transmitted data must be compatible across heterogeneous networks [15].

Machine learning functionalities require a systemic view, considering the sensing devices' resource constraints, communication costs to remote servers and the sampling rate required to detect and characterize interesting phenomena. There is also a high correlation between data accesses and user location. Because of the time and space dynamic nature of sensor densities, system performance depends on the mobility patterns of the sensing devices. Uniform coverage for a given area is hard to achieve, as sensors tend to visit zones in a given area in a non-uniform fashion. And as interesting events might be rare, sparse data models need to be considered. For such cases data-mining techniques can be applied. Another approach is to have actuated sensing devices, i.e. sensors that are tasked to visit uncovered areas [15].

Some authors have provided a systematic architecture that can be used as a viewpoint to face these issues, consisting of five layers: pervasive sensing, data anonymization, hybrid learning, semantic inference, and application [11]. The pervasive sensing layer involves the gathering of data from the different data sources (mobile devices, static sensors, social web); The data anonymization layer anonymizes sensed data, offering different anonymization algorithms that can be applied according to the nature of the requirements; The hybrid learning layer applies machine-learning and data mining algorithms to convert low-level single-modality sensing data into high-level features or micro-context. Its focus is to mine data patterns and derive behavior and single space context, before multimodal intelligence is extracted; The semantic inference layer is needed when different micro-contexts need to be aggregated. Its objective is to match the inputted micro-contexts with an expected high-level activity; The application layer provides a set of accessible services that are sustained on the other layers. Applications may be installed directly on a mobile sensing device or on remote servers, communicating with the sensors.

6 The Cloud and the Internet of Intelligent Things

Internet of Things will become a reality provided by the proliferation of WSN. For that end, backend platforms that interconnect WSNs, need to be seriously addressed.

6.1 Wireless Sensor Networks Backend Infrastructures

The research on WSN has been primarily focused upon providing in-network support for sensor network applications. As a result, a variety of special-purpose Operating Systems for sensor nodes has been developed, such as: TinyOS [46], Contiki [47], Squawk [48] and Lorien [49]. The research community has also been very active in the development of programming abstractions for wireless sensor networks resulting in specialized component models such as NesC [50], OpenCom [51] and LooCI [52] along with macro-programming approaches such as TinyDB [53] and Kairos [54].

A key shortcoming of current research efforts is a lack of consideration of the WSN backend. Since the nodes of a sensor network have very limited storage and processing capabilities, sensor networks rarely, if ever, operate in isolation and are usually connected to a back-end modeling infrastructure. Examples from real world sensor-network deployments include computational models of flooding [55], pollution [56] and hurricanes.

Even if the WSN is designed in a special-purposed way to allow some processing within the network, the results are usually stored outside the network where they can also be further analyzed. A recent literature [57] is an example of the previously mentioned type of network where the concept of collector nodes is introduced. These nodes are better equipped with processing, storage and battery capabilities compared to the ordinary sensors. The collector nodes receive the data from the spatially distributed sensors and perform complex in-network computations according to the application needs.

The integration of Sensor Networks with Internet is one of the open research issue in Wireless Sensor Networks. Although it makes a lot of sense to integrate sensor networks with the Internet, we need to consider that the number of sensor nodes could be so large that it becomes impossible to allocate a MAC address for each one of them. As a result, global IP address is unavailable for each sensor node.

Currently, developers implement and connect to backend computational facilities in an ad-hoc manner, with little development support available. This situation is further complicated by the necessity of elastic computation to deal with the dynamism of the events being monitored by the WSN or with an increase in the number of nodes of the WSN.

In the following sub-sections we depict four different technologies that can serve as backend infrastructures for WSN. We start with Cluster Computing, move on to Grid Computing, then talk about High-Performance Computing and end with the emerging Cloud Computing technology, which is the technology this work intends to focus on.

Cluster Computing

Cluster computing is a form of distributed computing where sets of machines are used to increase throughput for computations. The cluster is formed in such a way that the use of multiple computers, multiple storage devices, and redundant interconnections, form what appears to users as a single highly available system. As traffic or availability assurance increases, all or some parts of the cluster can be increased in size or number.

Clustering has been available since the 1980s. Sysplex, from IBM, is a cluster approach for a mainframe system. Cluster computing can also be used as a relatively low-cost form of parallel processing for scientific and other applications that lend themselves to parallel operations. An early and well-known example was the Beowulf project in which a number of off-the-shelf PCs were used to form a cluster for scientific applications.

The authors of [58] show the benefits of cluster and grid computing in solving several parallel problems. The results of their work prove a considerable amount of speed up in the problem resolution time directly proportional to the increasing number of CPUs used.

Despite the apparent benefits of cluster computing, of moving from mainframes to commodity clusters, it became also apparent that those clusters are quite expensive to operate.

Next we talk about grid computing, although similar, there are significant differences between cluster and grid computing. The biggest difference is that a cluster is homogenous while grids are heterogeneous. The computers that are part of a grid can run different operating systems and have different hardware whereas the cluster computers all have the same hardware and OS. A grid can make use of spare computing power on a desktop computer while the machines in a cluster are dedicated to work as a single unit and nothing else. Grids are inherently distributed by its nature over a LAN, metropolitan or WAN. On the other hand, the computers in the cluster are normally contained in a single location or complex. Finally, another difference lies in the way resources are handled. In case of Cluster, the whole system (all nodes) behaves like a single system view and resources are managed by centralized resource manager. In case of Grid Computing, every node is autonomous i.e. it has its own resource manager and behaves like an independent entity.

Grid Computing

Grid Computing refers therefore to the federation of computer resources from multiple administrative domains to reach a common goal. The grid can be thought of as a distributed system with non-interactive workloads that involve a great number of computer processing cycles or access to large amounts of data.

Grid Computing requires the use of software that can divide and farm out pieces of a program to as many as several thousand computers. A well-known example of grid computing in the public domain is the on-going SETI@Home project in which thousands of people share the unused processor cycles of their PCs in the vast search for signs of rational signals from outer space.

In recent years, there has been clear evidence that a new class of distributed applications is emerging, denoted Pervasive Management Support (PMS) systems. PMS combines sensor networks and grid technologies to support the ongoing management of evolving physical systems by providing what-if analysis to help people make faster and better real-world decisions.

Why has the Grid been the technology of choice for PMS systems? An alternative to support heavy computational requirements would be to have a dedicated computational facility, however, despite its benefits, the Grid offers a potentially more attractive solution. A Grid solution can be cheaper by employing pooled resources, it can be more flexible by scaling up the computational power on demand, and it can offer the needed redundancy to fulfill the PMS system requirements.

Hereafter it is described previous research work that integrate together sensor networks and Grid Computing.

FireGrid [59] integrates several technologies, two of which are Grid computing and wireless sensors, to create a system that helps predicting and managing fire threats. A typical FireGrid scenario could involve ten thousand sensors, each one tasked to monitor the environment and ensure that information on the environment is delivered where it is required. Grid Computing is considered vital to the success of the project considering the fundamentally distributed, heterogeneous and loosely coupled architecture of the FireGrid system. FireGrid requires significant computational power on demand, with little advance notice; it need to assimilate data from thousands of sources; and it needs to interactively communicate with building management, control systems and human beings, such as firefighters for example.

The work done in [60] addresses a growing problem in the UK, flooding. The authors designed a system that combines wireless sensor networks with grids for computationally intensive flood modeling purposes. This system uses the GridStix Platform to achieve a "local grid" allowing the sensors to perform local computations, and the GridKit Middleware Platform to interface the wireless sensor network with the Grid.

In [61] it is proposed a system architecture called SPRING, which extends the grid computing paradigm to the sharing of sensor resources in wireless sensor networks. The authors point out that one of the major design issue/challenge was the Grid's Application Programming Interfaces (APIs) for sensors. Although they believed that the natural choice would be to adopt the grid standards and APIs, the Open Grid Services Architecture (OGSA), due to the limited resources available on the sensors, such approach would be unfeasible. As a result, they chose to use the Globus Toolkit 3.2 to implement the Grid Interface layer.

The authors in [62] analyze key problems of combining WSN and grid computing and show a sensor grid architecture based on an embedded wireless gateway. They explain that in order to communicate with the Grid, instead of having a web service in each sensor, thus wasting the precious resources of the sensor, a wireless gateway was implemented, which could connect the WSN and the Grid conveniently.

The work in [63] proposes a solution to monitor cardiovascular diseases using a personal digital assistant (PDA) and applying Grid Computing as technology e-nabler. Medical staff can access and use the application in Software as a service (SaaS) basis.

Although the merging of wireless sensor networks with the Grid seems a natural path to follow, a major difficulty in developing applications for such system has been the lack of nonstandard interfaces and complex programming models. It seems that what is truly required to support WSN-Grid integration is a new mind-set, one that moves away from thinking of grids as large-scale "batch processors" and towards thinking of them as flexible, agile, autonomic computational resources. Such mind-set seems to be more closely related to the emerging distributed computing paradigm named Cloud Computing, which has been introduced in the previous sections and will be further analysed on Sect. 2.3.4.

High Performance Computing

High-Performance Computing (HPC) integrates systems administration (including network and security knowledge) and parallel programming into a multidisciplinary field that combines digital electronics, computer architecture, system software, programming languages, algorithms and computational techniques.

Traditionally, High-Performance Computing used supercomputers to solve advanced computation problems. However, with the changes to computing paradigms, many HPC systems have shifted from supercomputing to computing clusters and grids.

One should also keep in mind that most current applications are not designed for HPC technologies but are retrofitted, meaning they are not actually designed or tested for scaling to more powerful processors or machines. Since networking clusters and grids use multiple processors and computers, these scaling problems can cripple critical systems in future supercomputing systems.

The article in [64] describes a system for augmented testing of wireless networks that tries to overcome the existing problems in systems for testing of tactical hardware in complex battlefield scenarios due to the complex network interfaces and sensor fusion algorithms. The system design combines dedicated High Performance Computing resources with a scalable, high fidelity network emulation and a Computer Generated Forces model to virtually represent the tactical network, force movement, interactions and communication loads to systems under test. Though the paper describes performance results it does not address any implementation issues they might had to overcome regarding the flow of data between the sensors and the HPC resources, nor the flexibility or scalability of the system if sensor network surpasses the maximum expect of ten thousand nodes.

Hyperspectral remote sensing is concerned with the measurement, analysis and interpretation of the electromagnetic spectrum. In [65], the authors present the work they have done to bring the benefits of high performance and distributed computing to the hyperspectral imaging community. They have implemented several solutions using available techniques for HPC and review the obtained results. The work is however limited to hyperspectral techniques applied to HPC and not easily extended to a more general-purpose sensing application.

Cloud Computing

Cloud Computing is fundamentally an evolution of distributed systems technologies such as Cluster Computing and Grid Computing, relying heavily on the virtualization of resources.

For sensor networks, the implementation of backend modeling resources can be particularly problematic due to the tight-coupling of sensing applications with their environment. For instance, consider a flood modeling and warning application that makes use of live sensor data. During standard system operation, the timeliness of flood predictions may not be critical, however, when a flood occurs, the need for timely flood warnings increases. Using traditional computational technologies, extending computational capacity may require the manual allocation of additional resources and may even require the refactoring of the modeling application itself. In an elastic system such as the Cloud, it is possible to incrementally increase available computational power to meet application demands in a more fine-grained manner. Also, if changes need to be made to the modeling application, they can be easily achieved in the Cloud.

Next we describe cases where wireless sensor networks were paired up with Cloud Computing.

The authors of [66] propose a solution to automate vital data collection of patients by using sensors attached to existing medical equipment and afterwards delivering it to the Cloud. However the architecture suggested in this work has a single point of failure, it depends on The Exchange Service. The Exchange Service is responsible for receiving collected data from sensors and dispatching it to the storage service hosted on the Cloud.

The work done in [67] proposes a system for monitoring the parameters of an industrial application process, such as temperature, pressure and level. The authors describe how the sensors communicate between them and with an Integration Controller, which is, by turn, responsible for handling the interaction with the Cloud platform.

The authors in [68] were motivated by the need to integrate and process data from ground-based sensors with satellite data. The authors describe how they solved this problem by implementing a pipeline system for satellite data reprojection and reduction using Windows Azure. The pipeline consists of three main independent data-centric tasks. The first task is responsible for data collection. It gathers the necessary source files, either from a remote file transfer site or from Azure Blob Storage if previously cached, for the reprojection task; the second task uses the data collected in the previous task for one or more reprojection tasks. Once the reprojection computation is finished, the results are stored in Windows Azure Blobs; the third task, the reduction task, is optional and consists in executing an analysis algorithm on the results from the previous task, thus generating final data from which conclusions may be asserted. Their experiments showed that running a practical large-scale science data processing job in the pipeline using 150 moderately-sized Azure virtual machine instances, produced analytical results 90 times faster than was possible with a high-end desktop machine.

Melchor and Fukuda [69] mentions that the recent popularity of sensor networks and cloud computing brought new opportunities of sensor-cloud integration that facilitates users to immediately forecast the future status of what they are observing. The authors describe the main challenge to be the establishment of elastic data channels from sensors to users. To solve this problem a Connector data software toolkit is described, implemented and tested. The tool kit consists of a Connector-API, a Connector-GUI, a Web Server and a Sensor Server. The Connector-API is a program-side I/O and graphics package that allows cloud jobs to behave as various protocol clients to access remote data. The Connector-GUI is a user-side GUI that forwards keyboard/mouse inputs to a user program and also receives standard outputs from a remote user program. The Web Server is a web-based GUI that allows mobile users to access cloud-based data-analyzing jobs from web browsers (in case their mobile devices are not capable of running the Connector GUI). The Sensor Server is a sensor-side data publisher that provides two types of channels to allow interaction between sensors and users. One allows data to be retrieved from the sensors through the Connector-API, the other allows remote users to manage sensors trough the Connector-GUI.

The work done in [70] proposes a novel framework to integrate the Cloud Computing model with WSN. The framework is composed of: a Data Processing Unit, a Pub/Sub Broker, a Request Subscriber (RS), an Identity and Access Management Unit (IAMU), and a Data Repository (DR). The framework works as follows: users connect to the Cloud through the IAMU, which gives them access based on their user account. Once a user is authenticated he can issue data requests to the RS, which in turn creates a subscription and forwards it to the Pub/Sub Broker. Once the subscription is resolved, the resulting data is made available at the DR.

6.2 Wireless Sensor Networks and Health Care

Fox et al. [71] exploited the technology to support rapid analysis of genome sequencing and patient record problems that typify the high-end biomedical computation challenges of today, with focus on Cloud Computing. They show examples of data intensive science applications (such as the Smith-Waterman Computation) in the area of biology and health using several modern technologies, amongst which are Hadoop and Dryad, which in turn have many services that can be executed in virtual machine based clouds such as Azure (Microsoft) and EC2 (Amazon).

Despite traditional health care services being often provided on the clinical environment, nowadays, remote health monitoring (telemedicine [72]) is becoming increasingly common due to the advancements in short-range and wide area wireless technologies, which provide us with mobile, wearable and flexible health monitoring systems.

One important benefit is reducing the costs of disease treatment by increasing health monitoring and doctor-patient efficiency. Moreover, constant monitoring will increase early detection of adverse conditions and diseases. This ability will be a reality in a near future and it will be enabled by the incremental integration of wireless sensor networks, consumer electronics and the Cloud.

SC3 [73] is a health care monitoring system, which integrates WSN with the Cloud. In this work the authors point out that one of the advantages of using the Cloud is the fact that patient information and data can be accessed globally and resources can be shared by a group of hospitals rather than each hospital having a separate IT infrastructure.

The work done in [74] describes a series of prototypes for health. All the prototypes addressed in the paper use wireless sensors and associated monitoring applications. Furthermore, the authors defend that each of the technologies are generic with respect to their sensing, notification, and logging functions, and can easily be extended or adapted to support a variety of applications.

In [75], it is detailed an architecture for a mobile monitoring system capable of observing the health status of a patient on the move. The system is comprised of sensor nodes, a grid network and a middleware to manage and control the sensors.

While the work in [76] proposes a remote health care monitoring system that uses sensor networks and describes the general architecture, they focus on a Quality of Service (QoS) MAC protocol, MACH, exclusively designed for such networks. MACH enables emergency traffic to meet its QoS requirements. The authors went to prove that the proposed MAC gives preference to emergency traffic, which is crucial for medical applications.

SINDI [77] is a system designed to monitor the health of people while at their home. SINDI has the following components: a wireless sensor network for gathering data about the user and his or her environment, a Master Processor located at the home of the user for data storing and processing and a data server to store and manage the data of several homes. The authors choose to focus on the WSM and the Master Processor, whilst leaving the data server component out of the scope of the article.

The authors of [78] propose an architecture for a personal wireless sensor network for mobile health care monitoring. In their work they consider a wireless sensor network comprised of several smart sensor nodes and one control node responsible for the communication within the sensor network and the communication with a remote database. They suggest that the control node be executed by a PDA that will communicate with the remaining nodes through Bluetooth. This work is mainly focused on the WSN and their respective control node.

In [79] a system for home healthcare via a wireless biomedical sensor network is described. The authors describe the hardware implementation of the sensors as well as the implementation of the system. Similarly to [78], the authors defend the use of an intermediate actuator between the data gathered by the sensors and its propagation to the Internet. In [78], such intermediate was the Master Processor; in this case it is described as a home based monitoring system. Once the data gathered from the sensors is transmitted to the home based monitoring system, it is sent to the appropriate Hospital Health Monitoring Centre (HHMC), where it is integrated with the permanent medical data of the given patient.

6.3 Machine-to-Machine (M2M) Communications in the Cloud

M2M refers to technologies that empower devices with the ability to communicate with each other [77, 78]. In M2M communications, sensors are often employed to capture events that are relayed through a network to an application that translates them into meaningful information, such as beginning the shipment of some items that got out of stock. With the IoT, M2M communication has expanded beyond one-to-one communication and changed into a system of networks that transmits data to personal appliances [79].

Since low-cost sensors are becoming more and more common, what is holding the IoT and M2M back? On the one hand the lack of what might be called Sensor Commons. Sensor Commons is a future state whereby data will be available in real time, from a multitude of sensors that are relatively similar in design and method of data acquisition. This allows comparing data across multiple sensors (important for cooperative sensing). The devices do not need to be all calibrated together, however one should ensure that they are more or less recording the same thing with similar levels of precision. In order to be able to store, process and display the information gathered by the multitude of sensors that build a M2M system, significant computing power is required. With the emergence of Cloud Computing this problem can be more easily addressed as described hereafter.

Indeed, Cloud Computing allows for data collection, processing, interface and control to be separated and distributed to the most appropriate resources. Current M2M implementations combine the above mentioned aspects into one solution, either chips in sensor bodies or an onsite laptop or desktop PC tied within a local network perform the data processing and determine the control logic.

However, once data moves to the Cloud most current limitations disappear [42, 80–82]. Storing data in the cloud means that, from a storage point of view, data buffers within devices no longer matter. Cloud storage is near limitless and so historic data can be saved for as long as it is considered valuable. Applications are also not restricted by tiny processors, low-power consumption, and special purpose programming languages.

In addition, there is not only access to super-fast processors [83–85], but if there are server or storage bottlenecks, these can be quickly addressed by launching on-demand more servers or horizontally scaling storage. Lastly, interfaces can move to Web browsers, mobile phones, wallboards, and tablets, eliminating the need for having screens as a part of every device. Medical devices no longer have to come with their own monitors. Separating the data input from processing from display not only means lower costs (less components to the devices) but also easier upgrading of diagnostics and far better visualization capabilities.

Hence, separating the components and putting the data in the Cloud allows devices to keep getting smarter while not necessarily becoming more complex.

6.4 Cloud Robotics

Robotics has faced strong barriers in its evolution because there are inherent challenges to it that until recently technologies were unable to overcome. Researchers have been able to apply robotics to controlled environments because robots in factories have fixed behaviors, but for robotics to reach a level of per-vasiveness in which robots are present in our daily lives, robots need to adapt themselves to environments that can change very frequently and need to be able to respond to unexpected events.

Robots to reach this level of adaptation need to analyze their surroundings and to store information about it and to process this information [86]. Robots however usually do not have enough computational and battery power to process that information while still being mobile and with a small size form factor.

Cloud Robotics is a new paradigm aiming to solve some of these current robotics issues. Cloud robotics is a combination of cloud computing and robotics. Robots operating from a cloud can be more portable, less expensive and have access to more intelligence than an ordinary robot [86]; these robots could also offload Central Processing Unit's (CPU) heavy or energy demanding tasks to remote serves, relying on smaller and less power hungry onboard computers [87].

Robots in a cloud configuration can perform autonomous learning, obtain knowledge, share knowledge, and even evolve. Through the robotics cloud, robots can execute collaborative tasks and provide efficient services. Robots can upload and share their acquired knowledge on the servers, which are responsible for knowledge, storage and scheduling [86].

The cloud is promising for the evolution of robotics, but it also has its short-comings. Robots rely on sensors and feedback to accomplish a task, this might not be accomplished on the cloud because cloud based applications can be slow or difficult to run because of internet limitations [86], And robots usually have several closed feedback control loops between sensors and actuators demanding real-time action, for which network latency is an important concern.

Besides the introduction of the cloud for computing in robotics, there is a new paradigm in robotics, which is Robotics as a Service (RaaS). By using Service Oriented Architecture (SOA), the robots can publish their functionalities as services available to the clients and can extend these functionalities by searching for services available in the cloud [88]. RaaS provides some advantages like having a layer of common services with standard interfaces, and making the invocations of the services device-independent.

RoboEarth is an existing solution for Cloud Robotics. This platform lets robots collect, store and share information independently of specific robot hardware. Such capabilities enable robots to provide functions and behaviors that were not explicitly programmed, such as the ability to recognize certain objects. RoboEarth platform provides a repository in the cloud for robots and high-level components for interaction with the robots, all packaged in a REST-full API. It implements components for ROS (Robot Operating System) compatible operating systems, as well as components for robot-specific controllers accessible via a Hardware Abstraction Layer platform.

7 Conclusions

We discussed on this chapter the emerging of Intelligent Things on the internet with more power than mere sensors, such as AI processing and actuation capabilities. This enables a new Internet of Things: of intelligent agents, and eventually of living creatures on the future. Currently, smart devices, such as home appliances and light bulbs, are appearing allowing for some degree of actuation. In the short term, more intelligent things, like appliance robots (such as smart lamps, or the currently available smart vacuum cleaner) will also be communicating with each other, and with humans, through the Internet.

On one hand, typical robotics and artificial intelligence approaches face now new challenges. A robotic body is much more than a physical hardware platform encased on a metal structure: robots will have a large part of their sensors and actuators in the surrounding environment. The Internet of Things becomes part of a robot body. A robot's brain will be in the cloud, and its AI algorithms may control, or "incarnate", different physical robotic platforms according to needs. Hence, a robot brain is physically distributed on the Internet, and no more associated to a specific robotic body. We expect robots to be controlling one hardware platform as body, having acquired memory experiences on their learning processes from data perceived while occupying other robotic bodies.

On the other hand, all things are not only becoming connected to the Internet, but they are also increasingly become equipped, initially with sensors and later actuators and the processing required for closing the loop. Such capabilities will allow new forms of communication to things, especially with humans.

The IoIT is therefore closely interrelated with pervasive robotics [87], the embedding of actuation and processing capabilities (behind sensing) into everyday objects. Hence, IoIT goes further than IoT paradigm of connecting billions of things, into the paradigm of transforming everyday objects into intelligent things that communicate with each other, and with people. To enable IoIT, it is necessary ubiquitous middleware platforms that integrate in a seamless and transparent manner not only data from sensors, but also data into actuators, and that are able to transfer artificial intelligence processing among terminal devices and the cloud according to system needs.

Current software integration platforms, including robotics middleware such as ROS and Yet Another Robotic Platform (YARP), or smarthome middleware such as Cosm and OpenHAB, do not meet yet such requirements. With IoIT, the robots and the smarthome become the same entity, and this will require new middleware that integrates seamless communication services, data and the needed processing intelligence, for the Internet of Intelligent Things.

References

1. H.S. Woelffle, P. Guillemin, P. Friess, Sylvie, Vision and challenges for releasing the internet of things (Office of the European Union, 2010)
2. L. Atzoria, A. Ierab, G. Morabito, The internet of things: a survey. Comput. Netw. **54**(15) (2010)
3. C. Gouveia, J. Grilo, E. Nobre, A. Arsénio, Using energy harvesting for robots: review of the state of the art, in *International Workshop on Energy Efficiency for a More Sustainable, World* (2012)
4. A. Iera, C. Floerkemeier, J. Mitsugi, G. Morabito, The internet of things. Guest editorial. IEEE Wirel. Commun. 17(6) (2010)
5. D. Evans, The internet of things how the next evolution of the internet is changing everything, in *CISCO White Paper* (2011)
6. R. Brooks, *Cambrian Intelligence: The Early History of the New AI* (MIT Press, 1999)
7. K. Framling, J. Nyman, Information architecture for intelligent products in the internet of things, in *Beyond Business Logistics proceedings of NOFOMA* (2008), pp. 224–229
8. Y. Chen, H. Hu, Internet of intelligent things and robot as a service. Simul. Modell. Pract. Theory (Elsevier, Amsterdam, 2012)
9. D. Peebles, H. Lu, N. Lane, T. Choudhury, A. Campbell, Community-guided learning: exploiting mobile sensor users to model human behavior, in *Proceedings of 24th AAAI Conference on, Artificial Intelligence* (2010)
10. N. Lane, E. Miluzzo, H. Lu, D. Peebles, T. Choudhury, A. Campbell, D. College, Adhoc and sensor networks: a survey of mobile phone sensing. IEEE Commun.0 Mag. 140–150 (2010)
11. D. Zhang, B. Guo, B. Li, Z. Yu, Extracting social and community intelligence from digital footprints: an emerging research area. Ubiquitous Intell. Comput. 4–18 (2010)
12. H. Lu, W. Pan, N. Lane, T. Choudhury, A. Campbell, SoundSense: scalable sound sensing for people-centric applications on mobile phones, in *Proceedings of the 7th International Conference on Mobile Systems, Applications, and Services* (ACM, 2009), pp. 165–178
13. E. Miluzzo, N. Lane, K. Fodor, R. Peterson, H. Lu, M. Musolesi, S. Eisenman, X. Zheng, A. Campbell, Sensing meets mobile social networks: the design, implementation and evaluation of the cenceme application, in *Proceedings of the 6th ACM Conference on Embedded Network Sensor Systems* (ACM, 2008), pp. 337–350

14. A. Campbell, S. Eisenman, N. Lane, E. Miluzzo, R. Peterson, H. Lu, X. Zheng, M. Musolesi, K. Fodor, G.S. Ahn, The rise of people-centric sensing. Internet Comput. IEEE **12**(4), 12–21 (2008)

15. T. Abdelzaher, Y. Anokwa, P. Boda, J. Burke, D. Estrin, L. Guibas, A. Kansal, S. Madden, J. Reich, Mobiscopes for human spaces. IEEE Pervasive Comput. **6**(2), 20–29 (2007)

16. P. Fitzpatrick, A. Arsenio, E. Torres-Jara, Reinforcing robot perception of multi-modal events through repetition and redundancy and repetition and redundancy. Interact. Stud. **7**(2) (2006)

17. L. Atzori, A. Iera, G. Morabito, M. Nitti, The social internet of things (SIoT)—When social networks meet the internet of things: concept, architecture and network characterization. Comput. Netw. **56**(16), 3594–3608, 14 Nov (2012)

18. M. Kranz, L. Roalter, F. Michahelles, Things that Twitter: social networks and the Internet of things, in *What can the Internet of Things do for the Citizen (CIoT) Workshop at The Eighth International Conference on Pervasive Computing (Pervasive 2010)* (2010)

19. L. Atzori, A. Iera, G. Morabito, SIoT: giving a social structure to the internet of things. IEEE Commun. Lett. **15**(11), 1193–1195 (2011)

20. L. Caldas, *GENE_ARCH*: An evolution-based generative design system for sustainable architecture, in *Lecture Notes in Artificial Intelligence, Lecture Notes in Computer Science Series* ed. by I.F.C. Smith, EG-ICE, LNAI 4200 (Springer, Berlin Heidelberg, 2006), pp. 109–118

21. M. Mun, S. Reddy, K. Shilton, N. Yau, J. Burke, D. Estrin, M. Hansen, E. Howard, R. West, P. Boda, PEIR, the personal environmental impact report, as a platform for participatory sensing systems research, in *Proceedings of the 7th International Conference on Mobile Systems, Applications, and Services*, New York, USA, ACM (2009), pp. 55–68

22. S. Reddy, K. Shilton, G. Denisov, C. Cenizal, D. Estrin, M. Srivastava, Biketastic: sensing and mapping for better biking, in *ACM Conference on Human Factors in, Computing Systems (CHI)* (Apr 2010)

23. K. Finkenzeller, in *RFID Handbook: Fundamentals and Applications in Contactless Smart Cards and Identification*, 2nd edn (Wiley, New York, 2003)

24. S. Sarma, Towards the 5 cent tag. MIT Auto-ID Center, Tech. Rep., 11 (2001)

25. M. Musolesi, E. Miluzzo, N. Lane, S. Eisenman, T. Choudhury, A. Campbell, The second life of a sensor—Integrating real-world experience in virtual worlds using mobile phones, in *Proceedings of HotEmNets '08* (2008)

26. R. Kwok, Phoning in data. Nature **458** (April 2009)

27. Cosm Ltd.: Cosm—Internet of Things Platform Connecting Devices and Apps for Real-time Control and Data Storage, 15 Nov 2012. https://cosm.com

28. OpenRemote: OpenRomete | Open Remote Automation | OpenRemote; 15 Nov 2012. http://www.openremote.com

29. openhab; openhab—empowering the smart home—Google Project Hosting, 15 Nov 2012. http://code.google.com/p/openhab/

30. Ninja Blocks; Ninja Blocks—The API for Atoms, 15 Nov 2012. http://ninjablocks.com

31. IFTTT: IFTTT / About IFTTT, 15 Nov 2012. https://ifttt.com/wtf

32. Physical Graph Corporation: SmartThings—Make Your World Smarter, 15 Nov 2012, http://smartthings.com

33. N. Aharony, W. Pan, C. Ip, I. Khayal, A. Pentland, Social fMRI: Investigating and shaping social mechanisms in the real world. *Pervasive and Mobile Computing* (2011)

34. W. Grosky, A. Kansal, S. Nath, L. Jie, F. Zhao, SenseWeb: an infrastructure for shared sensing. IEEE MultiMedia **14**(4), 8–13, Oct–Dec (2007)

35. LIFX: LIFX; 15 Nov 2012. https://www.facebook.com/lifxlabs

36. Koninklijke Phillips Electronics N.V.; Philips hue, 15 Nov 2012. http://www.meethue.com/en-US

37. Unified Computer Intelligent Corporation: Ubi—The Ubiquitous Computer, 15 Nov 2012. http://theubi.com

38. A. Kansal, M. Goraczko, Building a sensor network of mobile phones, in *On Information Processing in Sensor*, pp. 547–548 (2007)
39. S. Consolvo, D. McDonald, T. Toscos, M. Chen, J. Froehlich, B. Harrison, P. Klasnja, A. LaMarca, L. LeGrand, R. Libby, Others: activity sensing, in *Conference on Human Factors in Computing Systems* (ACM, 2008), pp. 1797–1806
40. A. Kapadia, N. Triandopoulos, C. Cornelius, AnonySense: opportunistic and privacy-preserving context collection, in *Proceedings of the 6th International Conference on Pervasive Computing (Pervasive '08)*, ed. by Jadwiga Indulska, Donald J. Patterson, Tom Rodden, Max Ott (Springer, Berlin, Heidelberg, 2008), pp. 280–297
41. A. Arsenio, *Teaching a Robotic Child—Machine Learning Strategies for a Humanoid Robot from Social Interactions*, Book Chapter in Humanoid Robots (New Developments, 2007)
42. P. Parwekar, From Internet of Things towards cloud of things, in *2nd International Conference on Computer and Communication Technology (ICCCT)* (2011), pp. 329–333
43. S. Wenchong, L. Maohua, Tactics of handling data in internet of things, in *IEEE International Conference on Cloud Computing and Intelligence Systems (CCIS)* (2011), pp. 515–517
44. A. Metnitzer, Nokia Siemens Networks Machine2Machine Solution. MIPRO 2011. Opatija (2011), pp. 386–388
45. B. Longstaff, S. Reddy, D. Estrin, *Improving Activity Classification for Health Applications on Mobile Devices Using Active and Semi-Supervised Learning*, Proc. Pervasive Health (IEEE Press, 2010), pp. 1–7
46. P. Levis, S. Madden, J. Polastre, R. Szewczyk, A. Woo, D. Gay, J. Hill, M. Welsh, E. Brewer, D. Culler, *Tinyos: an operating system for sensor networks* (Springer, New York, 2004). In Ambient Intelligence
47. A. Dunkels, B. Gronvall, T. Voigt, *Contiki—a lightweight and flexible operating system for tiny networked sensors*, in Workshop on Embedded Networked Sensors (Nov 2004)
48. D. Simon, J. Daniels, C. Cifuentes, D. Cleal, D. White, in *The Squawk Java Virtual Machine* (Sun Microsystems, 2005)
49. B. Porter, G. Coulson, Lorien: a pure dynamic component based operating system for wireless sensor networks, in *MidSens '09: Proceedings of the 4th International Workshop on Middleware Tools, Services and Run-Time Support for Sensor, Networks* (2009)
50. D. Gay, P. Levis, R.V. Behren, M. Welsh, E. Brewer, D. Culler, The nesc language: A holistic approach to networked embedded systems, in *PLDI '03: Proceedings of the ACM SIGPLAN 2003 Conference on Programming Language Design and Implementation* (2003)
51. G. Coulson, G. Blair, P. Grace, F. Taiani, A. Joolia, K. Lee, J. Ueyama, T. Sivaharan, A generic component model for building systems software. ACM Trans. Comput. Syst. **26**(1) (2008)
52. D. Hughes, K. Thoelen, W. Horre, N. Matthys, S. Michiels, C. Huygens, W. Joosen, Looci: A loosely-coupled component infrastructure for networked embedded systems, in *7th International Conference on Advances in Mobile Computing & Multimedia (MoMM09)* (Dec 2008)
53. S. Madden, M. Franklin, J. Hellerstein, W. Hong. Tinydb: an acquisitional query processing system for sensor networks. ACM Trans. Database Syst. (2005)
54. R. Gummadi, N. Kothari, R. Govindan, T. Millstein, Kairos: a macro-programming system for wireless sensor networks, in *SOSP '05: Proceedings of the twentieth ACM symposium on Operating systems principles* (2005)
55. D. Hughes, P. Greenwood, G. Coulson, G. Blair, F. Pappenberger, P. Smith, K. Beven, An experiment with reflective middleware to support grid-based flood monitoring. Wiley Inter-Sci. J. Concurr. Comput. Pract. Experience **20**(11) (Nov 2007)
56. W. Tsujita, A. Yoshino, H. Ishida, T. Moriizumi, Gas sensor network for air-pollution monitoring. Chem. Sens. Actuators B (2005)
57. A. Giridhar, P. Kumar, Toward a theory of in-network computation in wireless sensor networks. IEEE Commun. Mag. **44** (Apr 2006)

58. K. Fujisawa, M. Kojima, A. Takeda, M. Yamashita, High performance grid and cluster computing for some optimization problems, in *SAINT-W' 04: Proceedings of 2004 Symposium on Applications and the Internet* (2004)
59. D. Berry, A. Usmani, J. Torero, A. Tate, S. McLaughlin, S. Potter, A. Trew, R. Baxter, M. Bull, M. Atkinson, FireGrid: Integrated emergency response and fire safety engineering for the future built environment, in *Proceedings of the UK eScience All Hands Meeting* (2005)
60. D. Hughes, P. Greenwood, G. Blair, G. Coulson, F. Pappenberger, P. Smith, K. Beven, An intelligent and adaptable grid-based flood monitoring and warning system, in *Proceedings of the UK eScience All Hands Meeting* (2006)
61. H. B. Lim, Y. M. Teo, P. Mukherjee, V. T. Lam, W. F. Wong, S. See, Sensor grid: integration of wireless sensor networks and the grid. Local Comput. Netw. (2005)
62. L. Mingming, L. Baiping, L. Wei, Information security wireless sensor grid, in *5th International Conference on Information Assurance and, Security* (2009)
63. K. Sundararaman, J. Parthasarathi, S. Rao, A. Rao, Hridaya A tele-medicine initiative for cardiovascular disease through convergence of grid, Web 2.0 and SaaS, in *Pervasive Computing Technologies for Healthcare* (2008)
64. K. LeSueur, E. Jovanov, Performance analysis of the augmented wireless sensor network testbed, in *41st Southeastern Symposium on System Theory University of Tennessee Space Institute Tullahoma* (March 2009)
65. K. Fujisawa, M. Kojima, A. Takeda, M. Yamashita, High performance grid and cluster computing for some optimization problems, in *Applications and the Internet Workshops* (2004)
66. C. Rolim, F. Koch, C. Westphall, J. Werner, A. Fracalossi, G. Salvador, A cloud computing solution for patient's data collection in health care institutions, in *Second International Conference on eHealth, Telemedicine, and Social Medicine* (2010)
67. V. Rajesh, J.M. Gnanasekar, R.S. Ponmagal, P. Anbalagan, Integration of wireless sensor network with cloud, in *International Conference on Recent Trends in Information, Telecommunication and Computing* (2010)
68. J. Li, M. Humphrey, D. Agarwal, K. Jackson, C. van Ingen, Y. Ryu, eSchience in the cloud: A MODIS satellite data reprojection and reduction pipeline in the Windows Azure platform, in *IEEE International Conference on E-Schience Workshops* (2009)
69. J. Melchor, M. Fukuda, A design of flexible data channels for sensor-cloud integration, in *Systems Engineering (ICSEng) 21st International Conference* (2011)
70. K. Ahmed, M. Gregory, Integrating wireless sensor networks with cloud computing, in *Mobile Ad-hoc and Sensor Networks (MSN) 2011 Seventh International Conference* (2011)
71. G. Fox, X. Qiu, S. Beason, J. Choi, J. Ekanayke, T. Gunarathne, M. Rho, H. Tang, N. Devadasan, G. Liu, Biomedical case studies in data intensive computing. Cloud Comput. **5931** (2009)
72. J. Craig, V. Patterson, Introduction to the practice of telemedicine. J. Telemed. Telecare **11**(1), 3–9 (2005)
73. X. Le, S. Lee1, P. Truc, L. Vinh, A. Khattak, M. Han, D. Hung, M. Hassan, M. Kim, K. Koo, Y. Lee, E. Huh, Secured WSN-integrated cloud computing for u-Life care, in *Consumer Communications and Networking Conference* (2010)
74. C. Baker, K. Armijo, S. Belka, M. Benhabib, V. Bhargava, N. Burkhart, A. Der Minassians, G. Dervisoglu, L. Gutnik, M. B. Haick, C. Ho, M. Koplow, J. Mangold, S. Robinson, M. Rosa, M. Schwartz, C. Sims, H. Stoffregen, A. Waterbury, E. S. Leland, T. Pering, P. K. Wright, Wireless sensor networks for home health care, in *AINAW 07': Proceedings of the 21st International Conference on Advanced Information Networking and Applications, Workshop* (2007)
75. S. Oh, C. Lee, u-healthcare sensorgrid gateway for connecting wireless sensor network and grid network. Adv. Commun. Technol. (2008)
76. D. Benhaddou, M. Balakrishnan, X. Yuan, Remote healthcare monitoring system architecture using sensor networks, in *Region 5 Conference* (2008)

77. D. Merico, A. Mileo, R. Bisiani, Pervasive wireless-sensor-networks for home healthcare need automatic reasoning, in *Pervasive Computing Technologies for Healthcare* (2009)
78. S. Krco, V. Delic, Personal wireless sensor network for mobile health care monitoring, in *Telecommunications in Modern Satellite, Cable and Broadcasting Service* (2003)
79. R.A. Rashid, S.H.S. Arifin, M.R.A. Rahim, M.A. Sarijari, N.H. Mahalin, Home healthcare via wireless biomedical sensor network, in *Proceedings of the IEEE International RF and Microwave Conference* (2008)
80. W. Shi, M. Liu, Tactics of handling data in internet of things, in *Cloud Computing and Intelligence Systems (CCIS)* (Sept 2011)
81. L. Wu, X. Ping, Y.G. Wang, J. Fu, Cloud storage as the infrastructure of cloud computing, in *Intelligent Computing and Cognitive Informatics (ICICCI)* (2010)
82. R. Xue, Z. Wu, A. Bai, Application of cloud storage in traffic video detection, in *Computational Intelligence and Security* (2011)
83. H. Wang, P. Zhu, Y. Yu, Q. Yuan, Cloud computing based on internet of things, in *Mechanic Automation and Control Engineering (MACE)* (July 2011)
84. Q. Li, T. Zhang, Y. Yu, Using cloud computing to process intensive floating car data for urban traffic surveillance, in *International Journal of Geographical Information Science* (2011)
85. M. Armbrust, A. Fox, R. Griffith, A. D. Joseph, R. Katz, A. Konwinski, G. Lee, D. Patterson, A. Rabkin, I. Stoica, M. Zaharia, Above the clouds: a Berkeley view of cloud computing, in *UC Berkeley Technical Report UCB/EES* (2009)
86. F. Ren, Robotics cloud and robotics school, in *7th IEEE International Conference on Natural Language Processing and Knowledge Engineering (NLP-KE)* (2011)
87. E. Guizzo, Robots with their heads in the clouds. Spectrum IEEE **48**(3), 16–18 (2011)
88. Y. Chen, D. Zhihui, M. García-Acosta, Robot as a service in cloud computing, in *5th IEEE International Symposium on Service Oriented System Engineering (SOSE)* (2010)
89. A. Arsenio, Towards pervasive robotics, in *Proceedings of the 18th international joint conference on Artificial intelligence (IJCAI'03)* (Morgan Kaufmann Publishers Inc., San Francisco, 2003), pp. 1553–1554

Cloud Computing: Paradigms and Technologies

Ahmed Shawish and Maria Salama

Abstract Cloud Computing has recently emerged as a compelling paradigm for managing and delivering services over the internet. It is rapidly changing the landscape of information technology, and ultimately turning the long-held promise of utility computing into a reality. With such speedy progressing and emerging, it becomes crucial to understand all aspects about this technology. This chapter provides a comprehensive overview on the Cloud's anatomy, definition, characteristic, affects, architecture, and core technology. It clearly classifies the Cloud's deployment and service models, providing a full description of the Cloud services vendors. The chapter also addresses the customer-related aspects such as the Service Level Agreement, service cost, and security issues. Finally, it covers detailed comparisons between the Cloud Computing paradigm and other existing ones in addition to its significant challenges. By that, the chapter provides a complete overview on the Cloud Computing and paves the way for further research in this area.

1 Introduction

Cloud Computing has recently emerged as a compelling paradigm for managing and delivering services over the internet. The rise of Cloud Computing is rapidly changing the landscape of information technology, and ultimately turning the

A. Shawish (✉)
Department of Scientific Computing, Faculty of Computer and Information Science,
Ain Shams University, Cairo, Egypt
e-mail: ahmed.gawish@fcis.asu.edu.eg

M. Salama
Department of Computer Science, Faculty of Informatics and Computer Science,
British University in Egypt, Cairo, Egypt
e-mail: maria.salama@bue.edu.eg

F. Xhafa and N. Bessis (eds.), *Inter-cooperative Collective Intelligence:*   39
Techniques and Applications, Studies in Computational Intelligence 495,
DOI: 10.1007/978-3-642-35016-0_2, © Springer-Verlag Berlin Heidelberg 2014

long-held promise of utility computing into a reality. The latest emergence of Cloud Computing is a significant step towards realizing this utility computing model since it is heavily driven by industry vendors. It attracts business owners due to its ability to eliminate the provisioning plan overhead, and allows enterprises to start from the small scale and dynamically increase their resources simultaneously with the increase of their service demand. Cloud Computing promises to deliver reliable services through next-generation data centers built on virtualized compute and storage technologies. Users will be able to access applications and data from a Cloud anywhere in the world following the pay-as-you-go financial model.

With such speedy progressing of the Cloud Computing and emerging in most of the enterprise business and scientific research areas, it becomes crucial to understand all aspects about this technology. The aim of this chapter is to provide a complete overview on the Cloud Computing through a comprehensive descriptions and discussion of all aspects of this technology. In this chapter, Cloud Computing anatomy is presented along with the essential definitions from different perspectives and prominent characteristics. The affects of Cloud Computing on organizations and enterprises were also addressed in terms of time, management and operational costs. The architecture design of Clouds has been discussed, as well as the key technologies behind it; such as virtualization and web services.

Clouds are classified according to their deployment models as private, community, public and hybrid Clouds. Clouds also offer different service models; software, platform and infrastructure as service. In such perspective, the chapter addresses the Cloud Computing services classification by clearly differentiating between deployment and service models in one hand, and beneath each of them on the other hand. The Cloud providers and vendors are also addressed and described.

As more providers offer computing services using Cloud infrastructure, the method of determining the right price for users become crucial for both providers and consumers. The Service Level Agreement that drives the relation between the provider and the consumer becomes also of a great significance. Similarly the security of the data across the Cloud is gaining a great interest due to its sensitivity. This chapter discusses the Cloud customer-related aspects in terms of the Service Level Agreements, service cost, service pricing, and security issue.

In fact, the Cloud Computing is built on top of several other technologies, for example Distributed Computing, Grid Computing, and Utility Computing. This chapter covers a comprehensive comparison between the Cloud Computing and preceding paradigms; in terms of architecture, resources sharing, QoS guarantees and security.

The challenges facing the new paradigm; such as security, availability and resources management; should be carefully considered in future research in order to guarantee the long-term success of Cloud Computing. The chapter tackles these challenges and paves the way for further research in this area.

2 Anatomy of Cloud Computing

This section presents a general overview of Cloud Computing; including its definitions, characteristics, and organizational affects. Clouds architecture is also addressed, as well as the key technologies on which Cloud Computing rely.

Cloud Computing is the new cost-efficient computing paradigm in which information and computer power can be accessed from a web browser by customers. Cloud Computing is the Internet-based development and use of computer technology. Loosely speaking, Cloud Computing is a style of computing paradigm in which typically real-time scalable resources such as files, data, programs, hardware, and third party services can be accessible from a Web browser via the Internet to users. These customers pay only for the used computer resources and services by means of customized Service Level Agreement (SLA), having no knowledge of how a service provider uses an underlying computer technological infrastructure to support them. The service load in Cloud Computing is dynamically changed upon end-users' service requests [1].

Cloud Computing shifts the computation from local, individual devices to distributed, virtual, and scalable resources, thereby enabling end-users to utilize the computation, storage, and other application resources, which forms the Cloud, on-demand [2].

2.1 Definition

Cloud Computing has been coined as an umbrella term to describe a category of sophisticated on-demand computing services initially offered by commercial providers. It denotes a model on which a computing infrastructure is viewed as a "Cloud", from which businesses and individuals access applications from anywhere in the world on demand [3]. The main principle behind this model is offering computing, storage, and software "as a service". There are many definitions of Cloud Computing, but they all seem to focus on just certain aspects of the technology [4].

An early definition of Cloud Computing has been proposed as follows: A computing Cloud is a set of network enabled services, providing scalable, Quality of Service (QoS) guaranteed, normally personalized, inexpensive computing platforms on demand, which could be accessed in a simple and pervasive way [5].

Markus Klems claims that immediate scalability and resources usage optimization are key elements for the Cloud. These are provided by increased monitoring, and automation of resources management in a dynamic environment. Other authors disagree that this is a requirement for an infrastructure to be considered as a Cloud [6].

According to a 2008 paper by Carl Hewitt called "ORGs for Scalable, Robust, Privacy-Friendly Client Cloud Computing" published by IEEE Internet Computing, "Cloud Computing is a paradigm in which information is permanently stored in servers on the internet and cached temporarily on clients that include desktops, entertainment centers, table computers, notebooks, wall computers, handhelds, sensors, monitors, etc." [7].

A Berkeley Report in February 2009 states "Cloud Computing, the long-held dream of computing as a utility has the potential to transform a large part of the IT industry, making software even more attractive as a service" [8].

From an economic perspective, Cloud Computing is defined as follows: "Building on compute and storage virtualization technologies, and leveraging the modern Web, Cloud Computing provides scalable and affordable compute utilities as on-demand services with variable pricing schemes, enabling a new consumer mass market" [9].

From the Quality of Service perspective, Clouds have been defined as a large pool of easily usable and accessible virtualized resources (such as hardware, development platforms and/or services). These resources can be dynamically reconfigured to optimum resource utilization. This pool of resources is typically exploited by a pay-per-use model in which guarantees are offered by the Infrastructure Provider by means of customized SLA's" [10]. The definition refers to a pay-per-use economic model taken from the paradigm of Utility Computing.

Some authors focus on the business model (collaboration and pay-as-you-go) and the reduction in capital expenditure by the realization of Utility Computing. Until recently, it was often confused with the Cloud itself, but it seems now agreed that it is just an element of the Cloud related to the business model. Another major principle for the Cloud is user-friendliness [6]. Buyya et al. [11] added that to reach commercial mainstream it is necessary to strengthen the role of SLAs between the service providers and the consumers of that service. McFedries [12] described the data center (conceived as a huge collection of clusters) as the basic unit of the Cloud offering huge amounts of computing power and storage by using spare resources. The role of virtualization in Clouds is also emphasized by identifying it as a key component [11]. Moreover, Clouds have been defined just as virtualized hardware and software plus the previous monitoring and provisioning technologies [6].

The US National Institute of Standards and Technology (NIST) has published a working definition [13] that seems to have captured the commonly agreed aspects of Cloud Computing. This definition describes Cloud Computing using:

• Five characteristics: on-demand self-service, broad network access, resource pooling, rapid elasticity, and measured service.
• Four deployment models: private Clouds, community Clouds, public Clouds, and hybrid Clouds.
• Three service models: Software as a Service (SaaS), Platform as a Service (PaaS), and Infrastructure as a Service (IaaS).

An encompassing definition of the Cloud taking into account Cloud features has been proposed as follows. Clouds are a large pool of easily usable and accessible virtualized resources (such as hardware, development platforms and/or services). These resources can be dynamically reconfigured to adjust to a variable load (scale), allowing also for an optimum resource utilization. This pool of resources is typically exploited by a pay-per-use model in which guarantees are offered by the Infrastructure Provider by means of customized SLAs. On the other hand, looking for the minimum common denominator would lead us to no definition as no single feature is proposed by all definitions. The set of features that most closely resemble this minimum definition would be scalability, pay-per-use utility model and virtualization [14].

Obviously, the Cloud concept is still changing and these definitions show how the Cloud is conceived today.

2.2 Characteristics

The features of Cloud Computing are that it offers enormous amounts of power in terms of computing and storage while offering improved scalability and elasticity. Moreover, with efficiency and economics of scale, Cloud Computing services are becoming not only a cheaper solution but a much greener one to build and deploy IT services [15].

The Cloud Computing distinguishes itself from other computing paradigms in the following aspects [5]:

- *On-demand service*: Computing Clouds provide resources and services for users on demand. Users can customize and personalize their computing environments later on, for example, software installation, network configuration, as users usually own administrative privileges.
- *QoS guaranteed offer*: The computing environments provided by computing Clouds can guarantee QoS for users, e.g., hardware performance like CPU speed, I/O bandwidth and memory size. The computing Cloud renders QoS in general by processing Service Level Agrement (SLA) with users.
- *Autonomous System*: The computing Cloud is an autonomous system and it is managed transparently to users. Hardware, software and data inside Clouds can be automatically reconfigured, orchestrated and consolidated to present a single platform image, finally rendered to users.
- *Scalability*: The scalability and flexibility are the most important features that drive the emergence of the Cloud Computing. Cloud services and computing platforms offered by computing Clouds could be scaled across various concerns; such as geographical locations, hardware performance, and software configurations. The computing platforms should be flexible to adapt to various requirements of a potentially large number of users.

2.3 Affects

Recently, there has been a great deal of hype about Cloud Computing. Cloud Computing is on the top of Gartner's list of the ten most disruptive technologies of the next years. Since Cloud Computing makes several promises, in terms of time , management and operational costs, enterprises need to understand the affects of Cloud Computing which are focused on the following specific topics [16]:

- *The organizational change brought about with Cloud Computing*: The type of organizational change that Cloud Computing results in can be demonstrated by taking a look at, for example, IT procurement within enterprise. Simplistically, procurement is based on obtaining estimates for things, then getting those estimates signed-off by management to allow the procurement to proceed. Capital and operational budgets are kept separate in this process, and it can take several months between the decision to procure hardware and the hardware being delivered, setup and ready to use. The use of Cloud Computing can greatly reduce this time period, but there is more significant change related to the empowerment of users and the diffusion of the IT department's authority. For example, a company's training coordinator who requires a few servers to run a week-long web-based training course can bypass their IT department and run the training course in the Cloud. They could pay their Cloud usage-bill using their personal credit card and charge back the amount as expenses to their employee.
- *The economic and organizational implications of the utility billing model in Cloud Computing*: New Cloud Computing pricing models based on market mechanisms are starting to emerge, but it is not yet clear how such models can be effectively used by enterprise. An example of such models is used by Amazon's Spot Instances, which allows users to bid for unused capacity in Amazon's data centers. Amazon runs the user's instances as long as the bid price is higher than the spot price, which is set by Amazon based on their data center utilization.
- *The security, legal and privacy issues that Cloud Computing raises*: Security, legal and privacy issues are widely acknowledged as being important in Cloud Computing. Most of the security and privacy issues in Cloud Computing are caused by users' lack of control over the physical infrastructure. This leads to legal issues that are affected by a Cloud's physical location, which determines its jurisdiction. Furthermore, multi-tenancy brings the need for new solutions towards security and privacy. For example, Denial of Service (DoS)[1] attacks were a common concern even before Cloud Computing became popular, but when an application is targeted by a DoS attack in the Cloud, the user or owner could actually end-up paying for the attack through their increased resource usage. This could be significantly higher than the peak usage of that application in an in-house data-center with limited resources.

---

[1] A Denial of Service attack (DoS attack) is an attempt to make a machine or network resource unavailable to its intended users.

2.4 Architecture

Many organizations and researchers have defined the architecture for Cloud Computing. Basically, the whole system can be divided into the core stack and the management. In the core stack, there are three layers: (1) Resource (2) Platform and (3) Application.

The resource layer is the infrastructure layer which is composed of physical and virtualized computing, storage and networking resources.

The platform layer is the most complex part which could be divided into many sub-layers; e.g. a computing framework manages the transaction dispatching and/ or task scheduling. A storage sub-layer provides unlimited storage and caching capability.

The application server and other components support the same general application logic as before with either on-demand capability or flexible management, such that no components will be the bottle neck of the whole system [17].

Based on the underlying resource and components, the application could support large and distributed transactions and management of huge volume of data. All the layers provide external service through web service or other open interfaces. Cloud Architecture is depicted in Fig. 1.

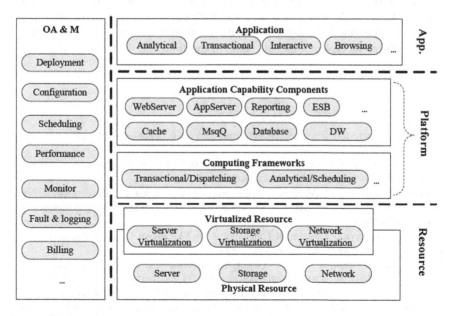

Fig. 1 Cloud architecture

2.5 Technologies behind Cloud Computing

A number of enabling technologies contribute to Cloud Computing, several state-of-the-art techniques are identified [5]:

- *Virtualization* technology: Virtualization technologies partition hardware and thus provide flexible and scalable computing platforms. Virtual machine techniques, such as VMware[2], and Xen[3] offer virtualized IT infrastructures on demand. Virtual network advances, such as Virtual Private Network[4] (VPN), support users with a customized network environment to access Cloud resources. Virtualization techniques are the bases of the Cloud Computing since they render flexible and scalable hardware services.
- *Orchestration of service flow and workflow*: Computing Clouds offer a complete set of service templates on demand, which could be composed by services inside the computing Cloud. Computing Clouds therefore should be able to automatically orchestrate services from different sources and of different types to form a service flow or a workflow transparently and dynamically for users.
- *Web service and SOA*: Computing Cloud services are normally exposed as Web services, which follow the industry standards such as WSDL[5], SOAP[6] and UDDI[7]. The services organization and orchestration inside Clouds could be managed in a Service Oriented Architecture (SOA). A set of Cloud services furthermore could be used in a SOA application environment, thus making them available on various distributed platforms and could be further accessed across the Internet.
- *Web 2.0*: Web 2.0 is an emerging technology describing the innovative trends of using World Wide Web technology and Web design that aims to enhance creativity, information sharing, collaboration and functionality of the Web. Web 2.0 applications typically include some of the following features/techniques:

---

[2] VMware software provides a completely virtualized set of hardware to the guest operating system.

[3] Xen is a Hypervisor providing services that allow multiple computer operating systems to execute on the same computer hardware concurrently.

[4] A virtual private network (VPN) is a private computer network that interconnects remote (and often geographically separate) networks through primarily public communication infrastructures such as the Internet.

[5] WSDL (Web Services Description Language) is an XML-based language that is used for describing the functionality offered by a Web service.

[6] SOAP, originally defined as Simple Object Access Protocol, is a protocol specification for exchanging structured information in the implementation of Web Services in computer networks.

[7] Universal Description, Discovery and Integration (UDDI) is a platform-independent, Extensible Markup Language (XML)-based registry by which businesses worldwide can list themselves on the Internet, and a mechanism to register and locate web service applications.

- Cascading Style Sheets[8] (CSS) to separate of presentation and content
- Folksonomies (collaborative tagging, social classification, indexing and social tagging).
- Semantic Web technologies.
- REST, XML and JSON-based APIs.
- Innovative Web development techniques such as Ajax.
- XHTML and HTML markup.
- Syndication, aggregation and notification of Web data with RSS or Atom feeds.
- Mashups, merging content from different sources, client- and server-side.
- Weblog publishing tools.
- Wiki to support user-generated content.
- Tools to manage users' privacy on the Internet.

The essential idea behind Web 2.0 is to improve the interconnectivity and interactivity of Web applications. The new paradigm to develop and access Web applications enables users access the Web more easily and efficiently. Cloud Computing services in nature are Web applications which render desirable computing services on demand. It is thus a natural technical evolution that the Cloud Computing adopts the Web 2.0 technique.

- *World-wide distributed storage system*: A Cloud storage model should foresee:

  - A network storage system, which is backed by distributed storage providers (e.g., data centers), offers storage capacity for users to lease. The data storage could be migrated, merged, and managed transparently to end users for whatever data formats. Examples are Google File System and Amazon S3.
  - A distributed data system which provides data sources accessed in a semantic way. Users could locate data sources in a large distributed environment by the logical name instead of physical locations. Virtual Data System (VDS) is good reference.

- *Programming model*: Users drive into the computing Cloud with data and applications. Some Cloud programming models should be proposed for users to adapt to the Cloud infrastructure. For the simplicity and easy access of Cloud services, the Cloud programming model, however, should not be too complex or too innovative for end users. The MapReduce is a programming model and an associated implementation for processing and generating large data sets across the Google worldwide infrastructures. The MapReduce and the Hadoop are adopted by recently created international Cloud Computing project of Yahoo!, Intel and HP.

---

[8] CSS is a style sheet language used for describing the presentation semantics (the look and formatting) of a document written in a markup language; most commonly to style web pages written in HTML and XHTML.

2.6 Advantages of the Use of Cloud Technology

The advantages of Cloud Computing is that it offers enormous amounts of power in terms of computing and storage while offering improved scalability and elasticity. Moreover, with efficiency and economics of scale, Cloud Computing services are becoming not only a cheaper solution but a much greener one to build and deploy IT services.

The Cloud Computing distinguishes itself from other computing paradigms in the following aspects [5]:

- *On-demand service provisioning*: Computing Clouds provide resources and services for users on demand. Users can customize and personalize their computing environments later on, for example, software installation, network configuration, as users usually own administrative privileges.
- *QoS guaranteed offer*: The computing environments provided by computing Clouds can guarantee QoS for users, e.g., hardware performance like CPU speed, I/O bandwidth and memory size. The computing Cloud renders QoS in general by processing Service Level Agreement (SLA) with users.
- *Autonomous System*: The computing Cloud is an autonomous system and it is managed transparently to users. Hardware, software and data inside Clouds can be automatically reconfigured, orchestrated and consolidated to present a single platform image, finally rendered to users.
- *Scalability and flexibility*: The scalability and flexibility are the most important features that drive the emergence of the Cloud Computing. Cloud services and computing platforms offered by computing Clouds could be scaled across various concerns, such as geographical locations, hardware performance, and software configurations. The computing platforms should be flexible to adapt to various requirements of a potentially large number of users.

3 Classifying Clouds and Vendors

In this section, different classifications of Clouds are presented, as well as vendors are nominated.

There are diverse dimensions to classify Cloud Computing, two commonly used categories are: service boundary and service type [17].

- From the service boundary's view, Cloud Computing can be classified as public Cloud, private Cloud and hybrid Cloud. The public Cloud refers to services provided to external parties. The enterprises build and operate private Cloud for themselves. Hybrid Cloud shares resources between public Cloud and private Cloud by a secure network. Virtual Private Cloud (VPC) services released by Google and Amazon are examples of Hybrid Cloud.

- From the service type's view, Cloud Computing can be classified as Infrastructure as a Service (IaaS), Platform as a Service (PaaS) and Software as a Service (SaaS). SaaS provide services to end users, while IaaS and PaaS provide services to ISV and developers - leaving a margin for 3rd party application developers.

3.1 Deployment Models

The Cloud model promotes four deployment models:

- *Private Cloud*: The Cloud infrastructure is operated solely for an organization. It may be managed by the organization or a third party and may exist on premise or off premise.
- *Community Cloud*: The Cloud infrastructure is shared by several organizations and supports a specific community that has shared concerns (e.g., mission, security requirements, policy, and compliance considerations). It may be managed by the organizations or a third party and may exist on premise or off premise.
- *Public Cloud*: The Cloud infrastructure is made available to the general public or a large industry group and is owned by an organization selling Cloud services.
- *Hybrid Cloud*: The Cloud infrastructure is a composition of two or more Clouds (private, community, or public) that remain unique entities but are bound together by standardized or proprietary technology that enables data and application portability (e.g., Cloud bursting for load-balancing between Clouds).

3.2 Service Models

Cloud Computing is gaining popularity to the extent that the new XaaS service category introduced will gradually take the place of many types of computational and storage resources used today [18]. Cloud Computing delivers infrastructure, platform, and software (application) as services, which are made available as subscription-based services in a pay-as-you-go model to consumers. These services in industry are respectively referred to as Infrastructure-as-a-Service (IaaS), Platform-as-a-Service (PaaS), and Software-as-a-Service (SaaS) [19]. Table 1 summarizes the nature of these categories and lists some major players in the field [20].

3.2.1 Infrastructure-as-a-Service

Infrastructure-as-a-Service, also called Hardware-as-a-Service was coined possibly in 2006. As the result of rapid advances in hardware virtualization, IT automation and usage metering& pricing, users could buy IT hardware, or even an entire data center, as a pay-as-you-go subscription service [5]. Infrastructure-as-a-Service

Table 1 Cloud computing services classification

Category	Characteristics	Product type	Vendors and products
SaaS	Customers are provided with applications that are accessible anytime and from anywhere	Web applications and services (Web 2.0)	SalesForce.com (CRM) Google documents Clarizen.com (Project management) Google mail (automation)
PaaS	Customers are provided with a platform for developing applications hosted in the Cloud	Programming APIs and frameworks; Deployment system	Google AppEngine Microsoft Azure Manjrasoft Aneka
IaaS/ HaaS	Customers are provided with virtualized hardware and storage on top of which they can build their infrastructure	Virtual machines management infrastructure, Storage management	Amazon EC2 and S3; GoGrid; Nirvanix

(IaaS) or Hardware-as-a-Service (HaaS) solutions deliver IT infrastructure based on virtual or physical resources as a commodity to customers. These resources meet the end user requirements in terms of memory, CPU type and power, storage, and, in most of the cases, operating system as well. Users are billed on a pay-per-use basis. They have to set up their applications on top of these resources that are hosted and managed in data centers owned by the vendor.

Amazon is one of the major players in providing IaaS solutions. Amazon Elastic Compute Cloud (EC2) provides a large computing infrastructure and a service based on hardware virtualization. By using Amazon Web Services, users can create Amazon Machine Images (AMIs) and save them as templates from which multiple instances can be run. It is possible to run either Windows or Linux virtual machines, for which the user is charged per hour for each of the instances running. Amazon also provides storage services with the Amazon Simple Storage Service (S3), users can use Amazon S3 to host large amount of data accessible from anywhere [20].

3.2.2 Platform-as-a-Service

Platform-as-a-Service solutions provide an application or development platform in which users can create their own application that will run on the Cloud. More precisely, they provide an application framework and a set of API that can be used by developers to program or compose applications for the Cloud. PaaS solutions often integrate an IT infrastructure on top of which applications will be executed. This is the case of Google AppEngine and Microsoft Azure, while other solutions, such as Manjrasoft Aneka, are purely PaaS implementations.

Google AppEngine is a platform for developing scalable web applications that run on top of data centers maintained by Google. It defines an application model and provides a set of APIs that allow developers to take advantage of additional services such as Mail, Datastore, Memcache, and others. AppEngine manages the execution of applications and automatically scales them up/down as required. Google provides a free but limited service, while utilizes daily and per minute quotas to meter and price applications requiring a professional service.

Azure is a Cloud service operating system that serves as the development, runtime, and control environment for the Azure Services Platform. By using the Microsoft Azure SDK, developers can create services that leverage the .NET Framework. These services have to be uploaded through the Microsoft Azure portal in order to be executed on top of Windows Azure. Additional services, such as workflow execution and management, web services orchestration, and access to SQL data stores, are provided to build enterprise applications.

Aneka, commercialized by Manjrasoft, is a pure PaaS implementation and provides end users and developers with a platform for developing distributed applications for the Cloud by using .NET technology. The core value of Aneka is a service oriented runtime environment that is deployed on both physical and virtual infrastructures and allows the execution of applications developed by means of various programming models. Aneka provides a Software Development Kit (SDK) helping developers to create applications and a set of tools for setting up and deploying Clouds on Windows and Linux based systems. Aneka does not provide an IT hardware infrastructure to build computing Clouds, but system administrators can easily set up Aneka Clouds by deploying Aneka containers on clusters, data centers, desktop PCs, or even bundled within Amazon Machine Images [20].

3.2.3 Software-as-a-Service

Software or an application is hosted as a service and provided to customers across the Internet. This mode eliminates the need to install and run the application on the customer's local computers. SaaS therefore alleviates the customer's burden of software maintenance, and reduces the expense of software purchases.

Software-as-a-Service solutions are at the top end of the Cloud Computing stack and they provide end users with an integrated service comprising hardware, development platforms, and applications. Users are not allowed to customize the service but get access to a specific application hosted in the Cloud. Examples of SaaS implementations are the services provided by Google for office automation, such as Google Mail, Google Documents, and Google Calendar, which are delivered for free to the Internet users and charged for professional quality services. Examples of commercial solutions are SalesForce.com and Clarizen.com, which provide online CRM (Customer Relationship Management) and project management services, respectively [20].

3.2.4 Data-as-a-Service

Data in various formats and from multiple sources could be accessed via services by users on the network. Users could, for example, manipulate the remote data just like operate on a local disk or access the data in a semantic way in the Internet. Amazon Simple Storage Service (S3) provides a simple Web services interface that can be used to store and retrieve, declared by Amazon, any amount of data, at any time, from anywhere on the Web. The DaaS could also be found at some popular IT services, e.g., Google Docs and Adobe Buzzword. ElasticDrive is a distributed remote storage application which allows users to mount a remote storage resource such as Amazon S3 as a local storage device.

3.3 Cloud Vendors

Commercial Cloud Computing providers offer their services according to three fundamental service models. Table 2 summarizes how these three classes virtualize computation, storage, and networking [8].

Amazon EC2 is at one end of the spectrum. An EC2 instance looks much like physical hardware, and users can control nearly the entire software stack, from the kernel upwards. At the other extreme of the spectrum are application domain-specific platforms; such as Google AppEngine and Force.com, the SalesForce business software development platform. AppEngine is targeted exclusively at traditional web applications, enforcing an application structure of clean separation between a stateless computation tier and a stateful storage tier. Similarly, Force.com is designed to support business applications that run against the salesforce.com database, and nothing else. Microsoft's Azure is an intermediate point on this spectrum of flexibility vs. programmer convenience. Azure applications are written using the .NET libraries, and compiled to the Common Language Runtime, a language independent managed environment.

Aside the commercial Cloud vendors, another implementation of clouds is taking place, especially in the research context. Three main Clouds could be nominated as Open Source Clouds. The first Cloud Architecture, Eucalyptus, is an IaaS system with the aim of creating an open-source infrastructure architected specifically to support Cloud Computing research and infrastructure development. The system combines a Cloud Controller responsible for processing incoming user requests, manipulating the Cloud fabric and processing SLAs in company with a Client Interface that utilizes Internet standard protocols for instance HTTP XML and SOAP. Another Cloud architecture named Enomalism is another IaaS system that presents an organization with the capability to manage virtual infrastructures (including networks), virtual machine images and fine grained security and group management, in addition to the creation of virtual machine images. A third Cloud architecture, named OpenNebula, based on the research being performed by the

Table 2 Cloud computing vendors and how each providers virtualized resources

	Amazon Web Services	Microsoft Azure	Google AppEngine
Computation model (VM)	x86 Instruction Set Architecture (ISA) via Xen VM computation elasticity allows scalability, but developer must build the machinery, or third party VAR such as rightScale must provide it	Microsoft Common Language Runtime (CLR) VM; common intermediate form executed in managed environment machines are provisioned based on declarative descriptions (e.g. which roles can be replicated); automatic load balancing	Predefined application structure and framework; programmer-provided handlers written in Python, all persistent state stored in MegaStore (outside Python code) Automatic scaling up and down of computation and storage; network and server failover; all consistent with 3-tier Web app structure
Storage model	Range of models from block store (EBS) to augmented key/blob store (SimpleDB) Automatic scaling varies from no scaling or sharing (EBS) to fully automatic (SimpleDB, S3), depending on which model used Consistency guarantees vary widely depending on which model used APIs vary from standardized (EBS) to proprietary	SQL data services (restricted view of SQL Server) Azure storage service	MegaStore/BigTable
Networking model	Declarative specification of IPlevel topology; internal placement details concealed Security Groups enable restricting which nodes may communicate Availability zones provide abstraction of independent network failure Elastic IP addresses provide persistently routable network name	Automatic based on programmer's declarative descriptions of app components (roles)	Fixed topology to accommodate 3-tier Web app structure Scaling up and down is automatic and programmer invisible

Reservoir Project, the European research initiative in virtualized infrastructure and Cloud Computing, combines features of IaaS and PaaS in one architecture [19].

4 Cloud Computing Aspects

In this section, customer-related aspects; such as security, service pricing, Service Level Agreements and security; are presented.

4.1 Cost and Pricing

Cloud Computing providers have detailed costing models and metrics that are used to bill users on a pay-per-use basis. This makes it easy for users to see the exact costs of running their applications in the Cloud and it could well be that the design of their system can have a significant effect on its running costs.

From a Consumer Perspective, the three pricing models that are used by Cloud service providers are namely tiered pricing, per-unit pricing and subscription-based pricing. Tiered pricing is where different tiers each with different specifications (e.g. CPU and RAM) are provided at a cost per unit time. An example is Amazon EC2, where a small tier virtual machine has the equivalent of a 1.0 GHz CPU, 1.7 GB of RAM with 160 GB of storage and costs $0.085 per hour at the time of writing, whereas as a large tier machine has the equivalent of four 1.0 GHz CPUs, 7.5 GB of RAM with 850 GB of storage and costs $0.34 per hour 5. Per-unit pricing is where the user pays for their exact resource usage, for example it costs $0.15 per GB per month to store data on Amazon's Simple Storage Service (S3). Subscription-based pricing is common in SaaS products; such as Salesforce's Enterprise Edition CRM that costs $125 per user per month. Elasticity, which is the ability to quickly scale up or down one's resource usage, is an important economic benefit of Cloud Computing as it transfers the costs of resource over-provisioning and the risks of under-provisioning to Cloud providers. An often cited real-world example of elasticity is Animoto.com whose active users grew from 25,000 to 250,000 in three days after they launched their application on Facebook.

From a Provider Perspective, the cost is mainly of building Cloud data centers, which is different from building enterprise data centers. The costs of Cloud data centers can be reduced by looking at the cost of servers, infrastructure, power, and networking. For example, running data centers at hotter temperatures can reduce cooling costs and building micro data centers near users can also reduce bandwidth costs. A tool to model the Total Cost of Ownership (TCO) of setting up and maintaining a Cloud by taking into account the costs of hardware, software, power, cooling, staff and real-estate has been developed. The tool would probably be useful for both large and small Cloud providers as the cost factors are common to both. Calculating the TCO of a Cloud starts by taking the required number of physical servers as an input. Also, a method for calculating a cloud's Utilization Cost has been provided. This allows cloud providers to calculate costs in a similar manner to TCO. However, rather than inputting the number of physical servers into the model, they can start by inputting the maximum number of virtual machines that they need in their cloud. The model would then calculate the required number of physical servers and racks depending on the virtual machine density that is defined by the user (this is the maximum number of virtual machines that can be deployed on a physical server). Modeling the Utilization Cost can be useful for cloud providers as it allows them to see and analyze a detailed breakdown of the total costs of building a cloud [16].

Currently, providers follow a fairly simple pricing scheme to charge users fixed prices based on various resource types. For processing power (as of 15 October 2008), Amazon charges $0.10 per virtual computer instance per hour (h), Sun Microsystems charges $1.00 per processor (CPU) per h , and Tsunamic Technologies charges $0.77 per CPU per h. Charging fixed prices is simple to understand and straightforward for users, but does not differentiate pricing to exploit different user requirements in order to maximize revenue. Moreover, in utility computing it is not fair to both the provider and users since different users have distinctive needs and demand specific QoS for various resource requests that can change anytime. Instead of charging fixed prices for these heavy users, charging variable prices has been advocated providing guaranteed QoS through the use of advanced reservations. Advance reservations are bookings made in advance to secure an available item in the future and are used in the airline, car rental, and hotel industries. In the context of utility computing, an advance reservation is a guarantee of access to a computing resource at a particular time in the future for a particular duration. Hence, implementing autonomic metered pricing for a utility computing service to self-adjust prices to increase revenue becomes important. In particular, autonomic metered pricing can also be straightforward for users through the use of advanced reservations. With advanced reservations, users can not only know the prices of their required resources in the future ahead, but are also able to guarantee access to future resources to better plan and manage their operations [21].

4.2 Service Level Agreement

The Cloud constitutes a single point of access for all services which are available anywhere in the world, on the basis of commercial contracts that guarantee satisfaction of the QoS requirements of customers according to specific Service Level Agreements (SLAs). The SLA is a contract negotiated and agreed between a customer and a service provider. That is, the service provider is required to execute service requests from a customer within negotiated quality of service (QoS) requirements for a given price [1]. The purpose of using SLAs is to define a formal basis for performance and availability the provider guarantees to deliver. SLA contracts record the level of service, specified by several attributes; such as availability, serviceability, performance, operation, billing or even penalties in the case of violation of the SLA. Also, a number of performance-related metrics are frequently used by Internet Service Providers (ISPs); such as service response time, data transfer rate, round-trip time, packet loss ratio and delay variance. Often, providers and customers negotiate utility-based SLAs that determine the cost and penalties based on the achieved performance level. A resource allocation management scheme is usually employed with a view to maximizing overall profit (utility) which includes the revenues and penalties incurred when QoS guarantees are satisfied or violated, respectively. Usually, step-wise utility functions are used where the revenue depends on the QoS levels in a discrete fashion [22].

The service providers and their customers negotiate utility based SLAs to determine costs and penalties based on the achieved performance level. The service provider needs to manage its resources to maximize its profits. Recently, the problem of maximization of SLA revenues in shared service center environments implementing autonomic self-managing techniques has attracted vast attention by the research community. The SLA maximization problem considers (1) the set of servers to be turned ON depending on the system load, (2) the application tiers to servers' assignment, (3) the request volumes at various servers, and (4) the scheduling policy at each server as joint control variables [23].

4.3 Security

Certainly as the Cloud Computing environment matures, security concerns are being addressed more consistently. In October 2008, Google unveiled a new service-level agreement (SLA) that guarantees 99.9 % system accessibility for users of its Google Apps Premier Edition, joining the Amazon S3 SLA that launched in 2007. Other companies are addressing related angles of security. For example, Boulder, Colo.-based Symplified offers a unified access management system built for the Cloud architectures of SaaS. The company's Identity-as-a-Service approach provides on-demand features; such as single sign-on, provisioning, de-provisioning, auditing and compliance, access and authorization, identity synchronization, strong authentication, and identity integration [7].

One of the most serious concerns is the possibility of confidentiality violations. Either maliciously or accidentally, Cloud provider's employees can tamper with or leak a company's data. Such actions can severely damage the reputation or finances of a company. In order to prevent confidentiality violations, Cloud services' customers might resort to encryption. While encryption is effective in securing data before it is stored at the provider, it cannot be applied in services where data is to be computed, since the unencrypted data must reside in the memory of the host running the computation [14].

The Cloud also offers security advantages. For example, intruders have no access to the source code, and providers often work hard to provide clean, unbreakable barriers between customers. Security can differ greatly from application to application, from platform to platform, and from provider to provider, however. Yet on the whole, the Cloud holds much promise for better security.

Beyond generic concerns regarding the Cloud approach, each of the three models has its own security concerns.

With SaaS, users must rely heavily on their Cloud providers for security. The provider must do the work to keep multiple companies or users from seeing each other's data without permission. In addition, the provider must protect the underlying infrastructure from break-ins and generally has responsibility for all authentication and encryption. It's difficult for the customer to get the details that help ensure that the right things are being done. Similarly, it's difficult to get assurance that the application will be available when needed.

With PaaS, the provider might give some control to the people building applications atop its platform. For example, developers might be able to craft their own authentication systems and data encryption, but any security below the application level will still be completely in the provider's hands. Usually, the provider will offer little or no visibility into its practices. Plus, the platform provider must be able to offer strong assurances that the data remains inaccessible between applications. Large banks don't want to run a program delivered in the Cloud those risks compromising their data through interaction with some other program.

With IaaS, the developer has much better control over the security environment, primarily because applications run on virtual machines separated from other virtual machines running on the same physical machine, as long as there is no gaping security hole in the virtual machine manager. This control makes it easier to ensure that developers properly address security and compliance concerns, with the downside that building the application can be more expensive and time-consuming.

Backing up data poses another concern. Even though some providers do their own backups for the customer, much can still go wrong. Maybe they increase their prices and make it difficult to get data off their network [24].

5 Cloud Computing and Distributed Computing Paradigms

The emergence of Cloud Computing continues the natural evolution of Distributed Systems to cater for changes in application domains and system requirements [19].

Cloud Computing builds on top of several other technologies, i.e. Distributed Computing, Grid Computing, Utility Computing and Autonomic Computing, and it can be envisaged as a natural step forward from the Grid-utility model. In the heart of Cloud Computing infrastructure, a group of reliable services are delivered through powerful data computing centers that are based on modern virtualization technologies and related concepts such as component-based system engineering, orchestration of different services through workflows and Service-Oriented Architectures (SOAs) [22].

Cloud Computing is based on the use of distributed computing resources that are easily allocated, de-allocated, migrated and possibly re-allocated on user request. As such, it relies heavily on the use of virtualization technologies, able to offer an almost unlimited amount of virtual computing resources. Virtualization controls the access to physical resources in a transparent way, it is possible to offer computational resources with full control, in that final users can configure them as administrators, without any restriction [25].

5.1 Utility Computing Versus Cloud Computing

Cloud services should be accessed with simple and pervasive methods. In fact, the Cloud Computing adopts the concept of Utility Computing. In other words, users obtain and employ computing platforms in computing Clouds as easily as they access a traditional public utility (such as electricity, water, natural gas, or telephone network) [5]. In detail, the Cloud services enjoy the following features:

- The Cloud interfaces do not force users to change their working habits and environments, e.g., programming language, compiler and operating system. This feature differentiates Cloud Computing from Grid computing as Grid users have to learn new Grid commands and APIs to access Grid resources and services.
- The Cloud client software which is required to be installed locally is light-weight. For example, the Nimbus Cloudkit client size is around 15 MB.
- Cloud interfaces are location independent and can be accessed by some well-established interfaces like; Web services framework and Internet browser.

5.2 Volunteer Computing Versus Cloud Computing

Both Cloud and volunteer computing have similar principles, such as transparency. On both platforms, one submits tasks without needing to know the exact resource on which it will execute. However, in practice, the Cloud Computing infrastructures differ from volunteer computing platforms throughout the hardware and software stack. From the perspective of the user, there are two main differences, namely configurability (and thus homogeneity), and quality-of-service. Clouds present a configurable environment in terms of the OS and software stack with the Xen virtual machine forming the basis of EC2. So, while Clouds can offer a homogeneous resource pool, the heterogeneity of VC hardware and operating system is not transparent to VC application developers.

Clouds also provide higher quality-of-service than VC systems. Cloud resources appear dedicated, and there is no risk of preemption. Many Cloud Computing platforms, such as Amazon's EC2, report several nine's in terms of reliability. Cloud infrastructures consist of large-scale centralized compute servers with network-attached storage at several international locations. The infrastructures are accessed through services such as S3 also provide high-level web services for data management [26].

5.3 Grid Computing Versus Cloud Computing

Cloud Computing, born in the e-business context, and Grid computing, originated in the e-science context, are two different but similar paradigms for managing large sets of distributed computing resources [25].

Grid computing can be seen as one of several predecessors to Cloud Computing. Grid computing is often about making large computations using large amounts of resources, while as Cloud Computing is more about making large amounts of resources available to many different applications over a longer period of time. Clouds leverage modern technologies such as virtualization to provide the infrastructure needed to deploy services as utilities. Still, Cloud Computing and Grid computing share a lot of the underlying technology and many concepts from Grid computing can be modified and made suitable for Cloud Computing as well [27].

A comparison between Grid and Cloud is shown hereunder in the following aspects [5]:

- *Definition*: Grid computing, oriented from high performance distributed computing, aims to share distributed computing resource for remote job execution and for large scale problem solving. The Grid computing emphasizes the resource side by making huge efforts to build an independent and complete distributed system. Cloud Computing provides user-centric functionalities and services for users to build customized computing environments. Cloud Computing, which is oriented towards the industry, follows an application-driven model.
- *Infrastructure*: Grid infrastructure enjoys the following features. Grid infrastructure in nature is a decentralized system, which spans across geographically distributed sites and lack central control. Grid infrastructure normally contains heterogeneous resources, such as hardware/software configurations, access interfaces and management policies. On the contrary, from the viewpoint of users, computing Clouds operate like a central compute server with single access point. Cloud infrastructures could span several computing centers, like Google and Amazon, and in general contain homogeneous resources, operated under central control.

- *Middleware*: Grid computing has brought up full-fledged middleware, for example, Unicore, Globus Toolkit and gLite. The Grid middleware could offer basic functionalities like resource management, security control and monitoring & discovery. The Grid community has established well-defined industry standards for Grid middleware, e.g.,WSRF. The middleware for Cloud Computing, or the Cloud operating system, is still underdeveloped and lacks of standards. A number of research issues remain unsolved, for example, distributed virtual machine management, Cloud service orchestration, and distributed storage management.
- *Accessibility and application*: Grid computing has the ambitious objective to offer dependable, consistent, pervasive, and inexpensive access to high-end computational capabilities. However inexperienced users still find difficulties to adapt their applications to Grid computing. Furthermore, it is not easy to get a performance guarantee from computational Grids. Cloud Computing, on the contrary, could offer customized, scalable and QoS guaranteed computing environments for users with an easy and pervasive access. Grid computing has gained numerous successful stories in many application fields. A recent famous example is LCG project to process huge data generated from the Large Hadron Collider (LHC) at CERN. Although Amazon EC2 has obtained a beautiful success, more killer applications are still needed to justify the Cloud Computing paradigm.

Based on the fore mentioned discussion, it could be concluded that Grid computing has established well organized infrastructures, middleware and application experience. Cloud Computing recently outpaces Grid computing because it could offer computing environments with desirable features, such as QoS guarantee and computing environment customization. It would be a good solution to build Cloud Computing services on Grid infrastructure and middleware, thus providing users with satisfying services.

Table 3 compares different features of Grid and Cloud highlighting the similarities and differences between both paradigms.

In fact, Cloud and Grid computing paradigms have many points in common: both adopt large datacenters, both offer resources to users, both aim at offering a common environment for distributed resources. At the state of the art, there are two main approaches for their integration:

- *Grid on Cloud*: a Cloud approach (i.e. adoption of virtual machines) is adopted in order to build up and to manage a flexible Grid system. As in this context the Grid middleware runs on a virtual machine, the main drawback of this approach is performance. Virtualization inevitably entails performance losses as compared to the direct use of physical resources.
- *Cloud on Grid*: the complex and stable Grid infrastructure is exploited to build up a Cloud environment. This is a more common solution. In this case, a set of Grid services is offered in order to manage virtual machines. The use of Globus workspaces, a set of Grid services for the Globus Toolkit 4, is the prominent

Table 3 Grid versus gloud characterstics

Feature	Grid	Cloud
Resource Sharing	Collaboration (VOs, fair share)	Assigned resources are not shared
Resource heterogeneity	Aggregation of heterogeneous resources	Aggregation of heterogeneous resources
Virtualization	Virtualization of data and computing resources	Virtualization of hardware and software platforms
Security	Security through credential delegations	Security through isolation
High level services	Plenty of high level services	No high level services defined yet
Architecture	Service orientated	User chosen architecture
Software dependencies	Application domain-dependent software	Application domain-independent software
Platform awareness	The client software must be grid-enabled	The SP software works on a customized environment
Software workflow	Applications require a predefined workflow of services	Workflow is not essential for most applications
Scalability	Nodes and sites scalability	Nodes, sites, and hardware scalability
Self-management	Re-configurability	Re-configurability, self-healing
Centralization degree	Decentralized control	Centralized control (until now)
Usability	Hard to manage	User friendliness
Standardization	Standardization and interoperability	Lack of standards for Clouds interoperability
User access	Access transparency for the end user	Access transparency for the end user
Payment Model	Rigid	Flexible
QoS Guarantees	Limited support, often best-effort only	Limited support, focused on availability and uptime

solution. It is also adopted in the Nimbus project, where a Cloud environment is built on top of a Grid.

Both Grid Computing and Cloud Computing implement the more general model of Utility Computing. However, the motivation and approaches to realize an infrastructure that serves compute utilities are considerably different. Grid Computing evolved in a scientific community as a means to solve computationally challenging tasks in science and industry. Although Clouds can also be used for large-scale batch jobs, they focus on very different applications. Emerging in a commercial environment, the main purpose of Cloud Computing is to provide a platform to develop, test and deploy Web-scale applications and services [9]. The motivation behind research into Grid Computing was initially the need to manage large scale resource intensive scientific applications across multiple administrative domains that require many more resources than that can be provided by a single computer. Cloud Computing shares this motivation, but within a new context oriented towards business rather than academic resource management, for the stipulation of reliable services rather than batch oriented scientific applications [19].

6 The Road Ahead

Despite the fact that Cloud Computing promises to bring sweeping changes to the way of using information technology, decisions-makers are still struggling for insights on the challenges associated with trying to exploit it.

In this section, we present the lessons learned so far, discuss the challenges that should be addressed in future research directions.

6.1 Lessons Learned in Cloud Adoption

No two organizations had adopted Cloud Computing in exactly the same way or over the same timeframe. So far, the journeys followed were determined by history and accidents (experiments, unplanned trials, recommendations from peers) as well as variations in the structure, industrial context and computing needs of individual organizations. Two main themes, however, did emerge as important, being broadly applicable to multiple cases. In one of these the journey was initiated by the adoption of a Cloud-based service for a single key enterprise application. Good results then led to the adoption of more Cloud-based software development to extend the functionality of standard applications. In the other, the journey began with efforts to improve the capacity utilization of computing resources inside the organization. Servers and computer storage were considered around fewer, larger facilities, and some software applications were virtualized, so they ran anywhere in these facilities rather than on dedicated machines. Good results saw this expanded into an organization-wide strategy, followed by a shift to service providers outside the organization.

There are also many other variations; in one case the journey began with adoption of open source software (installed on premises). The associated shift from up-front software purchasing to pay-as-you-go maintenance led to a broader interest in paying for computer resources using utility-based pricing. In another case, the adoption of web-based email and collaboration tools started the process. In third case, the journey began with a desire to remove computing equipment from hostile work environments. As with the journeys taken, there was a wide variation in the extent of adoption.

6.2 Challenges and Future Research Directions

With the flourish of the Cloud Computing and the wide availability of both pro-viders and tools, a significant number of challenges are facing such new paradigm. These challenges are the key points to be considered for future research directions. Challenges to be faced include user privacy, data security, data lock-in, avail-ability, disaster recovery, performance, scalability, energy-efficiency, and pro-grammability. Some of these challenges are further described as follows:

- *Security, Privacy, and Trust*: With the movement of the clients' data into the Cloud, providers may choose to locate them anywhere on the planet. The physical location of data centers determines the set of laws that can be applied to the management of data [4]. Security and privacy affect the entire Cloud Computing stack, since there is a massive use of third-party services and infrastructures that are used to host important data and to perform critical operations. In such scenarios, trusting the provider is a fundamental issue to ensure the desired level of privacy [28]. Accordingly, it becomes crucial to give deep attention to such legal and regulatory perspectives.
- *Data Lock-Inand Standardization*: A major concern of Cloud Computing users is about having their data locked-in by a certain provider. Users may want to move data and applications out from a provider that does not meet their requirements. However, in their current form, Cloud Computing infrastructures and platforms do not employ standard methods of storing users' data and applications. Hence, the interoperate option is not yet supported and users' data are not portable. This is why a lot of work is requested in terms of standardi-zation and supporting interoperation option.
- *Availability, Fault-Tolerance, and Disaster Recovery*: It is expected that users will have certain expectations about the service level to be provided once their

applications are moved to the Cloud. These expectations include availability of the service, its overall performance, and what measures are to be taken when something goes wrong in the system or its components. Generally, users seek for a warranty before they can comfortably move their business to the cloud that is Service Level Agreement (SLA). SLAs, which include QoS requirements, must be ideally settled to cover many issues such as the service availability, the fault-tolerance, and the disaster recovery. This is why this topic is still hot and in need for a lot of effort to be professionally covered.

- *Resource Managementand Energy-Efficiency*: One important challenge faced by providers of Cloud Computing services is the efficient management of virtualized resource pools. Physical resources such as CPU cores, disk space, and network bandwidth must be sliced and shared among virtual machines running potentially heterogeneous workloads [29]. In addition to optimize application performance, dynamic resource management can also improve utilization and consequently minimize energy consumption in data centers. This can be done by judiciously consolidating workload onto smaller number of servers and turning off idle resources. Researches in this area are endless and still in need for more.

7 Conclusion

This chapter presents an overview on the Cloud Computing paradigm. First, the Cloud Computing anatomy has been described then it has been defined from different perspectives; such as the service and business models as well as its economical and quality of service aspects. The Cloud Computing distinguishes itself from other computing paradigms, by its promising features including on-demand, flexible and QoS guaranteed provisioned service. The affects of Cloud Computing on organizations and enterprises were also addressed in terms of time, management and operational costs. The architecture of Clouds has been discussed; differentiating between the Cloud core stack and the management. The first has three layers in the core stack: resource, platform and application layers; while the later includes managerial tasks such as billing and scheduling. Enabling technologies contributing to Cloud Computing; such as virtualization and web 2.0, web services and SOA, were also clarified.

Second, the Cloud Computing classification is addressed, clearly differentiating between deployment and service models apart, and beneath each of them. That is Clouds are classified according to their deployment models as private, community, public and hybrid Clouds. Clouds are also classified according to their offered service models; infrastructure, platform and software as services. Examples of the Cloud providers and vendors are listed such as Amazon Web Services, Microsoft Azure and Google AppEngine provide the above mentioned services respectively.

Next, the customer-related aspects were discussed including the pricing and costing models with metrics that are used to bill users on a pay-per-use basis. The Service Level Agreement, which is a contract between the customer and the service provider, was also addressed as well as the data security perspectives.

Furthermore, a comprehensive comparison between Cloud Computing and utility, volunteer and grid paradigms has been conducted; in terms of architecture, resources sharing, QoS guarantees and security. The analysis of the Cloud reports that such technology is based on the use of distributed computing resources that are easily allocated, de-allocated, migrated and possibly re-allocated based on the user request.

Finally, to guarantee the long-term success of Cloud Computing, the chapter tackles some significant challenges that face the Cloud paradigm. Challenges that need to be carefully addressed for future research like; user privacy, data security, data lock-in, availability, disaster recovery, performance, scalability, energy-efficiency, and programmability.

By that, this chapter has provided a comprehensive overview on the newly emerged Cloud Computing paradigm and paved the way for further research in this area.

References

1. K. Xiong, H. Perros, Service performance and analysis in cloud computing, in *Proceedings of Congress on Services - I (SERVICES 09)*. IEEE Computer Society Washington 2009, 693–700
2. Y. Sun, J. White, J. Gray, A. Gokhale, D. Schmidt, Model-Driven Automated Error Recovery in Cloud Computing, in *Model-driven Analysis and Software Development: Architectures and Functions*, ed. by J. Osis, E. Asnina (IGI Global, Hershey, PA, USA, 2011), pp. 136–155
3. R. Buyya, C.S. Yeo, S. Venugopal, J. Broberg, I. Brandic, Cloud Computing and emerging IT platforms: Vision, hype, and reality for delivering computing as the 5th utility. Future Gener. Comput. Syst. **25**, 599–616 (2009)
4. W. Voorsluys, J. Broberg, and R. Buyya, Introduction to Cloud Computing, in Cloud Computing: Principles and Paradigms, ed. by R. Buyya, J. Broberg, A.Goscinski (New York: Wiley, 2011), pp. 1–41
5. L. Wang, G. Laszewski, Scientific cloud computing: Early definition and experience, in Proceedings of 10th IEEE International Conference on High Performance Computing and Communications (Dalian, China, 2008), pp. 825–830
6. J. Geelan, Twenty one experts define cloud computing. Virtualization Mag. (2010)
7. N. Kho, Content in the cloud. EContent Mag. (2009)
8. M. Armbrust, A. Fox, R. Griffith, A. Joseph, R. Katz, A. Konwinski, G. Lee, D. Patterson, A. Rabkin, I. Stoica, and M. Zaharia, Above the Clouds: A Berkeley View of Cloud Computing, University of California at Berkley, USA, Technical Rep UCB/EECS-2009-28, (2009)
9. M. Klems, J. Nimis, and S. Tai, Do Clouds Compute? A Framework for Estimating the Value of Cloud Computing", in Designing E-Business Systems: Markets, Services, and Networks, vol. 22, C. Weinhardt, S. Luckner, and J. Stober, Eds. Heidelberg: (Springer Berlin, 2009), part 4, pp. 110–123

10. A. Tikotekar, G. Vallee, T. Naughton, H. Ong, C. Engelmann, S. Scott, A. Filippi, *Effects of virtualization on a scientific application running a hyperspectral radiative transfer code on virtual machines, in Proceedings of 2nd Workshop on System-Level Virtualization for High Performance Computing (HPCVirt 08)* (Glasgow, UK, 2008), pp. 16–23
11. R. Buyya, C. Yeo, S. Venugopal, *Market-oriented cloud computing: Vision, hype, and reality for delivering it services as computing utilities, in Proceedings of 10th IEEE International Conference on High Performance Computing and Communications (HPCC 08)* (Dalian, China, 2008), pp. 5–13
12. P. McFedries, The cloud is the computer. IEEE Spectrum **45**(8), 20–20 (2008)
13. P. Mell, T. Grance, *The NIST Definition of Cloud Computing* (National Institute of Standards and Technology, Information technology laboratory, 2009)
14. N. Santos, K. Gummadi, R. Rodrigues, *Towards trusted cloud computing, in Proceedings of conference on Hot topics in Cloud Computing (HOTCLOUD 09)* (San Diego, USA, 2009)
15. J. Grimes, P. Jaeger, J. Lin, *Weathering the storm: The policy implications of cloud computing, in Proceedings of iConference, University of North Carolina* (Chapel Hill, USA, 2009)
16. A. Khajeh-Hosseini, I. Sommerville, and I. Sriram, Research challenges for enterprise cloud computing, CoRR, abs/1001.3257, 2010
17. L. Qian, Z. Luo, Y. Du, L. Guo, *Cloud computing: An overview, in Proceedings of 1st International Conference on Cloud Computing* (Beijing, China, 2009), pp. 626–631
18. F. Aymerich, G. Fenu, S. Surcis, *A real time financial system based on grid and cloud computing, in Proceedings of ACM symposium on Applied Computing* (Honolulu, Hawaii, USA, 2009), pp. 1219–1220
19. R. Buyya, R. Ranjan, R. Calheiros, *InterCloud: Utility-oriented federation of cloud computing environments for scaling of application services, in Proceedings of 10th International Conference on Algorithms and Architectures for Parallel Processing (ICA3PP* (Busan, Korea, 2010), pp. 13–31
20. R. Buyya, S. Pandey, C. Vecchiola, *Cloudbus toolkit for market-oriented cloud computing, in Proceedings of 1st International Conference on Cloud Computing (CLOUDCOM 09)* (Beijing, China, 2009), pp. 24–44
21. C. Yeoa, S. Venugopalb, X. Chua, R. Buyya, Autonomic metered pricing for a utility computing service. J. Future Gener. Comput. Syst. **26**(8), 1368–1380 (2010)
22. C. Yfoulis, A. Gounaris, *Honoring SLAs on cloud computing services: a control perspective, in Proceedings of European Control Conference* (Budapest, Hungary, 2009), pp. 184–189
23. D. Ardagna, M. Trubianb, L. Zhangc, SLA based resource allocation policies in autonomic environments. J. Parallel Distrib. Comput. **67**(3), 259–270 (2007)
24. J. Viega, Cloud Comput. Common Man. J. Comput. 42(8), 106–108 (2009). IEEE Computer Society
25. E. Mancini, M. Rak, U. Villano, PerfCloud: GRID services for performance-oriented development of cloud computing applications", in Proceedings 18th IEEE International Workshops on Enabling Technologies: Infrastructures for Collaborative Enterprises (WETICE 09). (Groningen, The Netherlands, 2009), 201–206 (2009)
26. D. Kondo, B. Javadi, P. Malecot, F. Cappello, amd D. Anderson, Cost-benefit analysis of cloud computing versus desktop grids, in Proceedings of IEEE International Symposium on Parallel and Distributed Processing (IPDPS 09), IEEE Computer Society Washington, 2009, pp. 1–12
27. E. Elmroth, F. M'arquezy, D. Henriksson, D. Ferrera, *Accounting and billing for federated cloud infrastructures, in Proceedings of Eighth International Conference on Grid and Cooperative Computing (GCC 09)* (Lanzhou, China, 2009), pp. 268–275
28. R. Buyya, S. Pandey, C. Vecchiola, Cloudbus toolkit for market-oriented cloud computing, in Proceedings of 1st International Conference on Cloud Computing (CLOUDCOM 09), (Beijing, 2009), pp. 24–44

29. W. Voorsluys, J. Broberg, S. Venugopal, and R. Buyya, Cost of virtual machine live migration in clouds: A performance evaluation, in Proceedings of 1st International Conference on Cloud Computing, (Beijing, 2009), pp. 254–265
30. L. Vaquero, L. Rodero-Merino, J. Caceres, M. Lindner, A break in the clouds: Towards a cloud definition. ACM SIGCOMM Comput. Commun. Rev. 9(1), 50–55 (2009)
31. D. Armstrong, and K. Djemame, Towards quality of service in the cloud, in Proceedings of 25th UK Performance Engineering Workshop, (Leeds, 2009)

A Comparison of Two Different Approaches to Cloud Monitoring

Beniamino Di Martino, Salvatore Venticinque, Dimosthenis Kyriazis and Spyridon V. Gogouvitis

Abstract Monitoring is a relevant issues above all for customers of Public Clouds. In fact they need to detect under-utilization, overload conditions, and check SLA fulfillment. However they cannot trust twice the provider for the SLA and for its checking, and have limited knowledge about the infrastructure to understand the reasons of real bottlenecks. They can only access the virtual resources. In this Chapter we provide a comparison of two different approaches to the monitoring of Clouds Infrastructure, developed within the research activities of two European FP7-ICT projects. The first one has been developed by the VISION Cloud project and it is based on asynchronous message delivery mechanism for the collection, propagation and delivery of all events to their respective recipients. The second one is based on , the mobile agents framework of the mOSAIC project, for provisioning and managements of heterogeneous Cloud resources.

1 Introduction

Cloud was born as an emerging computing paradigm that aims at turning the computing resources into any other utilities, which are sold and delivered as a service by the Internet. For a business exploitation of such computer utility the best

B. Di Martino · S. Venticinque (✉)
Department of Industrial and Information Engineering, via roma 29 81031 Aversa, Italy
e-mail: salvatore.venticinque@unina2.it

B. Di Martino
e-mail: beniamino.dimartino@unina.it

D. Kyriazis · S. V. Gogouvitis
National Technical University of Athens, Iroon Polytechniou 9, 15773 Athens, Greece
e-mail: dimos@mail.ntua.gr

S. V. Gogouvitis
e-mail: spyrosg@mail.ntua.gr

F. Xhafa and N. Bessis (eds.), *Inter-cooperative Collective Intelligence:*
Techniques and Applications, Studies in Computational Intelligence 495,
DOI: 10.1007/978-3-642-35016-0_3, © Springer-Verlag Berlin Heidelberg 2014

effort policy cannot be used and a more suitable service provisioning, based on a Service Level Agreement (SLA), is required. In this context the monitoring that is offered as a service by the provider itself is not enough [1]. The user cannot trust twice the provider for the SLA and for its checking. On the other hand some parameters depend by the application and can be monitored only by instrumenting the application itself. That is why a relevant problem to be addressed concerns how and by whom monitoring should be performed. Monitoring and tools are available for real and virtualized resources, but different scenarios have to be dealt with. In private Cloud all is available to the user, but in a public Cloud customers can only access the virtual resources, meanwhile the provider manages hardware and hypervisor. This means that monitoring information is accessible only to providers for administration purposes, and eventually shared to the customers.

However Cloud customers cannot check the compliance of the Service Level Agreement (SLA) trusting the monitoring service of their own same provider. In fact the Cloud provider has a conflicting interest ensuring the guarantees on service levels it provides. GoGrid SLA [2] is very detailed if compared with others, but its customers are obliged to trust GoGrid also for the monitoring: *individual servers will deliver 100% uptime as monitored within the GoGrid network by GoGrid monitoring systems (Server Uptime Guarantee). Only failures due to known Go-Grid problems in the hardware and hypervisor layers delivering individual servers constitute Failures and so only those Failures are covered by this Server Uptime Guarantee and this SLA.*

Besides Cloud customers need to detect under-utilization and overload conditions, to take decisions about load balancing and resource reconfiguration. In both the cases it is necessary to dimension the Cloud resource to avoid useless expenses and to not fail to satisfy the service requirements when workloads change dynamically. In this chapter we introduce general requirements for monitoring of IaaS Clouds and present how they have been addressed by the research activities of two relevant European project: VISION Cloud[1] and mOSAIC[2]. In the next section we present what and why it needs to monitor Clouds at infrastructure level. In Sect. 3 related work is discussed. mOSAIC and VISION projects are introduced in Sect. 4. In the rest of the chapter we focus on the monitoring solution developed within these projects. Section 5 refers to the VISION platform an asynchronous message delivery mechanism for the collection, propagation and delivery of all events to their respective recipients. Section 6 describes the monitoring services. Finally conclusions are due.

---

[1] http://www.visionCloud.eu

[2] http://www.mosaic-Cloud.eu

2 Monitoring IaaS Clouds

Different requirements were identified for Cloud applications, both data- and computing-intensive applications. These developed usage patterns are related with the *runtime maintenance use cases*, as described in [3], and cover different use cases like 'monitoring of sensor data and sensor diagnosis', 'monitoring the status of an intelligent maintenance system', 'monitoring simulations in a model exploration system', or 'resource monitoring at infrastructure level [4, 5] and for security purposes [6].

Performance, availability and utilization of services monitoring are interesting both for governance and management level, and fit with the specific requirements following first three use cases mentioned above. However monitoring at service level becomes critical because of the potential conflicts that might occur between provider and customers in case of outage, for example. It will be necessary to grant that vendor and consumer meet the conditions of the achieved agreement, which can have financial or legal consequences, with impact at governance level.

As specified in Cloud Computing Use Cases whitepaper [7], "*all Cloud services must be metered and monitored for cost control, chargebacks and provisioning*. On provider side monitoring is necessary for effective utilization of Computing resources, to check service levels, which have been agreed with the customers, to lead negotiation strategies, such as overbooking, in order to maximize the total incoming. At this level the monitoring addresses the full range of resources, both hardware and virtual.

Different requirements are based on consumers' needs, like monitoring their own resources for checking that service level agreements are continuously respected and for detecting under-utilization or overloading conditions of their resources in order to renegotiate the agreement if it is necessary.

Monitoring is required at each Cloud deployment level, including the infrastructure level (IaaS), the platform level (PaaS) and the application-level (SaaS).

IaaS delivers a computing infrastructure over the Internet and is enabled to split, assign and dynamically resize these resources to build custom infrastructures, just as demanded by customers.

Cloud IaaS SLAs are similar to SLAs for network services, hosting and data center outsourcing. The main issues at IaaS concerns the mapping of high level application requirements on monitored parameters which measure low level service levels.

For IaaS there are many standardization efforts like OCCI [3] (Open Cloud Computing Interface), SOCCI[4] (Service-Oriented Cloud-Computing Infrastructure) by The Open Group, the ISO Study Group on Cloud Computing[5] (SGCC),

---

[3] http://occi-wg.org/

[4] https://collaboration.opengroup.org/projects/soa-soi/

[5] http://www.iso.org/iso/jtc1_sc38_home

Table 1 Different monitored resources at infrastructure (IaaS) level

Service	Monitored Parameters and Performance Indexes
Compute	Cumulative Availability (connectivity uptime). *E.g. 99.5%, 99%*
	Role instance uptime. *E.g.* 99.9 %, 99 %
	Outage length. *E.g. Monthly time*
	Server reboot time. *E.g.* 15 minutes
	Availability. *E.g.* 100 %
Network	Packet loss. *E.g.* 0.1 % per month
	Latency within the Cloud network or within the same VLAN. *E.g. 1 ms*
	Mean jitter, maximum jitter. *E.g. 0.5 ms within any 15 minute period*
	Bandwidth. *E.g. Mb/s, Gb/s*
	IOPS. *E.g. I/O per second.*
Storage	Latency with the internal compute resource. *E.g.* < 1 ms
	Bandwidth with the internal compute resource. *E.g. Gb/s*
	Processing time (it does not include transferring). *E.g. Blob copy*< 60 s,
	time to process data< 2 s x amount of MB,
	list tables returns or continues within 10 s
	Maximum restore time. *E.g. Number of minutes*
	Uptime. *E.g.* If< 99.9% refund 10% credits,
	Uptime. *If* < 99% refund 25% credits

and there is a de facto industry alignment on IaaS service level objectives, warranties, guarantees, performance metrics, etc.

Table 1 captures the most important monitored resources at infrastructure (IaaS) level and associated monitored parameters, together with their performance indexes.

Compute, network and storage are the main abstractions of services at this level, as captured in Table 1. Different monitored parameters for compute, like cumulative availability, outage length, server reboot time, are used to check the service level objectives. Usually there are no service level agreements established for compute performance: customers are simply guaranteed to have the compute for which they paid, with technical enforcement at the hypervisor level. However performance information at hypervisor level are always available to the vendors and sometimes to the customers.

As providers usually offer a network SLA, some support for monitoring information of the Cloud provider's data center connectivity to the Internet is provided. This information can be monitored and provided also for internal communication between different resources (devices) within provider's data center or, if there are multiple data centers, between those data centers. Values of key performance indicators are usually related to geographical locations.

3 Related Work

The increasing use of Cloud Computing to manage applications, services and resources puts even more emphasis on the necessity of monitoring the QoS parameters in order to have the desired performances [8]. Furthermore the resources control is very important when a customer wants to check the compliance of the provider offered services with the ones that he/she has signed into the SLA. Infrastructure-level resource monitoring [8] aims at the measurement and reporting of system parameters related to real or virtual infrastructure services offered to the user (e.g. CPU, RAM, or data storage parameters). Traditional monitoring technologies for single machines or Clusters are restricted to locality and homogeneity of monitored objects and, therefore, cannot be applied in the Cloud in an appropriate manner [9].

Many available tools need to run on the physical machine and use facilities of the hypervisor. In a Private Cloud they can be used for management purpose, but in a Public Cloud they are controlled by the Cloud vendor. Cloud vendors share a subset of information to the customers, monitoring services and supporting different performance measurements, depending on the monitoring technology. At the state of the art there are many tools which provide Cloud monitoring facilities, like Cloudkick,[6] Nimsoft Monitor,[7] Monitis,[8] Opnet,[9] RevealCloud.[10] All of them are proprietary solutions and do not aim at defining a standard for monitoring interoperability .

The Open Grid Forum has proposed Open Cloud Computing Interface (OCCI [10]) as a standard for the Cloud infrastructures level. In particular, an extension of OCCI to support mechanisms for Monitoring and SLA agreement negotiation is described in [11].

As resource usage and the service workload can change very frequently, a continuous and updated answer must be provided to the user by efficiently monitoring the environment (e.g., a change in the answer must be detected as soon as possible while minimizing the communication cost). Furthermore the Cloud elasticity allows for the possibility to change dynamically over time the system configuration, and so the distributed monitoring must adapt itself quickly to the new requirements. That is why we propose an agents based monitoring service. In [12] authors claims that an approach based on software agents is a natural way to tackle the monitoring tasks in the aforementioned distributed environments. Agents move and distribute themselves to perform their monitoring tasks. In [13] an optimal control of mobile monitoring agents in artificial-immune-system-based (AIS-based) monitoring networks has been studied. We integrate here an agents based

---

[6] https://www.Cloudkick.com/

[7] http://www.nimsoft.com/solutions/nimsoft-monitor

[8] http://portal.monitis.com

[9] http://www.opnet.com/

[10] http://copperegg.com/

monitoring architecture with some technologies for monitoring network and host like s-flow, which have been extended to support the transport of monitoring information of virtualized resources. For example host-sflow[11] exports physical and virtual server performance metrics using the sFlow protocol. Ganglia [14] and other collectors of performance measures are already compliant with its specification.

General monitoring tools also include Nagios [15], an open source network monitoring system used to monitor services and hosts, and MonALISA [16], a Jini-based monitoring framework. These approaches can be considered as metering tools that do not provide a lot in the direction of advanced monitoring functionality.

In the field of Grid computing, the Globus Toolkit Monitoring and Discovery System (MDS4) [17] provides mechanisms for monitoring and discovering resources and services. MDS lacks metering tools of each own and relies on external information providers. GridICE [18] is based on Nagios and follows a centralized approach thus requiring multiple Nagios servers in order to facilitate a large distributed environment.

Monitoring solutions in Clouds are usually tailored to specific vendors. For example Amazon CloudWatch [19] provides the functionality to receive information about specific metrics, such as the number of read bytes, from EBS services. Windows Azure [20] provides monitoring capabilities through the Azure Diagnostic Manager. An open source software targeting Clouds is the Lattice Monitoring Framework [8], which follows the notion of producers and consumers of information in trying to minimize network utilization for monitoring information.

4 Cloud Experience in European Research Projects

Both VISION Cloud and mOSAIC projects address the interoperability issues for an integrated utilization of current heterogeneous Cloud technologies, to overcome the vendor lock-in problem that prevents to use different providers at the same time, or change the first choice when original business condition or user requirements are not valid anymore. VISIONCloud focuses on data-intensive storage services, but mOSAIC provides a Cloud platform to develop applications, which needs a complete IaaS infrastructure where to run.

4.1 VISION Cloud

VISIONCloud is an FP7-ICT project, which is developing a storage platform addressing the challenge of providing data-intensive storage Cloud services. Five key enablers have been identified that help in meeting today's challenges of storage Clouds.

---

[11] http://host-sflow.sourceforge.net/

1. *Raising the Abstraction Level of Storage.* Content in VISION Cloud is stored through a new data model that fits the scale of the Cloud, through relaxed semantics and consistency, yet provides rich metadata capabilities and a strong security model.
2. *Data Mobility and Federation.* Users today are not able to easily migrate their data between providers, suffering for what is known as *data lock-in.* VISION Cloud is based on the Cloud Data Management Interface (CDMI) proposed by the Storage Networking Industry Association (SNIA) which is an open interface rather that providing a proprietary interface. Moreover, the platform incorporates the needed functionality for federation scenarios to take place, allowing users to have a consistent view of their stored content, irrespective of what providers are being used.
3. *Computational Storage.* In most Cloud scenarios compute and storage resources are considered separate entities, frequently provided through different services. In VISION Cloud we are developing a new construct, called *storlet,* which provides the means to the user to define computations on stored data objects that ensure computations will be executed closed to the storage location, avoiding unnecessary data movements.
4. *Content-Centric Access to Storage.* To access a traditional storage system, the user must know the location of the system and the place of the artifact or file in a storage hierarchy. Current storage systems for Cloud applications overcome these limitations, but they are tailored to a specific domain, for example, Google books, Flickr and YouTube. In contrast, we are developing a generic Cloud storage system with methods to facilitate access to storage by its content and relationships, rather than details of the underlying storage containers. These methods support any domain by allowing the definition of domain-specific storage optimizations.
5. *Advanced Capabilities for Cloud-Based Storage.* The QoS mechanisms and SLAs offered by current Cloud storage providers are rather rudimentary. They guarantee only service availability and give service credit or refunds for lack thereof. In our approach, models, requirements and SLA Schemas are expressed not only on storage resources and services, but also on the content descriptions for the underlying storage objects, in support of content centric storage as required for business critical and sensitive applications.

4.2 mOSAIC

The *Open source API and Platform for multiple Clouds* (mOSAIC) is an FP7-ICT project, which is developing a platform that promotes an open-source Cloud application programming interface (API) and a platform targeted for developing multi-Cloud oriented applications. Its goal is to provide enough freedom both at resource and programming level such that Cloud-based services can be easily developed and deployed.

With a flexible architecture of the platform, designed around the use of open and standard interfaces, its main goal is to provide a unified Cloud programming interface which enables the flexibility needed to build interoperable applications across different Cloud providers [21, 22].

The main use cases related to the Cloud application life-cycle are : (1) the *development*, that involves writing new code, configuration files, and everything that is needed to build up the Cloud application; (2) the *deployment*, that consists in the acquisition of resources from Cloud providers, needed for execution of the Cloud application; and (3) the *execution*, that involves the installation, configuration and execution of all the software components as well as the monitoring and management of the Cloud application. In order to sustain such use cases, mOSAIC identifies the following needs:

– *APIs and programming paradigms*: a model to develop Cloud applications, which focuses on distributed aspects in order to enhance solution scalability and fault tolerance;
– *provisioning system*: a system able to acquire resources from Cloud providers in an homogeneous way and provide them to Cloud applications;
– *execution platform*: a flexible environment that can be started over provisioned resources and enables the execution of applications developed by the Cloud APIs;
– *middleware hiding the complexity and heterogeneity*: there is a need for finding a way to reason on Cloud resources and features in order to manage the complexity of different solutions available in the Cloud environment, and automatize as much as possible the execution processes.

The mOSAIC solution is an open-source software platform and framework for developing and deploying distributed applications on the Cloud. mOSAIC provides a set of facilities and APIs that support the developer through the whole life-cycle of a Cloud Application. In particular, mOSAIC API introduces supplementary layers of abstractions that offer uniform access to Cloud Resource, independently from the Cloud provider and the technologies it supports, allowing applications to work with abstract Cloud resources. Being mOSAIC a deployable Cloud PaaS , Cloud applications runs consuming resources obtained from Cloud providers.

Cloud Agency is the provisioning subsystem of the mOSAIC framework that is responsible for brokering, monitoring and autonomic reconfiguration of IaaS resources which are necessary for the execution of Cloud applications [23–25].

The Cloud Agency services are easily accessible to the mOSAIC platform through a REST interface.

5 Monitoring in the VISION Cloud platform

The main responsibility of the monitoring system within VISION Cloud is the collection, propagation and delivery of all events to their respective recipients. To this end, the system employs an asynchronous message delivery mechanism, and on top of it, a simple distributed rule engine to decide where each event should be transmitted and whether it should be aggregated with other events first.

5.1 Overall Architecture and Requirements

The infrastructure that is considered by the platform is based on inexpensive commodity hardware following current trends in data intensive application design and Bid Data analytics [26, 27]. The architecture is composed of multiple data centers, each one composed of clusters of nodes. Each node in the systems hosts a number of direct attached disks. Given today's hardware, the initial hardware configuration for the nodes could be 4 or 8 way multiprocessors (taking multi-core into account) with 12 to 16 GB of RAM. Each node could have 12 to 24 high capacity direct attached disks (e.g., 2TB SATA drives). The architecture, design and implementation should support a system with hundreds of storage clusters, where each storage cluster can have several hundred nodes and the storage clusters are spread out over dozens of data centers.

The targeted sizes of the platform include accommodating 10^5 tenants each having up to 10^7 individual users with an available storage of 10^3 petabytes per tenant.

The platform provides a RESTful interface based on CDMI that allows both data related operations, such as storing and retrieving objects, as well as management operations, such as creating users or negotiating SLAs. In VISION Cloud's data model the main building block is the data object which contains data of arbitrary size and type. Data objects are "grouped" in containers which are the unit of placement. Data objects and containers have metadata associated with them that can be both system and user generated. A tenant is the entity that subscribes to VISION Cloud's services by signing different SLA's according to the level of service desired and is billed accordingly. A tenant has users, who actually consume the services and are the owners of data objects.

The monitoring mechanism is essential in the functioning of the platform. Monitoring information are needed for the proper management of SLAs by the platform as well as needed to provide insights on the performance of the Cloud such as notifying the tenant of service failures or SLA violations. Moreover, a number of management operations require an effective monitoring mechanism in order to function properly, such as automatic billing mechanisms, placement decisions and analytics to forecast future system behavior. The monitoring system also needs to be able to handle the scale of the platform, in terms of tenants, users and concurrent operations and needs to be efficient, so as to not become the bottleneck in the operation of the system.

The main requirements of the monitoring system of VISION Cloud follows:

- Aggregation. The monitoring system should go beyond a metering tool, to provide advanced monitoring functions by aggregation of numerous information sources.
- Scalability. The system must be able to scale to sizes typical of Cloud set-ups, such as multiple datacenters hosting hundreds of clusters.
- Low-overhead. The system needs to be able to handle a large volume of events with the minimal network and processing overhead.
- Extensibility. The system needs to be able to accommodate new information sources and aggregation rules without any downtime.
- Federation. The system should be able to provide an open interface to allow federation scenarios to take place.

5.2 Monitoring Architecture

The main aim of the monitoring solution of VISION Cloud is to go beyond being a metering tool but instead to provide aggregation functions on the collected events at various levels.

The approach taken in the VISION Cloud monitoring component is based on the natural structure of the Cloud. Monitoring instances are deployed and run on all machines and clusters, since, potentially, any piece of the infrastructure could be a source of events. The running instances are responsible for collecting low-level monitoring information and events from each of the nodes they reside on, before filtering/aggregating and distributing them. Other Cloud subsystems that reside on the same host and are interested in generating or consuming events use the corresponding libraries (described below) to interact with the locally running instance.

Producers of monitoring information are typically at the node level, while consumers can be at any level. An initial design decision was to perform as much aggregation as possible on the node level before moving to the cluster and Cloud levels. The reasoning behind this design lies in the fact that communication costs rise while moving from the node to the Cloud levels. Therefore, by aggregating information at the node level the communication overhead is minimized. Therefore the system distinguishes between a number of different aggregation levels. More specifically a rule can be defined at a per node, per cluster, per datacenter or per Cloud level combined with a time frame. An aggregation at a per Cloud level - for example - means that information needs to be collected and aggregated at the Cloud level as a user can issue request on any host of the Cloud. Still, partial aggregation can be conducted at a lower level. Therefore, in the case of a Cloud level aggregation event, aggregation is first conducted at the lower levels to minimize communication cost. For example in the case of a rule stating *"report the number of read operations of a user per 30 minutes"* each node that receives a read request from a particular user will issue an event to the local monitoring instance. The monitoring instance will aggregate these events and will only report them to the higher level (in this case the cluster level) at specific time intervals. As a consequence the number of monitoring events flowing through the system can be kept at the lowest needed level.

5.3 Implementation Details

An instance of the monitoring mechanism resides on every node of the Cloud and is responsible for collection events from that node. It consists of the following sub-modules:

- Vismo-Core. This is the main monitoring instance. There is a unique instance that runs on each node and its main purpose is to coordinate with the various other modules. In the most basic terms, it acts as a conductor of events and as such, can be seen as the backbone of the system, receiving events from the event producers and distributing them to the event consumers. Moreover, it is responsible for collecting locally produced events, performing partial (node-level) aggregation and pushing the events to consumers.
- Vismo-Dispatch. The sole purpose of this library is to connect a producer to the locally running monitoring instance. In doing so, events generated in the producer are passed in instantly to the monitoring process. Under the hood the open source zmq library is used, in a pull/push fashion.
- Vismo-Notify. The library is used by the various consumers to declare interest in one or more group of events, called topics. A client of this library can register to one or more topics. Upon registration, the library is responsible for notifying the client of the arrival of new events. The notification happens in an asynchronous fashion to the main client program, in another thread. Here, also, the 0MQ library [28] is employed, using a PUBSUB mechanism.
- Rule system. A basic rule system is used to evaluate every event received and trigger different processing actions, such as partial aggregation or immediate dispatching according to rules. A rule consists of two parts, a "when" part and a "then" part. The former is used to filter/match events based on the event properties (such as, for example, the name of the container, in a read operation), while the latter is used, upon successful matching, to perform some kind of computation, like summing over one or more event keys for a period of time, or dispatching the event to a specified consumer.
- Aggregator. This module is used to generate new events which are the result of an aggregation method upon a list of raw events. Typically, the aggregation happens over events of the same type (e.g, all read events in the last ten minutes are aggregated on the number of the response size in bytes). Another option is to

collect a number of raw events and group them by a given property field (e.g, all the read events in the last minute, issued to the container "bar").

- Web Interface. The web interface is used by the Cloud administrators (and other clients) to check what rules are executing and update or delete the system with new rules, at runtime. A new rule is passed down to all the monitoring instances of the Cloud; a rule to be deleted is removed from all instances (and any partial aggregation result is discarded).
- Vismo-Probes. These constitute various low level probes that are external to the main instance and collect data about CPU and memory usage, network load, etc., on a per node basis.

The modules are developed in JAVA, while the networking between consumers, producers and the monitoring modules is achieved through the use of 0MQ [28]. 0MQ supports many communication patterns and various transports like in-process, inter-process, TCP and multicast depending on the location of the receiver of the message. It also allows for control of many communication policies and QoS parameters, such as timeouts and high water marks. An important aspect of the library is that there is no explicit need for brokers, allowing for direct communication between peers, thus reducing maintenance costs. Moreover, the library has a very small memory footprint.

Hardware metrics are collected through probes that are configured to pass the relevant information to the local representative of the monitoring subsystem. The modularity of the implementation also allows for other metering systems to be deployed and to be used as sources of information, as long as the specified event format is followed. Services residing on the same host can also utilize the interfaces the monitoring system exposes to propagate events to it using the same format. Consumers of information are able to subscribe to specific topics and receive new events. Each topic represents a specific rule.

5.4 Event and Rule Format

Produced events follow a simple format that contains some static fields as well as fields that are specific to the event generated. The fields that each message must contain are as follows:

- IP: This field holds the originating IP of the event.
- Service: This field holds the name of the producer of the event.
- Timestamp: The library used by the producer of the event adds a timestamp to each message.
- ID: The library used to send the event to the monitoring instance adds a unique id to the event.
- Type: This field holds the type of the generated event.

Each producer of events can add fields to the events produced as he deems necessary. The rules that can be supplied to the monitoring mechanism to specify how each event should be handled are divided in three sections. In the first section, the topic name is defined which will be the id used to register to the specific queue and must be unique within the Cloud. In the same section the aggregation level and period (in minutes) are also defined. The *"when"* (or predicate) section of the rule provides basic boolean predicates on the fields of the events. If there is a field specified in a rule and the field does not exist in some event, the event will be discarded. Since the rules engine might be matching a number of rules per event, all applicable rules will be processed (and multiple actions taken). The *"then"* (or action) section (which can be empty) will apply a function over the specified event fields and pass the results to a queue. If the action is empty, the event is pushed as it is.

As an example the following rule specifies that the sum of read bytes from the users of tenant NTUA are needed to be reported every 10 minutes, if the reads have come from the object service and the # bytes read are larger than 1000. The aggregation result will be queued under topic reads_ntua which the interested consumers will need to read.

```
topic=reads_ntua
agg_level=Cloud
agg_period=10

when
(e.type = Bytesread) AND (e.tenant = ntua) AND
(e.producer = object_service) AND (e. bytes >= 1000)

then
SUM e.bytes
```

6 Cloud Agency

Cloud Agency [29] is a multi agent system (MAS) that accesses, on behalf of the user, the utility market of Cloud computing to maintain always the best resources configuration that satisfies the application requirements. This system is being designed and developed within the research activities of the FP7 mOSAIC project [22]. It supplements the common management functionalities, which are currently provided by IaaS Private and Public infrastructure, with new advanced services, by implementing transparent layer to IaaS Private and Public Cloud services. offers four main services in order to manage the Cloud resources: provisioning, management, monitoring and reconfiguration. Agents' execution can be

82 B. Di Martino et al.

orchestrated by invoking the service requests, which are offered by through an OCCI compliant RESTFull interface. The OCCI interface implements two kind of requests. Asynchronous Service Requests (ASR) are used to ask for something to be executed. For example to start a Negotiation, to accept or to refuse an SLA, to change a Policy, etc.. Requests, as any other actions are not-blocking. It means that execution is started remotely, but the client can continue to run. On the other hand, requests will generate future events, which have to be handled by the requester. Synchronous Service Requests (SSR) are used to get information. For example clients can ask for reading an SLA, the status of a negotiation, to get the list of vendors, or the list of resources. Queries are synchronous, they return immediately the response if it is available, an exception otherwise. The OCCI-MTP allows the communication between the client and the Cloud Agency, it has to manage both asynchronous and synchronous requests. Notification (CAN) are sent as asynchronous responses after an event raised by a an ASR or by the execution of an internal workflow.

In Fig. 1 the component diagram of Cloud Agency is shown.

The monitoring service is implemented by Meters and by an Archiver which are responsible for low level performance measurement and for collection and management of monitoring information. The Resources Monitoring Service is used to get an up to date knowledge of the performance figures of the Cloud Infrastructure, and an history of the Cloud behavior. The mOSAIC user can start performance meters which are running in the Cloud itself. In order to easily use the exposed REST services and to integrate these ones into a user application, it has been developed a set of JAVA APIs (CA-Client). According to the event-driven interaction model the APIs allow sending asynchronous and synchronous requests and handling asynchronous notification by implementing callbacks.

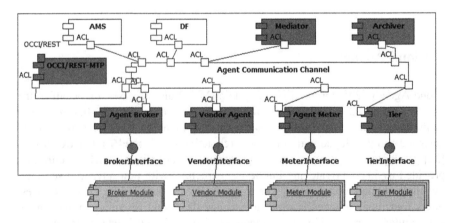

Fig. 1 Architecture

6.1 Monitoring Architecture

The monitoring architecture has been designed by decoupling the monitoring problem in two part:

- metering the Cloud, by installing and configuring a set of probes on the Cloud resources to measure low level performance indexes.
- monitoring Service Levels, by collecting, analyzing and using the produced information to compute and checking the service level performance indexes.

6.2 Metering the Cloud

In order to evaluate high level parameters either for checking the fulfillment of SLA, or to evaluate system workload, we have to provide a wide set of measures. Our probes are implemented by mobile agents, which are dispatched to Computing resources, from which sites they communicate their measures to a centralized collector. Each agent is configured with a Measuring Period (Tm), which is the period for taking a new measure, and a Communication Period (Tc), which is the period to communicate the last available set of measures. Of course the values of two parameters have to be chosen as a trade off between overhead and resolution of information. Time Periods work as heartbeats, which allow to evaluate the availability of both the computing resource and of the internal network. In fact when the communication fails the measure is stored locally, and it will sent by the next retry. The failed transmissions allow for the evaluation of the internal connectivity, while the lack of measures allows to detect a server downtime. Beyond this common set-up different agents have been designed and developed to measure and notify different parameters.

6.3 Monitoring the Cloud

The main issue here concerns the mapping of low level measures, which can be obtained by installing probes distributed across the computing infrastructure, to high level parameters, which describe service levels. In our solution a centralized Archiver collects low level metrics and use them as an input to perform the computation of high level metrics. The Archiver agent makes it possible to compute indexes for evaluating availability and reliability using missing heartbeats. Let us define Tfi as the time of a failure. The Archiver cannot distinguish between a network failure and a server failure if an heartbeat is missing, till a new measure has been notified. However after a new notification the history of server failures and missing notification can be recovered. We can evaluate the Mean Time Between Failures (MTBF) for the Network and the Server downtime by the identification of consecutive failures of communication, and of missing measures in the history. Of course the specific formula to compare the availability/reliability is eventually adapted according to the SLA of the specific provider. For example, in the case of Azure, the availability is defined in terms of time to recover. It means the elapsed time between a VM failure and the start of a new instance. Furthermore in the case of Azure this period will contribute to estimate the availability only if it lasts for more than 5 minutes. Again low level metrics are useful to evaluate the performance for a distributed Cloud infrastructure. In fact the response time, the throughput, the reliability of a service that uses a distributed Cloud infrastructure depends on the number of resources and on their interconnection. A three tier application that uses a web server, an application server and a database, running each one on a different machine can increase performance, but does not improve reliability. A load balancer with all tiers replicated on each virtual machine can increase service availability and reliability, but it does not improve the service response time.

6.4 Implementation Details

For metering the Cloud, agents move to the target computing resource, download, install and run Host-sFlow to profile the host performance, NetPerf for evaluation the internal network, bootchart for the boot time. Finally a traceroute to a list of public web sites is used to monitor the uptime of the external network. In Table 2 the set of provided low level metrics are listed. Currently is possible to choose Host-sFlow based metrics, that provides scalable, multi-vendor, multi-OS performance monitoring with minimal impact on the systems being monitored. NetPerf is used by Meter agents that are enabled to evaluate performance of the internal network. Other agents use UNIX net tools to evaluate performance of external network.

By the CA monitoring service it is possible to select a set of metrics, each one representing a Meter Agent that is able to perform the chosen measurements. Once the user has chosen the metric, the Archiver agent is responsible to create the user selected Meter Agent. This one retrieves from the Archiver Agent the location on

Table 2 Cloud Agency metrics

Meter type	Low level metrics
common meter	VM heartbeat, Network heartbeat
host meter	Boot-time, bytes in, bytes out, cpu aidle, cpu idle, cpu nice, cpu num, cpu speed, cpu system, cpu user, cpu wio, disk free, disk total, gexec, load fifteen, load five, load one, machine type, mem buffers, mem cached, mem free, mem shared, mem total, os name, os release, part max used, pkts in, pkts out, proc run, proc total, swap free, swap total
network meter	min latency, max latency, mean latency, stddev latency, packet loss, bandwidth (for external and internal network)

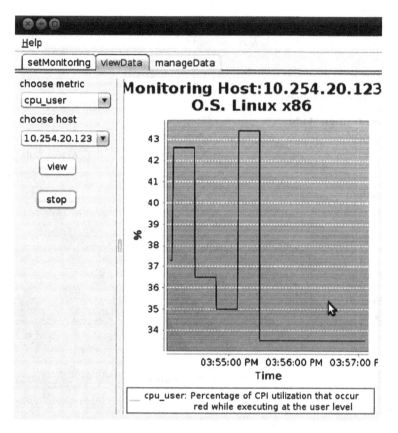

Fig. 2 Monitoring Dashboard

which the measurements have to be executed and migrates on the resource carrying on its configuration. After the migration, the Meter Agent downloads the required libraries to perform the measures, installs the measurement libraries on the resource and starts its own job. The archiver collects the measures in order to compute analysis on them in the future. The user can take under control the Cloud infrastructure performances by using the monitoring dashboard that can be started by the CA-Client GUI. It allows the visualization of a list of available measures or setting up the computation of some metrics about performance indexes (Fig. 2).

The user can create a trigger to be notified asynchronously according to a specific time period or when a critical condition is verified. An example of available metric is the average value of a measure that is periodically computed and that is notified when it is out of a certain range. To do this, he/she can open the observer tool via the CA-Client GUI .

It is necessary to choose the parameter on which it needs to create the trigger or to delete existing ones (Fig. 3). After choosing the parameter, the user can add a

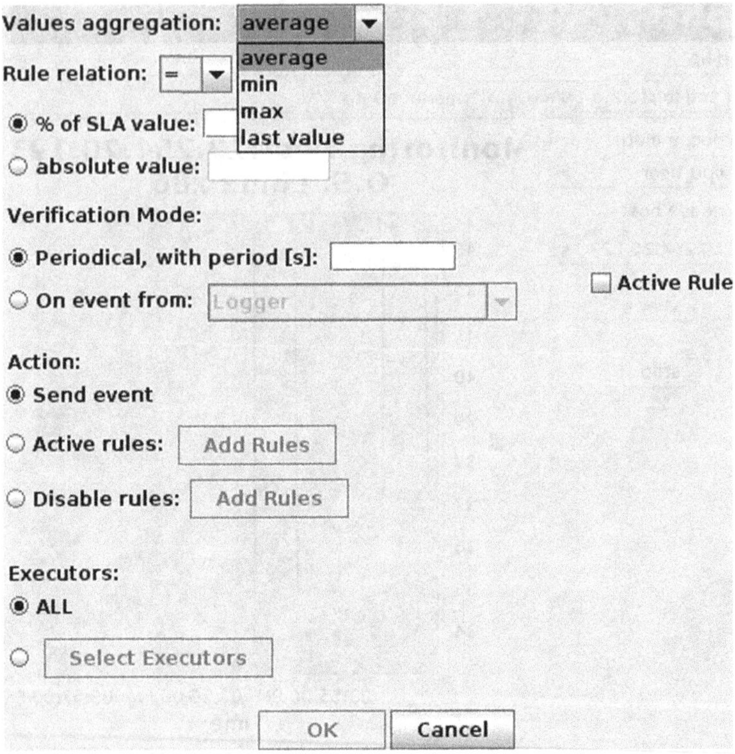

Fig. 3 Monitoring Observer

triggering rule: first of all it is possible to define the kind of monitoring results aggregation by choosing from different operations, such as average value, last received value and so on. It is also possible to determine the criterion by which the rule is verified by acting on two parameters: the comparison value and the order relation. As regards the former, it is possible to choose a percentage of the SLA value related to the considered parameter or an absolute value chosen by the user. The order relation can be selected by a number of possible values, for instance equality, diversity, strict inferiority, etc.

7 Critical Analysis

The research work described in this chapter have been developed within the activities of two different research projects which aim at providing APIs and services at platform level for development of Cloud applications. A first difference between the projects deals with the family of Cloud application they address.

VISION focuses on *data-intensive* applications and for this reason it aims at providing monitoring services for Cloud storages. On the other hand mOSAIC pursues a more general objective that is the development of a vendor agnostic APIs and a deployable platform. For this reason the monitoring is related above all to the Cloud infrastructure where the framework is running. For this reason in VISION much attention focuses on possibility to rise the abstraction level of storage resources to allow for a content centric programming paradigm. On the other hand a hard requirements becomes the network overhead. mOSAIC does not provide an abstraction of the Cloud infrastructure, that means the possibility to have a fine grained control over it, but it does not simplifies the task of its users. However both the project aim at using heterogeneous Cloud and this common goal requires to monitor service levels for checking the fulfillment of SLA by the providers they use. VISION Cloud is built on a Cloud federation and its monitoring facilities proved its users with single view of their distributed storage. mOSAIC manages a multi-Cloud infrastructure composed of many computer resources. Users can look at their computing infrastructure both as a collection of Cloud resources and as a collection of SLAs. Anyway both the solutions aim at providing a uniform and vendor agnostic interface to their services. VISION Cloud adopts the CDMI standard for its services. mOSAIC exposes a OCCI compliant interface. In the first case providers are federated if they are compliant with CDMI. Cloud Agency can support all the providers for which the related driver (vendor agent) is available.

The two approach propose a similar design for the monitoring services. We can see a distributed metering infrastructure composed of probes which are deployed on each node, and a facility for data aggregation that allows for the definition and the computation of metrics at higher level.

VISION Cloud provides metering solution that aims able to handle a large set of events with a minimal network and processing overhead, based on a message queue paradigm. However the strength of the solution is a rule based distributed mechanism that provides advanced monitoring functions by aggregation of numerous information sources.

Cloud Agency use mobile agents to dynamically deploy on computing nodes meters, which perform locally the necessary information at node level, and cooperate with other agents to make distributed measures. This high flexibility is not available at higher level for the aggregation of the information. In fact agent meters communicate the collected information to a centralized archiver that is responsible for the aggregation of the information and the notification of relevant events to the registered consumers.

In both cases the monitoring infrastructure is intended to run in the Cloud, but the persistence of the collected monitoring information is assured by the usage of persistent storages for VISION, and by a persistence service provided by Cloud Agency, which is able to store the centralized archive. All the temporary data are lost according the session-less model of the Cloud.

Regarding the technological choices we can observe that in both the projects code mobility has been exploited to move the computation close to data. In VISION it has been done for aggregation and filtering for reducing the network overhead. mOSAIC adopts the mobile agents paradigm to dynamically build the full monitoring infrastructure. In both the current implementation interoperability is provided by adopting a standard protocol (CDMI and OCCI) over the HTTP RESTfull transport layer. In this context this choice allows to mOSAIC delivering of agent based application according to the Cloud *As a Service* model. A performance evaluation of Cloud Agency service interface is presented in [30].

Both the monitoring solutions are mainly targeted to Cloud developers. The first considered user profile is interested to use APIs for building his/her own monitoring applications. For this reason a RESTfull interface can be easily integrated into any other applications. Besides both the projects developed language specific APIs ready to be used into their Cloud programming platform for programming *storlets* and *Cloudlets*. VISION Cloud, addressing big data analyst, provide a rule based paradigm for defining aggregation and filtering models. In the case of mOSAIC a relevant user profile is represented by Cloud deployers, who are interested to configure their monitoring infrastructure and use their ready facilities. For these target users both the solutions provide client applications with friendly graphic interface, which are web based, standalone clients or SDK plugins.

We can conclude that, at the state of the art, we have two complementary solutions regarding both the target resources to be monitored (storage and compute) and the architecture of the monitoring infrastructure (metering and aggregation).

We have found into two independent Cloud projects, focusing on different applications domains (data-intensive vs compute-intensive), common monitoring requirements that were born because of the choice to develop and run applications in an inter-cooperative Cloud context. It means the need of a monitoring service for supporting the decision makers (autonomic applications or human users) about the usage of the specific Cloud provider. The trustiness of the Cloud provider itself for the agreed service levels using its own monitoring service is obviously not feasible. Furthermore the usage of a third party monitoring service could be in some case expensive and not flexible. For this reason a solution that allows developers to integrated the monitoring services into their own application, or allows to administrators to build their own monitoring infrastructure is a promising approach.

8 Conclusion

We conclude the chapter by summarizing a comparison between the two monitoring approaches which provide many similarities even if they address two complementary Cloud scenarios. In fact first difference between the VISION Cloud approach and the mOSAIC one is related to the kind of Cloud resources to be managed. VISION Cloud focuses on storage services, meanwhile needs, above all, to monitor computational resources. In both case they aim at defining an uniform and standard interface to access heterogeneous Cloud vendors. CDMI is the VISION Cloud solution. A RESTfull OCCI compliant interface is provided by mOSAIC to solve the interoperability issue and to grant the sufficient transparency respect to the specific IaaS Cloud technology. Both the monitoring mechanisms collect low level performance indexes at node level, but this is done executing on physical infrastructures in the case of VISION Cloud and inside the Virtual machine for Cloud Agency. Both the solutions are distributed for what regards the collection of monitoring information and use an asynchronous message passing mechanism for communication. However in the case of VISION Cloud there are many distributed data aggregators and recipients, but mOSAIC uses a centralized archiver to get a complete view and to compute high level metrics which are available for applications and final users. A basic rule system is used to evaluate every event received and trigger different processing actions in VISION Cloud, but a similar functionality will be provided in mOSAIC by a complementary service out of the monitoring solution. Where VISION Cloud distributes the functionality of the monitoring system between components, there mOSAIC uses specialized agents. To address portability the technological solutions have been both implemented mainly in JAVA, by the distributed component paradigm in the case of VISION Cloud, and by the mobile agent based programming paradigm for Cloud Agency.

Acknowledgments The work described in this chapter has been partially supported by the FP7-ICT-2009-5-256910 (mOSAIC) and FP7-ICT-2009-5-257019 (VISION Cloud) EU projects.

References

1. R. Massimiliano, S. Venticinque, T. Mahr, G. Echevarria, G. Esnal, Cloud Computing Technology and Science, IEEE International Conference on, chap. Cloud Application Monitoring: The mOSAIC Approach, (IEEE Computer Society Press, Los Alamitos, CA, USA - NAZ), pp. 758–763. (2011)
2. The gogrid service level agreement, http://www.gogrid.com/legal/service-level-agreement-sla
3. DMTF: Architecture for Managing Clouds (June 2010), http://dmtf.org/sites/default/files/standards/documents/DSP-IS0102_1.0.0.pdf
4. T.F. Fortiş, G. Esnal, I. Padillo, T. Mahr, G. Ferschl : Cloud patterns for mosaic-enabled scientific applications. In: Euro-Par 2011 Workshops, Part I, Lecture Notes in Computer Science, vol. 7155, (Springer, Heidelberg) pp. 83–92. (2012)
5. D. Petcu, in Identifying cloud computing usage patterns. Cluster Computing Workshops and Posters (CLUSTER WORKSHOPS), 2010 IEEE International Conference on. pp. 1–8 (sept 2010)
6. M, F. S. V. B., D. M, On the Move to Meaningful Internet Systems, OTM 2012, vol. 7566, chap. mOSAIC-Based Intrusion Detection Framework for Cloud Computing, (Springer, Heidelberg - DEU), pp. 628–644. (2012)

7. Cloud Computing Use Cases Group: Cloud computing use cases white paper (July 2010), http://opencloudmanifesto.org/Cloud_Computing_Use_Cases_Whitepaper-4_0.pdf
8. S. Clayman, A. Galis, C. Chapman, G. Toffetti, L. Rodero-Merino, L. M. Vaquero, K. Nagin, B. Rochwerger, Monitoring service clouds in the future internet. in G. Tselentis, A. Galis, A. Gavras, S. Krco, V. Lotz, E.P.B. Simperl, B. Stiller, T. Zahariadis, eds. Future Internet Assembly, (IOS Press), pp. 115–126. (2010)
9. V.C. Emeakaroha, I. Brandic, M. Maurer, S. Dustdar, Low level metrics to high level slas - lom2his framework: Bridging the gap between monitored metrics and sla parameters in cloud environments. in W.W. Smari, J.P. McIntire, eds. HPCS. pp. 48–54. IEEE (2010)
10. T. Metsch, A. Edmonds et al. Open cloud computing interface - core and models. Standards Track, no. GFD-R in The Open Grid Forum Document Series (2011)
11. A. Edmonds, M.T. A. P., Open Cloud Computing Interface in Data Management-Related Setups, p. pp. 127. Springer (2011)
12. S. Ilarri, E. Mena, A. Illarramendi, Using cooperative mobile agents to monitor distributed and dynamic environments. Inf. Sci. **178**(9), 2105–2127 (2008)
13. W. Liu, B. Chen, Optimal control of mobile monitoring agents in immune-inspired wireless monitoring networks. J. Netw. Comput. Appli. **34**(6), 1818–1826 (2011)
14. M.L. Massie, B.N. Chun, D.E. Culler, The ganglia distributed monitoring system: design, implementation, and experience. Parallel Comput. **30**(7), 817–840 (2004)
15. W. Barth, Nagios: System and Network Monitoring. No Starch Press (2008)
16. I. Legrand, H. Newman, R. Voicu, C. Cirstoiu, C. Grigoras, C. Dobre, A. Muraru, A. Costan, M. Dediu, C. Stratan, MonALISA: An agent based, dynamic service system to monitor, control and optimize distributed systems. Comput. Phys. Commun. **180**(12), 2472–2498 (2009)
17. J.M. Schopf, L. Pearlman, N. Miller, C. Kesselman, I. Foster, M. D'Arcy, A. Chervenak, Monitoring the grid with the Globus Toolkit MDS4. J. Phys. Conf. Ser. **46**(1), 521 (2006)
18. S. Andreozzi, N.D. Bortoli, S. Fantinel, A. Ghiselli, G.L. Rubini, G. Tortone, M.C. Vistoli, GridICE: a monitoring service for Grid systems. Future Gener. Comput. Syst. **21**(4), 559–571 (2005)
19. Amazon CloudWatch, https://aws.amazon.com/cloudwatch/
20. Windows Azure, https://www.windowsazure.com/en-us/
21. D. Petcu, S. Panica, M. Neagul, From grid computing towards sky computing. case study for earth observation. pp. 11–20. Proceedings Cracow Grid Workshop 2010, Academic Computer Center, Poland (2010)
22. D. Petcu, C. Crciun, M. Neagul, S. Panica, B. Di Martino, S. Venticinque, M. Rak, R. Aversa, Towards a Service-Based Internet. ServiceWave 2010 Workshops, vol. 6569, chap. Architecturing a sky computing platform, (Springer, Heidelberg), pp. 1–13. Berlin - DEU (2011)
23. S. Venticinque, R. Aversa, B. Di Martino, M. Rak, D. Petcu, A cloud agency for SLA negotiation and management. in Proceedings of the 2010 conference on Parallel processing. (Springer, Heidelberg), pp. 587–594. Euro-Par 2010, (2011)
24. S. Venticinque, R. Aversa, B.D. Martino, D. Petcu, Agent based Cloud Provisioning and Management - Design and Prototypal Implementation. In: Closer 2010. pp. 184–191 (2011)
25. R. Aversa, B. Di Martino, M. Rak, S. Venticinque, Cloud Agency A Mobile Agent Based Cloud System. in Proceedings of the 2010 International Conference on Complex, Intelligent and Software Intensive Systems. pp. 132–137. CISIS '10, IEEE Computer Society, Washington, DC, USA (2010)
26. S. Ghemawat, H. Gobioff, S.T. Leung, The Google file system. SIGOPS Oper. Syst. Rev. **37**(5), 29–43 (Oct 2003)
27. R. Kouzes, G. Anderson, S. Elbert, I. Gorton, D. Gracio, The Changing Paradigm of Data-Intensive Computing. Computer **42**(1), 26–34 (Jan 2009)
28. Zeromq, http://www.zeromq.org/
29. R. Aversa, B. Di Martino, S. Venticinque, CISIS 2010 - International Conference on Intelligent, Complex and Intensive Systems, chap. Cloud Agency: A Mobile Agent Based

Cloud System, pp. 132–137. IEEE Computer Society, Washington D.C. - USA (2010), http://dx.medra.org/10.1109/CISIS.2010.143

30. S, V. L. T. B. D. M. Agents based cloud computing interface for resource provisioning and management. in 6th International Conference on Complex, Intelligent and Software Intensive Systems. pp. 249–256. IEEE Computer Society Conference Publishing Servic, USA (4–6/7/2012) http://dx.medra.org/10.1109/CISIS.2012.139

Intuitive User Interfaces to Help Boost Adoption of Internet-of-Things and Internet-of-Content Services for All

D. Tektonidis, C. Karagiannidis, C. Kouroupetroglou
and A. Koumpis

Abstract The idea we promote in the chapter is to provide better support to users with disabilities and impairments from the comfort of their home by means of providing them with a set of scalable services which can be either offered for free or purchased through some central form of a marketplace repository.

1 Introduction

A beer company in Argentina has produced a vending machine for drinkers that needs to be 'tackled' before it'll dispense a drink. The machine, called Rugbeer, is being used in a marketing campaign aimed at beer-loving rugby fans in the north of the country, a region where the sport is even more popular than soccer. According to the beer company, the unique machine has helped boost sales of Cerveza Salta by 25 % in bars where it was installed.[1]

---

[1] http://www.digitaltrends.com/cool-tech/rugbeer-the-vending-machine-you-have-to-tackle-to-get-your-beer

---

D. Tektonidis (✉) · C. Kouroupetroglou · A. Koumpis
Research Programmes Division, M.Kalou 6, GR – 54629, Thessaloniki, Greece
e-mail: dte@altec.gr

C. Kouroupetroglou
e-mail: chris.kourou@gmail.com

A. Koumpis
e-mail: akou@altec.gr

C. Karagiannidis
Special Education Department, Argonauton and Filellinon, University of Thessaly,
GR38221 Volos, Greece
e-mail: karagian@uth.gr

F. Xhafa and N. Bessis (eds.), *Inter-cooperative Collective Intelligence:*
Techniques and Applications, Studies in Computational Intelligence 495,
DOI: 10.1007/978-3-642-35016-0_4, © Springer-Verlag Berlin Heidelberg 2014

Intuitive User Interfaces (IUI) have long been considered as the holy grail of the HCI community, since they are assumed to facilitate the most natural, and thus successful form of communication. This is reflected by the fact that IUI is the only separate subsection term appearing in Wikipedia's definition section on usability.

At the same time, IUI have also stimulated some very fruitful debates and criticisms since intuition is a *personal* trait affected by various (cultural, usage, etc.) aspects and is dynamically modified (e.g. through experience), therefore it may be a moving target for UI design leading to inconsistent solutions.

Most work on IUI has been rather technology-driven so far, aiming to investigate how user intuition is related and adapted to the mainstream interaction technologies and metaphors available at different technology generations (window systems, tablets, mobile devices, etc.). Recent advances in motion sensing (e.g. MS Kinect), gesture recognition (e.g. SoftKinetic DepthSense Camera), face recognition (e.g. Technest SureMatch 3DT Suite), etc., facilitate radical new interactions through user gestures, voices and expressions, especially for people with special needs. People with disabilities are also being benefit from recent developments in software such as the Siri for iPhone and the Speech Recognition system for Android devices which bring speech recognition as another popular modality of interacting with machines.

In the present chapter we investigate these new opportunities, through the provision of accessible and ed services which understand and react to social context of usage by enabling each individual user to use her/his preferred (/natural) gestures, voices and expressions for interacting with each device in a smart environment (e.g. home, office, conference room, etc.) in a variety of social circumstances.

Let's look at a scenario that might shed light regarding how the future might look like:

Giorgio is living in Rome with his wife Maria and his blind son Matias and their grandma Estella. The family lives in a smart home with the 3 TV's, the electric kitchen and the washing machine connected to a network and being manipulated by all 4 of them.

All appliances adapt their interface to each user separately. Matias for example can control all of them using dialog interfaces. However, since he is not using the kitchen so often as his mother the interface adapts to a simple question answering interface asking him to specify how hot does he want his food to be and reminding him to avoid placing tupperware in the oven.

On the other hand when his mother Maria (who loves cooking) uses the kitchen the interface adapts to a more complex one allowing her to adjust all kind of parameters and timers controlling multiple parts (oven, grill, kitchen eyes) at the same time mainly using speech interface. When Giorgio comes to help her and tries to input commends the kitchen always asks him if he checked with his wife before overriding her instructions (Figs. 1 and 2).

The 3 TV sets are also adapting to users as well and in many cases they are used as control points for other devices in the home. The one in Estella's room knows her favorite programs and notifies her when they are on air. It can also check

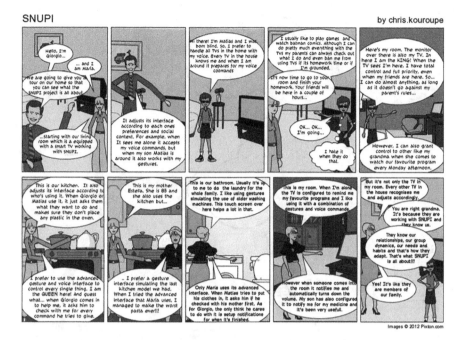

Fig. 1 How the future would look like in a home with 4 people all of them with different needs and capabilities operating multiple devices which are accessed by same modalities

whether the TV in the living room is free from the Giorgio who loves watching football games and Matias who enjoys playing games with his friends.

All TV sets are sharing the family's profiles and adapt their interface to who is in the room. Matias for example has priority in controlling the TV of his room but when he sits in the living room priority goes to his father and mother. Same also stands for Estella. However, the couple has the option to check and intervene on what their son is doing on the TV without even going in his room and sometimes they forbid content that they judge as inappropriate. Such rules cannot be over-ruled in any of the other 2 TV sets when Matias tried to use them. Rules like that also apply when Matias is working on his homework for school.

The washing machine can also be operated by all members of the family. Giorgio and his son are always using a simple question answering interface that asks them what kind of clothes they put in and also makes sure they didn't place clothes that cannot be washed together. In fact, when Matias is placing in his clothes the machine asks him to show her the articles one my one and make sure they match in colors. It then selects its program automatically making sure that they placed the appropriate amount of washing powder in the holders.

When Maria is using it, she prefers a complex interface combining gestures and speech to select the exact program and get notified in specific stages of the process. Estella is usually using the washing machine with an interface simulating the

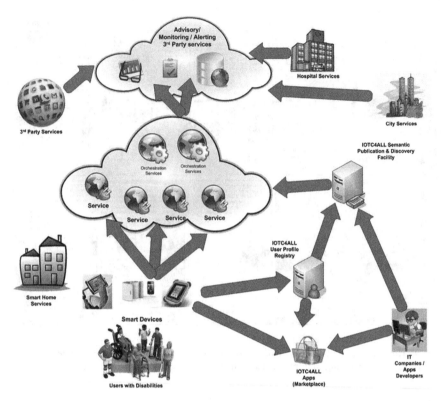

Fig. 2 Overall conceptualisation to support the IOTC4ALL vision

buttons interface she was using on her older washing machine. When she turns her
right hand she selects the appropriate program and her left hand the degrees.

Maria tried once to teach her about the complex interface she is using but to no
avail she would not remember every little detail and meshed things up a couple of
times.

2 The Technology Supply Quest

The arrival of Smartphones and Tablet PCs has created the Post-PC era and led to
a development of IOS and IOT. However people with disabilities cannot fully take
advantage of these benefits. Our vision is to provide better support to users with
disabilities and impairments from the comfort of their home by providing:

- **Advisory services** to help manage their health and social care needs and ensure
  they can effectively access relevant services.

- **Monitoring services** that help users manage on their own and remind them to about essential aspects such as taking their medication and monitoring their condition and generally look after themselves properly.
- **Alerting services** when the users situation suddenly deteriorates and urgent help is needed.
- **Dependable services** that will check whether services are functioning correctly and notify service provider when the connection to special care users is lost unexpectedly.
- **Accessible Interfaces** using suitable interface technology and user interfaces that are appropriate to the special needs of users.
- **Smart Interfaces** that are sensitive to the context and current needs and that ensure appropriate services are made use of.
- **Orchestration Services** that make sure that services are orchestrated effectively where needed to deliver a seamless service at the point of service delivery.
- **Third party services** when the users may want to use services such as online grocery shopping or public transport or use social networking facilities.

The concept is built on the fundamental principle that users should be supported as much as possible by accessible services to maintain independence through monitoring and advice and involve available services when needed to address emerging needs.

The services can be provided by both public and private providers (and even include wireless connection to devices such as blood pressure monitors, flow meters, glucose level monitors and the like) to ensure that the user is appropriately supported in managing by themselves while ensuring they receive emergency services if needed. This should include other services required for daily living and to facilitate the user managing on their own and giving them the freedom of independent living. It is also envisaged that access to social networking services be provided to mitigate social exclusion suffered by housebound users who live on their own.

We believe that making this vision (we call it "IOTC4ALL" – _I_nternet _of_ _T_hings and _C_ontent _4_ (: for) _ALL_) reality will have a significant impact on social inclusion and supporting disadvantaged users to lead independent lives with the assurance of support where needed. The vision is to develop an **open service model** that allows new service providers to enter and which can easily be extended to different member states independent of the technology used.

The aim is to enable to a large variety and diversity of users with special needs the usage of complex services through devices that although are considered smart they are difficult in terms of use for several people. Also a key aspect of the aspired architecture is the inclusion of people with disabilities to execute these services: though it is about services, the architecture requires an enriched Service Oriented Architecture by components that will enable and facilitate the discovery and execution of services from smart and mobile devices.

In practical terms we need to offer a platform where service providers can:

- Design services
- Implement Services
- Deploy
- Publish

And where the providers of devices and third party service providers can:

- Identify relevant services
- Compose and configure services for the specific user
- Configure interface devices (tablet, PDA, smartphones, web cams)
- Add further home-based healthcare devices where required (blood pressure meters, glucose meters etc).

To achieve these goals, we have to deal with the following issues:

- **Service Orientation**: Given that we need to get away from the website orientation to a service orientation how services can be exposed so that they can be accessed by special devices for impaired and disabled users that aim to make them accessible to those with special needs (and potentially combine them with other related services to provide a user centric overall service).
- **Discoverability**: Given a variety of services from a variety of providers and the needs and entitlements of individual users, how the services a user can access be discoverable by assistive devices or special interfaces users will use.
- **Service Entitlement**: Given that services are selective in terms of users meeting certain criteria it should be possible to determine the subset the user is entitled to given their personal circumstances and not to confuse users with irrelevant services and rather work with those they can use at that point in time.
- **Service Composition and Integration**: Services should be fashioned around the user rather than the user having to adapt to the services. So for the user the division between services should be transparent; some services may be provided in collaboration or based on joint or shared responsibilities by different organisations (healthcare, public services etc) and need to integrate seamlessly at the point of delivery to the user. This will require suitable approaches for service composition and integration.
- **Context sensitivity**: Given that the user will only requires some of the available services at any particular point in time it will be important to focus on the context and the immediate needs of the user to increase the assistive powers and not overburden the user who may be bewildered by the sheer amount of services available and what to use in any given situation given on the stated or inferred needs (the latter if some home based measuring devices are used for example)
- **Accessibility**: Given that users will not be IT literate or have special requirements with respect to interfaces in term of presentation, navigation and interface type (especially for blind and with motor disabilities) so that users can

effectively use the services. This may also have to be combined with a number of special devices for input/data collection or as output devices.

We envisage three main provider communities regarding the proposed service execution platform:

1. *Service providers* (public authorities and public services) aiming to manage their services (healthcare, social care and public services) delivery to the service users (disabled or aged) integrate them for a seamless service provision to the end user,
2. *Services consumers* will be (disabled or aged) customers that use services through apps from their home using mobile devices (e.g. smart phones, tablets etc.) or stationary devices (PC or PC enabled TV) directed at IT illiterate or inexperienced individuals without any technical knowledge.
3. *Third Party Providers* - IT solutions and value added service who are involved in the service design, composition (Service designers) and IT service provision. If the original service providers are unable to provide interfaces third party service providers will take on the role of generating devices and interfaces and additional services.

In the next section we present a set of informative scenarios of use that can be easily adapted and customised to fit the realities faced by a broad category of consumers and citizens today. Though emphasis is given to categories facing the risk of exclusion due to age or some type of disability, it is obvious that the same scenarios are applicable to the average European citizen and service customer.

3 Accessibility Matters: Opening Service Access to All

Accessibility is increasingly becoming an issue as well as assistive technology for both the able bodied and impaired users. Anti-discrimination legislation now requires businesses and public authorities alike to support disabled or disadvantaged users alongside their able-bodied counterparts. This is starting to attract attention also from the commercial community and with patent applications in these areas becoming more frequent, but where the EU is decidedly lagging behind as compared to the US as can readily be seen from the statistics readily available from ISO and EPO. A number of EU projects have begun to address aspects of accessibility, where movement and mental disabilities were more important, and to a lesser degree the sensory ones (including visual and hearing). Current EU-wide disability and accessibility-applicable research address Web, Sensors, Robotics and Brain-Machine Interfaces, as well as motor-related, mental-, and cognitive disabilities.

In a recent survey carried out by the authors it came to light concerning the user needs, that respondents rated cognitive, mental and multiple disabilities as important and that health and employment are areas where new eAccessibility solutions will be needed most. An initial Trend Analysis revealed that standard technology for devices and connectivity were less of an issue, as these are now relatively widely available, while virtualisation, suitable HCI Interfaces, Avatars and Robots as companions were identified as potential weak areas in need of further work. The time horizon for those is considerably longer making them less of an easy picking for researchers and industry alike. IOTC4ALL will be addressing healthcare needs of users with disabilities and making them substantially more accessible than what is available at present through much improved HCI interfaces.

Among the areas identified in need of development in the survey are solutions for increased independence of older people and active ageing with expectations from the community to meet the rights to participation of people with disabilities. With growing health care expenditure, Growing expenditure on social protection benefits, countermeasures to address fragmentation of (ICT) and develop affordable solutions with significant benefits are needed.

What is needed and where IOTC4ALL will make a concrete contribution to novelty will be by developing new models where users have more control over services they consume, new model for evaluating and implementing accessibility. This will promote the business value of investing in accessibility and raise awareness of accessibility issues in the industry.

Among the needs that can be identified in this area and which IOTC4ALL will be supporting:

- **Advanced active interfaces**: Dialog interfaces are a reality in many situations nowadays. However, they usually work with specific limited vocabulary and in a specific context. We aim to develop proactive interfaces that do not necessarily wait for the user to start the interaction but to prompt and remind users based on an analysis of context. This will be achieved by leveraging semantic technologies for assessment of context and accessing relevant services at that point in time so that the user receives focused assistance. IOTC4ALL will create innovative solutions here that are currently not available in the market nor have been reported in the academic literature.

- **Automatic adaptation of multimodal interfaces**: Modern devices such as tablets and smart phones can contain a lot of information about users, such as documents, music, preferences for various applications and so on. So, having a personal profile on a mobile device seems quite natural for the future of ICT. In addition, mobile phones are enhanced with connectivity sensors which allow them to interact with other machines through WiFi, Bluetooth, NFC to include additional monitoring devices such as blood pressure meters, glucose meters and pulse meters. Interfaces should be smart and work out the context and adapt to the needs and preferences of the user and who can interact with remote services to support the user. This goes in tandem with advanced active interfaces and will generate interfaces that adapt to the users disability problems and then actively seek to assist the user to assess their current needs and locate and access relevant services making also use of local devices where appropriate to take readings. Again there are no solutions of this

type currently available that cover all these aspects even if devices exist we can
draw on to develop the proposed innovative service.

- **Promoting discoverable and integratable services**: Ubiquitous connectivity
  and network services together with cloud based services and technologies mean
  that services can be accessed remotely but the way services are made available
  (usually using standard websites and applications) which are largely provider
  centric and the way they are presented often make it difficult to consume and
  integrate and orchestrate them so that users with disabilities can be readily and
  effectively be supported. A new service model is needed that does allow them to
  be discovered and accessed as a service so that appropriate smart interfaces can
  connect to them remotely and make the services available alongside each other
  and combinable and "orchestratable" at the interface level. We create innovation and make a concrete contribution in this area.

4 Setting up a Solution

The proposed Service Execution Environment based on the vision and the scenarios previously presented aims to provide a new approach that will improve the
use of services from user groups that have accessibility problems. The definition of
a service will be enriched with all the required features and parameters that are
related to the accessibility of the service. Therefore the proposed solution will
investigate and develop the following:

1. An enhanced **Service Model** that will include additional specifications related
   to the e-accessibility issues.
2. **Semantics** that will enrich existing services in order to increase their accessibility to groups of users that this is required.
3. A **PersonalisedApplication and Service Marketplace** where developers will
   be able to publish their applications and users will be able to find services and
   applications based on their profile.

4.1 The IOTC4ALL Service Model

The provision of the services for IOTC4ALL, will follow the life-event model that
has been implemented successfully in the past for e-Government Services.[2] The
life-event approach facilitates discovery of the service because it bases the services
discovery in the user profile and on WHAT the user wants to do ("I want to go to

---

[2] OneStopGov project http://islab.uom.gr/onestopgov/

the hospital", "I want to buy groceries"). The execution of these services will be based on the user profile based on his/her needs. Services execution will be differentiated based on the disabilities of the user ("What is the preferable means of transport for a visual impaired person", "How can he/she input the necessary data for executing the service").

The user will be able to download the application to his/her device from the IOT4CALL marketplace according to his/her needs. The services will be installed as Apps (Mobile Applications) to his/her mobile device (Tablet/Smartphone). The installed services will use information from the IOTC4ALL User Profile Registry. The execution of the service will involve the IOT4CALL Semantic Discovery Facility that will be able to locate and retrieve information from available information links. We use the term "information link" as in Linkeddata[3] to define every available system or service that can be accessed through the web.

The IOT4CALL Marketplace will provide a variety of applications that will make accessible a large variety of services capable to cover the scenarios and improving the lifestyle of the people with disabilities.

4.2 Enhancing Services Components to Enrich Their Accessibility

Today as the number of available services grows the percentage of people with disabilities that can actually access these services is reducing. Web Accessibilities guidelines are applied to many web portals (especially governmental). However these guidelines cannot be applied to mobile applications. Therefore the design for the usability of such applications is limited to the specifications of the mobile device (i.e size to the screen) and not to who is going to use is. The problem is not only to the user interface but also to the service that the mobile application is using. Of course, the cost for reengineering the existing service is huge therefore IOTC4ALL propose an approach that will enhance the existing creating wrappers that will include all the additional features that are related with accessibility.

For IOTC4ALL the usage of a service breaks down into components that are independent and can be enriched without changing its structure and operation. To facilitate the enrichment of the services components, IOTC4ALL will provide to the service providers semantic descriptions to enrich their service. Following the approach of Linkeddata where software systems are defined as Information Links, IOTC4ALL will use Semantic Web Technologies to define the requirements of enhancing existing services without changing their structure.

---

[3] http://www.linkeddata.org

4.3 The IOTC4ALL Marketplace

Marketplaces are a common practice from mobile applications and in the future for software in general. Existing Marketplaces (iTunes, Android Marketplaces etc) gather information from the user like the device he/she is using, location and other to suggest applications that may be in his/hers interest. For the applications developers, marketplaces enable them not only to sell their application but also to evolve providing new versions and communicating with their customers.

In IOTC4ALL we can make use of a Marketplace that will also gather information based on accessibility providing intelligent filtering for the available apps. In addition it will provide all the necessary infrastructure to facilitate both the users (customers) and the developers.

4.4 Smart Devices UI

IOTC4ALL' goal is to enable to a large variety and diversity of people with special needs the usage of complex services through devices that although are considered smart they are difficult in terms of use for several people. To give better supports to users with disabilities and impairments from the comfort of their home or in mobility it is crucial make the above devices fully accessible by means of the definition and implementation of novel interface solutions which are appropriate to the special needs of people.

In particular, a holistic approach in defining an adaptive smart interface should be adopted to pursue a global optimization by considering not only the different kind of devices but also their sensitiveness to the context and people needs, in so ensuring services are appropriately used.

In order to be useful, the design of user interfaces should consider many aspects of human behaviours and needs. The complexity of the extent of the involvement of a human in interaction with a machine is sometimes invisible compared to the simplicity of the interaction method itself.

The existing physical technologies for Human Computer Interaction (HCI) essentially can be categorized by the relative human sense that the device is designed for. These devices are basically relying on three human senses: vision, audition, and touch [1].

Input devices that rely on vision are the most used kind and are commonly either switch-based or pointing devices [2]. The switch-based devices are any kind of interface that uses buttons and switches like a (virtual or physical) keyboard [3]. The pointing devices examples are mice, joysticks, touch screens, graphic tablets and pen-based input [4]. The output devices can be any kind of visual display or printing device [5].

The devices that rely on audition are more advanced devices that usually need some kind of speech recognition [6]. Nowadays, all kind of non-speech and speech signals [7] and messages are produced by machines as output signals. Beeps, alarms, and turn-by-turn navigation commands of a GPS device are simple examples.

Recent methods and technologies in HCI are trying to combine former methods of interaction together and with other advancing technologies such as networking and animation. These new advances can be categorized in three sections: wearable devices [8], wireless devices [9]. The technology is improving so fast that even the borders between these new technologies are fading away and they are getting mixed together. Few examples of these devices are: radio frequency identification (RFID) products, personal digital assistants (PDA), and virtual tour for real estate business. Current directions and advances of research in user interface design include intelligent and adaptive interfaces as well as ubiquitous computing.

4.4.1 Intelligent and Adaptive HCI

As mentioned before, it is economically and technologically crucial to make HCI designs that provide easier and more pleasurable experience for the users. To realize this goal, the interfaces are getting more natural to use every day. Evolution of interfaces in note-taking tools is a good example. First there were typewriters, then keyboards and now touch screen tablet PCs that you can write on using your own handwriting and they recognize it change it to text and if not already made, tools that transcript whatever you say automatically so you do not need to write at all.

One important factor in new generation of interfaces is to differentiate between using intelligence in the making of the interface (Intelligent HCI) [10] or in the way that the interface interacts with users (Adaptive HCI) [11].

Intelligent HCI designs are interfaces that incorporate at least some kind of intelligence in perception from and/or response to users. A few examples are speech enabled interfaces that use natural language to interact with user and devices that visually track user's movements or gaze and respond accordingly.

Adaptive HCI designs, on the other hand, may not use intelligence in the creation of interface but use it in the way they continue to interact with users. An adaptive HCI might be a website using regular GUI for selling various products. This website would be adaptive -to some extent- if it has the ability to recognize the user and keeps a memory of his searches and purchases and intelligently search, find, and suggest products on sale that it thinks user might need. Most of these kinds of adaptation are the ones that deal with cognitive and affective levels of user activity.

Another example that uses both intelligent and adaptive interface is a tablet PC that has the handwriting recognition ability and it can adapt to the handwriting of the logged in user so to improve its performance by remembering the corrections that the user made to the recognised text.

Finally, another factor to be considered about intelligent interfaces is that most non-intelligent HCI design are passive in nature i.e. they only respond whenever invoked by user while ultimate intelligent and adaptive interfaces tend to be active interfaces. The example is smart billboards or advertisements that present themselves according to users' taste.

4.4.2 Ubiquitous Computing and Ambient Intelligence

Ubiquitous computing (Ubicomp), often used interchangeably by ambient intelligence and pervasive computing, refers to the ultimate methods of human-computer interaction that is the deletion of a desktop and embedding of the computer in the environment so that it becomes invisible to humans while surrounding them everywhere hence the term ambient.

The idea of ubiquitous computing was first introduced by Mark Weiser during his tenure as chief technologist at Computer Science Lab in Xerox PARC in 1998. His idea was to embed computers everywhere in the environment and everyday objects so that people could interact with many computers at the same time while they are invisible to them and wirelessly communicating with each other [12].

Ubicomp has also been named the Third Wave of computing. The First Wave was the mainframe era, many people one computer. Then it was the Second Wave, one person one computer which was called PC era and now Ubicomp introduces many computers one person era.

4.4.3 From Unimodal to Multimodal Interfaces

An interface mostly relies on number and diversity of its inputs and outputs which are communication channels that enable users to interact with computer via this interface. Each of the different independent single channels is called a modality [13]. A system that is based on only one modality is called unimodal. According to the nature of different modalities, they can be divided into three categories:

1. *Visual-Based*: Facial Expression Analysis, Body Movement Tracking (Large-scale), Gesture Recognition, Gaze Detection (Eyes Movement Tracking)
2. *Audio-Based*: Speech Recognition, Speaker Recognition, Auditory Emotion Analysis, Human-Made Noise/Sign Detections (Gasp, Sigh, Laugh, Cry, etc.), Musical Interaction
3. *Sensor-Based*: Pen-Based Interaction, Mouse and Keyboard, Joysticks, Motion Tracking Sensors and Digitizers, Haptic Sensors, Pressure Sensors, Taste/Smell Sensors.

Basic visual user interfaces in smart devices are the standard graphic based WIMP (window, icon, menu, and pointing device). Smart devices also have multiple modalities and are able to interpret information from various sensory and communication channels. For instance, smartphones uses touch based interactions and some (e.g.LG Optimus) allow users to use personalized gestures as shortcuts to commands. Multimodal interfaces process two or more combined user input modes (such as speech, pen, touch, manual gesture, gaze, and head and body movements) in a coordinated manner with multimedia system output [14]. They are a new class of interfaces that aim to recognize naturally occurring in the forms of human language and behavior, and which incorporate one or more recognition-based technologies, e.g., speech, pen, vision. Thus, they can provide richer and more natural methods of interactions, e.g., gesture, speech, and even all the five senses.

The term multimodal refers to combination of multiple modalities. In MMHCI systems, these modalities mostly refer to the ways that the system responds to the inputs, i.e. communication channels. The definition of these channels is inherited from human types of communication which are basically his senses: Sight, Hearing, Touch, Smell, and Taste. The possibilities for interaction with a machine include but are not limited to these types.

Therefore, a multimodal interface acts as a facilitator of human-computer interaction via two or more modes of input that go beyond the traditional keyboard and mouse. The exact number of supported input modes, their types and the way in which they work together may vary widely from one multimodal system to another. Multimodal interfaces incorporate different combinations of speech, gesture, gaze, facial expressions and other non-conventional modes of input. One of the most commonly supported combinations of input methods is that of gesture and speech.

An interesting aspect of multimodality is the collaboration of different modalities to assist the recognitions. For example, lip movement tracking (visual-based) can help speech recognition methods (audio-based) and speech recognition methods (audio-based) can assist command acquisition in gesture recognition (visual-based).

Some examples of applications of multimodal systems are Smart Video Conferencing, Intelligent Homes/Offices, Driver Monitoring, Intelligent Games, E-Commerce and of special interest for our project Helping People with Disabilities. A good example of Multimodal System for disable people is to address and assist disabled people (as persons with hands disabilities), which need other kinds of interfaces than ordinary people. In such systems, disabled users can perform work on the device by interacting using voice and head movements.

4.4.4 Smart Devices Today

In the past mobile devices were difficult to use and had limited user input methods (number keyboard or stylus) with small displays. Thus advanced usage was limited to business applications and platforms like Symbian and Blackberry. With the release of the iPhone this suddenly changed. Multi-touch interaction on a large colourful display enabled a world of new mobile applications affordable und manageable for everyone. The new interaction paradigm was adopted by all players in the market including Google, Nokia and also Microsoft. A few years later with Apple's iPad this concept has evolved into a new kind of personal computer, the tablet PC. While UMPCs and Netbooks appeared too bulky and complicated, the tablet PC brings the simplified mobile touch interaction of the iPhone into new interaction situations: touchscreen and overall user interface design are intuitive and visually accessible. The much larger screen should improve accessibility for people with low vision and other partial vision impairment.

Furthermore, the capacitive touchscreen is more accessible to those with dexterity limitations by introducing an electronic stylus; while those with limited finger movement and/or sensation in their fingertips have greater success using resistive touchscreen devices because of the ability to use a stylus. But, the electronic stylus for the capacitive touchscreen interfaces levels the playing field.

Cameras are increasingly used in conjunction with GPS functionality by software applications for augmented reality, mixed reality, and location-based services. Additionally, cameras are becoming an essential feature for devices used in the growing area of remote health monitoring. Being able to take a picture of a wound, infection or other visible condition will enhance the ability of people (with or without disabilities) to avoid the disruption and considerable effort of going to the doctor's office.

In addition, easy input devices have been developed, such as predictive text, restricted options, voice command, point and click.

5 Conclusions and Future Work

Smart house—smart hospital—smart city: are there any common threads amongst them? One may think that each one of these application areas is worth a dedicated examination. In a **house** we live for years so it is easy to see why a person may need to invest on making it (more) accessible. While in a **hospital** people stay (fortunately) only for short (and usually unexpected) periods of their life—and again there smartness expected from a hospital may be of totally different nature than the smartness we expect from a home to exhibit to its inhabitants. Finally, for the smart **city** case, the requirements are also different: people may need to "connect" to different types of information provision services and infrastructures and two basic service types are: (a) what services are offered to me **<here>** and (b) how can I get **<there>**.

What we can consider as common denominator of all these are:

- The users have access to their mobile phone (smartphone, Android or iphone) so the unique access point to information from the environment (house, hospital or city) comes through their mobile phone.
- It is through their mobile phone that they may recognize the various addressable entities i.e. the "things" (of the *Internet of Things*) that are surrounding them and which can be tagged with QR codes / RFID tags or be accessible through NFC, and
- It is *again* through their mobile phone that they will be informed about the availability of the various content elements (of the *Internet of Content*) that are retrievable and which can be tagged with QR codes / RFID tags or be accessible through NFC

What is our understanding of a service in the context of the IOTC4ALL vision? This comes up to a series of highly connected links and ideas:

- Any retrieval of location-based (house / hospital / city) content is a service (though a basic one—could we come up with more sophisticated ideas?)
- Any access to a "thing" for getting information about its status or for manipulating is also regarded as a service (again this may be regarded as basic ones).
- More sophisticated or complicated services may base on the composition of more than one basic services—this way we may support different *families of* scenarios to be realized.
- What we promote as an idea and as a role is this of industrial enterprises (like the ones participating in the project) which will invest on offering such types of composite services to people. So given one house / hospital / city setting, user X may decide to buy the service bundle by value added service provider A, or B or C, or, in case s/he wants may still access 'manually' each of the offered 'content' and 'things' manually. More specifically:

  – User X may decide to buy a basic service from value added service provider A that enables organized scenario selection according to a (simple) thematic organization.
  – Or spend some more money and buy the more expensive service B, where there is also increased intelligence i.e. the selection of a particular scenario follows the mood of the user and the user does not have to choose one. This case is of interest to categories like older people or bipolar patients who may be unable to make a decision on their own and so this extra degree of automation helps their independent living. Finally,
  – User X may choose to buy a premium service from provider C that offers him/ her with an advanced level of customer service and care that extends beyond basic service activities offered at level A and slightly more sophisticated or automated of level B, to include a more personalized and customized approach that create a customer-centric atmosphere. In this case, there may be increased interoperability offered for services to bridge gaps between the various environments that the user is changing—i.e. from the home through the city to the office and then back to home. This extra level of semantic processing that helps information flow between the various smart settings (house / city / hospital) is offered if the customer agrees to pay extra money.

Below we briefly summarise each of the supported scenario clusters:

Smart home scenario : A smart home in the context of this proposal is considered as an augmented environment with the ability to consolidate information ("content") embedded in appliances, as well as micro-systems and multi-modal sensors to offer users enhanced access to information and functionality and an ability to control and manage the environment at a higher level of abstraction. So why shouldn't the home take initiative to either understand (proactively) the user's mood and offer the most appropriate setting adjustment or react to user's wishes (user chooses to watch TV, light settings are adapted and phone calls, etc.)

Smart hospital scenario : What is our idea for a smart hospital? In the context of this proposal it is mainly related with enhancing accessibility of to its patients (as well as the medical staff working). The users in this pilot may be able-bodied patients as well as visually / cognitive / age challenged persons. The idea of accessibility in the hospital context relates with helping patients take their drugs on time (so all blister packs and other drugs' packages may contain RFID labels) and facilitating their access to medical equipment that embeds RFID tags. The doctors, nurses, caregivers and other staff members wear a "smart badge" storing information (besides their employee ID number) that may help the patient ask for advice or guidance. Same way as doctors, on arrival, each patient receives a wristband with an embedded RFID tag storing a unique identifier, and some information about his / her profile that besides medical information may include also information significant for the types of interaction that the particular patient is capable to perform with services / "content" / "things" available within the hospital environment.

Smart city scenario (mobile phone as a "white cane"): Consider a visually impaired person travelling through a city. Similar to a foreigner in a hostile place feeling lost and requiring guidance, the use of a mobile phone to provide such assistance is ideal. The primary aim of this scenario is to enable visually / cognitive / age challenged persons to acquire the necessary guidance and assistance by capturing geo-tagged information using their mobile phone, and return any information, such as a specific pharmacy, the local police station, or even points of interest. Additionally, the GPS capability of the phone is exploited to enhance the locally available services recognition process (service spots).

As far as the links with the other two scenarios are concerned we should notice that with the use of the mobile phone, the same way that a visually / cognitive / dementia challenged person is enabled to get guidance and assistance for the urban environment for short-range navigation and exploration in the home and hospital settings.

References

1. D. Te'eni, J. Carey, P. Zhang, *Human Computer Interaction: Developing Effective Organizational Information Systems* (John Wiley & Sons, Hoboken, 2007)
2. B.A. Myers, A brief history of human-computer interaction technology. ACM Interact. **5**(2), 44–54 (1998)
3. B. Shneiderman, *Designing the User Interface: Strategies for Effective Human-Computer Interaction*, 3rd edn, (Addison Wesley Longman, Reading 1998)
4. A. Murata, An experimental evaluation of mouse, joystick, joycard, lightpen, trackball and touchscreen for Pointing: Basic Study on Human Interface Design, in *Proceedings of the Fourth International Conference on Human-Computer Interaction 1991*, pp. 123–127 (1991)
5. J. Nielsen, *Usability Engineering* (Morgan Kaufman, San Francisco, 1994)
6. L.R. Rabiner, *Fundamentals of Speech Recognition* (Prentice Hall, Englewood Cliffs, 1993)
7. S. Brewster, Non speech auditory output, in *The Human-Computer Interaction Handbook: Fundamentals, Evolving Technologies, and Emerging Application*, ed. by J.A. Jacko, A. Sears (Lawrence Erlbaum Associates, Mahwah, 2003)
8. W. Barfield, T. Caudell, *Fundamentals of Wearable Computers and Augmented Reality* (Lawrence Erlbaum Associates, Mahwah, 2001)

9. M.D. Yacoub, *Wireless Technology: Protocols, Standards, and Techniques*, (CRC Press, London 2002) and virtual devices (K. McMenemy and S. Ferguson, A Hitchhiker's Guide to Virtual Reality, A K Peters, Wellesley (2007)
10. M.T. Maybury, W. Wahlster, *Readings in Intelligent User Interfaces* (Morgan Kaufmann Press, San Francisco, 1998)
11. A. Kirlik, *Perspectives on Human-Technology Interaction* (Oxford University Press, Oxford, 2006)
12. G. Riva, F. Vatalaro, F. Davide, M. Alaniz, *Ambient Intelligence: The Evolution of Technology* (IOS Press, Fairfax, Communication and Cognition towards the Future of HCI, 2005)
13. A. Jaimes, N. Sebe, Multimodal human computer interaction: a survey, Computer Vision and Image Understanding. Computer Vis. Image Underst. **108**(1–2), 116–134 (2007)
14. S.L. Oviatt, Advances in robust multimodal interface design. IEEE Comput. Graphics Appl. **23**, 52–68 (2003)

Intelligent Context: The Realization
of Decision Support Under Uncertainty

Philip Moore

Abstract Context is inherently complex and highly dynamic requiring intelligent context processing to address the diverse domains to which context has been employed and the broad range of available contextual information. The primary function of context-aware systems is targeting service provision; this requires the identification and selection of individuals based on their profile (a context). This identifies context-aware systems as decision-support systems where decisions must be arrived at under uncertainty. This chapter provides an overview of context and intelligent context processing. The context processing algorithm and extended context processing algorithm is introduced with an example implementation. An overview of fuzzy systems design as it relates to intelligent context processing is presented. The fuzzy variable and distribution function (often termed a membership function in the literature) are discussed with an analysis and conclusions relating to fuzzy variable and distribution function design. The chapter concludes with results, a discussion, conclusions, and open research questions.

1 Introduction

Context-aware systems are inherently complex and dynamic [15, 20]. Historically, the generality of contextual information used has been restricted to location, identity, and proximate data [16]. There is however an increasing research focus on the use of a broad and diverse range of domains and related contextual information in research projects and less so in commercial applications.

Addressing the range of available contextual information (if data can be captured, digitized, and codified it can be viewed as contextual information) [20] requires the application of computational intelligence implemented in intelligent

P. Moore (✉)
Faculty of Technology, Engineering and the Environment, School of Computing,
Telecommunications and Networks, Birmingham City University, Birmingham, UK
e-mail: philip.moore@bcu.ac.uk

F. Xhafa and N. Bessis (eds.), *Inter-cooperative Collective Intelligence:*
Techniques and Applications, Studies in Computational Intelligence 495,
DOI: 10.1007/978-3-642-35016-0_5, © Springer-Verlag Berlin Heidelberg 2014

context-aware systems [18]. A central function in context processing is *context matching* (CM) with *partial matching* [20]; this function is implemented in the *Context Processing Algorithm* (CPA) which, along with the design of the fuzzy *distribution function* (often termed a *membership function* in the literature) [2], forms this chapter's central contribution. The CPA presented in this chapter extends the *Context Processing Algorithm* (CPA) presented in [18]. The extended CPA (ECPA) implements additional functions to address uncertainty and the *Decision Boundary Proximity Issue* (DBPI) as discussed in [20, 21] and this chapter.

This chapter considers the complexity of context and sets out an overview of intelligent context processing. The ECPA is presented with an example implementation and conclusions related to the operation. The result derived from the ECPA is the *resultant value* (*rv*) (converted in to a semantic representation with an adjustment to address the DBPI (the (q) context matching metric) which represents the degree to which a user is a suitably qualified recipient for service. These metrics while interesting are not particularly useful in decision support; to effectively implement the ECPA a distribution function is required. The fuzzy variable and distribution function are discussed with an analysis and conclusions relating to fuzzy variable and distribution function design. The chapter concludes with a consideration of the DBPI, context processing under uncertainty with consideration of proposed further extensions to the ECPA and a discussion with open research questions.

2 The Complexity of Context

Context and context-aware systems (across a broad and diverse range of domains) are characterised by inherent complexity and uncertainty. Context-aware systems are predicated on the matching of an *entities* context with that of a resource [16]. An entity has been defined in [4] as: *"a person, place, or physical or computational object"*. Whilst predicated on contextual information (data) context-aware systems are generally characterised by the lack of *a-priori* knowledge related to the *input*(s) [in the case of resource allocation the resource (the *input*) being distributed] and the potential recipient(s) (the *output*). A discussion on this aspect of context-aware systems is presented in subsequent sections of this chapter.

Complexity theory addresses the study of complex and chaotic systems and how order, pattern, and structure can arise from them and the theory that processes having a large number of seemingly independent agents can spontaneously order themselves into a coherent system. To illustrate the point, consider context-aware systems in a hospital setting where context(s) must relate to individuals (patients / staff) and resources (e.g., X-Ray facilities: availability, workload, scheduling, etc).

The focal point of any healthcare system in a hospital setting is the patient. From this perspective there are a number of dynamic issues and challenges including: (1) a hospitalised patient is generally very ill and there may be unexpected change in a patient's 'state' (or *context* in computational terms), and (2) there is a high probability that the number of patients may change as in, for example, an influx of emergence cases in the accident and emergency department which can disrupt scheduled and planned medical procedures. The hospital environment is therefore subject to consistent dynamic complexity and uncertainty. It is clear that patient management in hospital environments requires the provision of 'real-time' information relating to patients' in terms of their 'numbers', current prevailing 'locations', and prevailing 'states' (or *context*). These demands can be viewed in terms of complexity theory (the processing of a large number of seemingly independent but inter-related entities in an integrated system) where the context-aware system is attempting to mitigate the impact of the dynamic change.

Additionally, there are potential issues for context-aware systems where the *scale* and *scope* of the system. This relates to the number and type of entities that must be accommodated and their related contextual information processed to achieve predictable decision support whilst accommodating constraint satisfaction and preference compliance as discussed in subsequent sections of this article.

This article introduces the concept of intelligent context processing to address the requirements identified and address the often conflicting challenges implicit in addressing the complexity and uncertainty that characterises context-aware systems along with the scale and scope of such systems.

3 Intelligent Context Processing

Context-awareness describes a concept in which the profile of an entity is defined by its context. The aim of context-aware systems is personalization and the targeting of service provision based on a current context which reflects the environmental, social, and infrastructure related contextual information As such context is highly dynamic and domain and application specific requiring the identification of domain specific function(s) and properties that relate to the domain of interest [16, 18, 20].

A context must reflect a user's current dynamic state [15, 20]. Location is central to context; context however includes more than just location [10, 16, 23]. As discussed in [20] a broad and diverse range of context factors combine to form a context definition, in fact, almost any information available at the time of an interaction can be viewed as contextual information.

Context-aware systems are intimately connected with the concept of *uncertainty*. Uncertainty shares with context domain specificicity and can be defined as: a state of having limited knowledge where it is impossible to exactly describe existing state or future outcome and may involve more than one possible outcome [2, 6, 12]. A fundamental factor in this connection is that uncertainty (in a problem solving situation) results from some information deficiency including: *incomplete, imprecise, fragmentary, vague,* or *contradictory*, or *deficient* in some way [12].

The various information deficiencies may result in differing types of uncertainty [12]:

- *Non-specificity* (or imprecision) which relates to sizes (cardinalities) of relevant sets of alternatives
- *Fuzziness* (or vagueness) which is a result of imprecise boundaries of fuzzy sets
- *Strife* (or discord) which expresses conflict(s) amongst various sets of alternatives.

Context processing and context matching (as implemented in the context processing algorithm as discussed in this section) has a clear connection with *fuzziness* in that context matching algorithm produces a resultant value (the rv and q metrics) which represents the degree to which an *output* context is a match for an *input* context as discussed in this chapter.

There are potential issues where fuzzy theories are applied in context processing and matching. For example, there is the possibility of ambiguity in locating an individual (more accurately an individual's device) geospatially. This may become apparent in the hospital scenario considered in Sect. 2 where for example a patient trolley (assigned to a patient with the associated porter) may be RFID tagged and the patient may also be RFID tagged. Where the patient, trolley, and the porter are separated (for operational reasons) ambiguity may result. The CPA, in applying prioritization using *weighting* of context properties is designed to address (or at least mitigate) such ambiguities and provide a location for the patient.

Central in the proposed approach to the processing of contextual information is the *Context Processing Algorithm* (CPA) [20, 21]. The CPA implements *Context Matching* (CM) (which extends the concept of data fusion) which provides a basis upon which contextual information can be processed in an intelligent context-aware system that enables *Partial matching* (PM), *Constraint Satisfaction*, *Preference Compliance* (CS) [20], and predictable decision support under uncertainty while accommodating the inherent complexity of context.

Context processing is designed to create the *input* context and access the *output* context(s) definitions and using the CPA to determine if the *output* (solution) context (properties) is an acceptable match with the *input* (problem) context (properties) [20, 21]. The likelihood of there being a perfect match is remote therefore CM must accommodate PM [20]; Fig. 1 graphically models CM, shown is the PM issue with a DB distribution function (a threshold) used in the CPA as discussed in [20, 21] and in subsequent sections of this chapter.

Essentially, the context-matching process is one of reaching a Boolean decision as to the suitability of a specific individual based on his or her context [20, 21]. The CPA is designed to address the PM issue as discussed in the following sections. As discussed in [17] accommodating CS represents a significant issue in intelligent context-aware systems; to mitigate CS violations the CPA algorithm implements a prioritizing bias in the CP and CM process [20, 21].

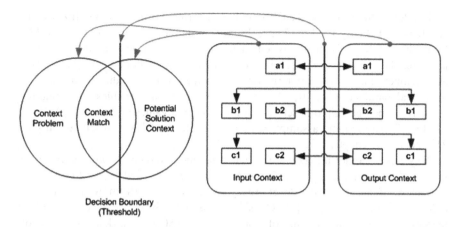

Fig. 1 The *context processing model*: Shown is the PM problem with the DB (threshold) used in the CPA. Note: the proposed approach enables multiple thresholds [20] in for example context-aware health monitoring systems where multiple decisions (prognoses) under uncertainty must be accommodated

The issues in personalization, intelligent context and its inherent complexity, CP, CM with PM, and CS have been explored in [15–20]. The CPA is addressed in [20, 21] and in the following sections.

Computational complexity can be viewed in terms of a measure of the degree of difficulty in solving a problem as measured by the number of steps in an algorithm. Computational complexity overhead is potentially an issue where, for example, Bayesian approaches, the C4.5 algorithm, or inference and reasoning approaches are adopted in context matching as discussed in [18–20]. In considering the CPA, while the complexity of the CM problem is handled with CS accommodated (or at least mitigated) and predictable decision support achieved. The computational complexity and the computational overhead in the CPA is low as the number of steps in the algorithm is clear with integrated security checking and stopping criteria. This is demonstrated by the results achieved as shown in Sect. 3.2.1 where the steps and the related results are set out. The algorithm has been tested on low powered devices including a Samsung N150 Plus 'netbook' with a 1.7 GHz processor and 1 GB physical memory; the running time for each context match is < 1 s.

3.1 The Context-Processing Algorithm

The CPA is predicated on *Event:Condition:Action* (ECA) rules concept as discussed in [1, 3, 7, 19, 20]. In the CPA the <*condition*> component of an ECA rule (the *antecedent*) employs an IF-THEN rule strategy where the IF component evaluates the rule <*condition*> resulting in an <*action*>. The <*action*> in the proposed approach can be either: (1) a Boolean decision, or (2) the firing of another rule or rule-set [19, 20].

To address the PM issue the CPA applies the principles identified in fuzzy logic
and fuzzy set theory [2, 12, 13, 24, 25] with a defined distribution function [2].
Conventional logic is generally characterized using notions based on a clear
numerical bound (the *crisp* case); i.e., an element is (or alternatively is not) defined
as a member of a set in binary terms according to a bivalent condition expressed as
numerical parameters {1, 0} [12, 13]. Fuzzy set theory enables a continuous
measure of membership of a set based on normalized values in the range [0, 1] [2,
12, 13]. These mapping assumptions are central to the CPA as discussed in [20,
21] and in this chapter.

A system becomes a fuzzy system when its operations are: "*entirely or partly
governed by fuzzy logic or are based on fuzzy sets*" [2] and "once fuzziness is
characterized at reasonable level, fuzzy systems can perform well within an
expected precision range" [2]. Consider a use-case where a matched context
mapped to a normalised value of, for example [0:80], has a defined degree of
membership. This measure, while interesting, is not in itself useful when used in a
decision-support system; in the CPA this is addressed using a decision boundary
solution to implement the essential process of *defuzzification* as discussed in [2]
and as it applies to the CPA in [20, 21] and in Sect. 4.

CM with PM imposes issues similar to those encountered in decision support
under uncertainty [6, 12], which is possibly the most important category of
decision problem [12] and represents a fundamental issue for decision-support
[22]. A discussion on the topic is beyond the scope of this chapter however a
detailed exploration of fuzzy sets and fuzzy logic can be found in [2, 6, 12, 22]
with an exposition on "Decision, Order and Time" presented in [22]. A comprehensive discussion on fuzzy system design principles can be found in [2] where
a number of classes of decision problem are identified and discussed. In summary,
Fuzzy Rule Based Systems have been shown to provide the ability to arrive at
decisions under uncertainty with high levels of predictability.

A discussion on the CPA, logic systems, and rule strategies with the related
conditional relationships for intelligent context-aware systems can be found in [19,
20]. For an in depth discussion on earlier CPA with CM and PM (as modeled in
Figs. 1 and 2) with example implementations, proof-of-concept, and a dataset
evaluation see [18, 20, 21]. The CPA is as follows where:

- $e := \{0, 1\}$—the numerical (Boolean) evaluation for each context property
  match {*true* = [1]} or {*false* = [0]}
- $w := \{0.10...1.00\}$—a weight applied to each context property to reflect its
  relative priority
- $av := \prod(e * w)$—the *Actual Value* (av) for each context property match evaluation in the range [0.00...1.00] following the application of the weight (w)
- $sav := \sum(av\ (a_1...a_n$—the *sum* of the *Actual Values* (**av**)
- $mpv := \sum(av\ (a_1...a_n)$—the *Maximum potential Value* (mpv). This represents a
  state in which all context property matches (av $(a_1...a_n)$) are *true* [1]. The mpv
  assumes that a perfect context match has been identified.

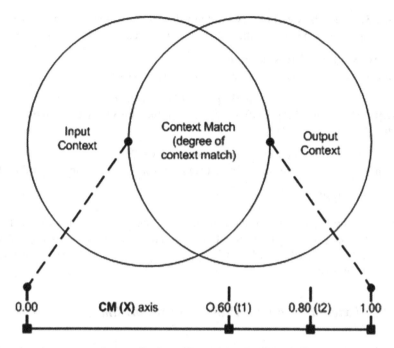

Fig. 2 The *extended context processing model*: Shown is the PM problem with the DB (threshold) used in the CPA as modeled in Fig. 1 with the context match mapped onto the solution space the CM(X) axis with the decision boundaries: (*t1*) (*t2*) (*t3*)

- **rv** := (**sav** / **mpv**)—the *resultant value* (rv) represents the degree to which the overall context match is true in the range [0.00...1.00].
- **t1** := the {0.60}decision boundary on the CM axis.
- **t2** := the {0.80} decision boundary on the CM axis.
- **t3** := the {1.00} decision boundary on the CM (X) axis. Note: the (**t3**) DB is a special case as there can be no instance where (**rv**) can exceed [1.0].

Figure 2 shows a mapping of CM and PM to the solution space with the decision boundaries (t1), (t2), and (t3). The CPA is as follows, shown are steps 1 to 6:

1. (**step 1**) Evaluate the context match {1, 0} for each individual context property, for example:

IF (**a1**(*input*)) **.equalTo** (**a1**(*output*)) THEN **e** = {1}
IF (**a1**(*input*)) **.notEqualTo** (**a1**(*output*)) THEN **e** = {0}
2. (**step 2**) Obtain the pre-defined property weighting (**w**) for each context property in the range [0.1, 1.0]:

w = **a1** [0.1, 1.0]

3. (**step 3**) Apply the weighting (**w**) to the value as derived from step 2 (note: the **w** is applied irrespective of the value of **e**. Thus retaining the result for **e**):

IF **e(a1)** = {1, 0} THEN **av** = (**e** * **w**)

4. (**step 4**) Sum the values derived from the CM process:

sav = $\sum$(**av**(a1) + **av**(b1) + **av**(b2) + **av**(c1) + av(c2))

5. (**step 5**) Compute the potential maximum value (**mpv**) for the context properties {a1, b1, b2, c1, c2}:

mpv = $\sum$(**w**(a1) + **w**(b1) + **w**(b2) + **w**(c1) + **w**(c2))

6. (**step 6**)Compute the resultant value (**rv**) for testing against threshold value (**t**):

rv = (**sav** / **mpv**)

Step 6 in the CPA results in a quantitative measure of the degree to which an input context is a match for an output context. This is expressed as the *resultant value* (*rv*). As discussed the (*t3*) DB is a special case as there can be no instance where (*rv*) can exceed [1.0].

3.2 The Extended Context Processing Algorithm

The CPA results in a quantitative representation (*rv*) of the CM. The *extended CPA (ECPA)* [21] adds an additional step to implement: (1) a semantic conversion of the (*rv*) metric—the (*q*) metric, and (2) the application of *Euclidean distance (ED)* as discussed in [20] to address the perceived *Decision Boundary Proximity Issue* (DBPI) as discussed in Sect. 4.2. The additional step (step 7) is as follows where: (*q*) is the semantic conversion of the (*rv*), and (*XC*) relates to the Euclidean Distance (ED) [5] currently set at [0.01] or 1 % of the normalized CM (*rv* + *xc*) from a DB.

(**Step 7**): Using the heuristically defined DB's (*t1* and *t2*) with the crisp DB (*t3*) determine if the output (potential solution) context definition is a suitably qualified match with the input context. Derive the semantic measures of the degrees of "qualifiedness": {"LQ", "GQ", "HQ"}. Note: as shown the value of **q** := (**rv** + **xc**).

IF ((*t1*) <(*rv* + *xc*)) THEN *q* = *"LQ"* (*Low Qualifiedness*)
IF ((*t1*) ≥ (*rv* + *xc*) < (*t2*)) THEN *q* = "GQ" (*Good Qualifiedness*)
IF ((*t2*) ≥ (*rv* + *xc*) ≤ (*t3*)) THEN *q* = "HQ" (*High Qualifiedness*)

An analysis of the ECPA supports a number of conclusions. The result of *step 7* is a semantic conversion (*q*) of the (*rv*); this retains the Boolean result obtained in step 1 of the CPA and achieves a semantic representation: {"LQ", "GQ", and "HQ"} of the CM. The semantic terms used are illustrative; the actual terms used in 'real-world' applications will be domain specific to reflect the fuzzy variable and uncertainty that characterizes a domain of interest.

Table 1 Expressed user preference / semantic context match relationships	LQ	GQ	HQ
GP	[0]	[1]	[1]
HO	[0]	[0]	[1]

There is a direct relationship between the {"LQ", "GQ", and "HQ"} semantic representation of the CM and an expressed user preference relating to the degree of membership derived from CM the user is happy with in, for example, an information system (see Table 1). The semantic measure may for example represent a failed (or *false*) context match and the {"GQ", "HQ"} metrics may relate to the well understood *precision* and *recall* metrics. A user may opt for a lower level of CM (higher recall) and specify in a semantic context modeling ontology [18]: "GP" (general precision) which relates to the "GQ" metric. Similarly, the "HP" (high precision) preference relates to the "HQ" level in the CM process. In the ECPA the results utilize the semantic metrics for the degrees of *"qualifiedness"*. Table 1 sets out the relationship between the semantic CM conversion (q) and the user expressed preference level. For example, where the actual result of the context processing is "GQ" and the user's expressed preference is "HP" the result of the CM is [0] as the CM result is less than the expressed preference.

Adopting this approach provides a basis upon which the use of distance can be extended to other relationships between a decision boundary and data points (the context match) with CM achieved using other methodologies such as machine learning techniques. Additionally, the use of ED [5] provides an effective basis upon which a relationship with DB's can be extended as considered in the discussion. This result, whilst interesting, is not in itself useful in implementing decision support; this is addressed in the implementation where domain specific rules are applied to interpret the CM result in terms of a Boolean decision {1, 0} (*true* or *false*) respectively, measured against the decision boundaries (t1) and (t2). Implementation demands that Defuzzification is implemented; this requires the design of a distribution function [2]; the following section addresses the design of a distribution function with an analysis and conclusions derived from the analysis.

3.2.1 Implementation

This section demonstrates the implementation of the ECPA; shown is an example program run:

```
run:
Begin Program Run
INFO [main] (SetupTDB.java:678) - Statistics-based BGP optimizer
*Error: null
*InData.inResType: 2
*loginAction: 2
*cp_R4: 2
```

```
*begin context matching
*eval_CoSub(iCS): cug1 (oCS): cug1
*(e): 1 *(W): 0.9 *(av): 0.9 *(sv): 0.9 *(mv): 0.9
*eval_MoSub(iMS): m1cug1 (oMS): m1cug1
*(e): 1 *(W): 0.8 *(av): 0.8 *(sv): 1.7 *(mv): 1.7
*eval_AcadQual(iAQ): aq2 (oAQ): aq1
*(e): 0 *(W): 0.8 *(av): 0.0 *(sv): 1.7 *(mv): 2.5
*eval_VocQual(iVQ): vq1 (oVQ): vq1
*(e): 1 *(W): 0.5 *(av): 0.5 *(sv): 2.2 *(mv): 3.0
*eval_AcadExp(iAE): ae1 (oAE): ae1
*(e): 1 *(W): 0.7 *(av): 0.7 *(sv): 2.9 *(mv): 3.7
*eval_VocExp(iVE): ve1 (oVE): ve1
*(e): 1 *(W): 0.6 *(av): 0.6 *(sv): 3.5 *(mv): 4.3
*eval_AcadInt(iAI): ai2 (oAI): ai1
*(e): 0 *(W): 0.7 *(av): 0.0 *(sv): 3.5 *(mv): 5.0
*eval_VocInt(iVI): vi1 (oVI): vi1
*(e): 1 *(W): 0.4 *(av): 0.4 *(sv): 3.9 *(mv): 5.4
*eval_FHEQ(resFHEQ): 6 (outFHEQ): 6
*(e): 1 *(W): 1.0 *(av): 1.0 *(sv): 4.9 *(mv): 6.4
*getResultantValue (sv / mv): 0.765625
*semanticMatch.outPrecision: GQ
*semanticMatch: 0.765625
*semanticMatch (q): 0.01
*semanticMatch: GQ
*applyPrecision(semPrec)(outPrec): GQ : GQ
*CM result: true
```

The output shown is taken directly from the terminal and shows the output resulting from an actual program run. Demonstrated is:

- The accessing of the data structure (a Jena TDB dataset).
- The control functions where the input type (*InData.inResType: 2) and the login action (*log inaction: 2) are tested for validity.
- The accessing of a context processing rule (*cp_R4: 2) in which the input data values are used in the context processing.
- The (9) context properties being tested and the overall context match computed.
- The comparative relationships between the *input* and *output* literal values.
- The resultant Boolean decision.
- The program output(s) demonstrate the use of the String functions: *#String-Tokenizer*, *#Trim*, and *#ToLowerCase* Java methods to ensure consistent representation of the literal String values.
- The resultant (**rv**) value (*getResultantValue (sv / mv): 0.765625).
- The DBPI adjustment is implemented in the (**q**) {*semanticMatch (q): 0.01} result.
- The user expressed preference as shown in Table 1.

The ECPA has been designed to implement error checking with stopping criteria to catch exceptions generated by errors in data entry.

4 The Distribution Function

The *distribution function* (DF), in association with fuzzy variable design, are important steps in the design of a fuzzy rule-based system [2]. Context and context-aware systems are domain specific requiring domain specific design; this characteristic is shared with the design of fuzzy rule-based decision support systems, the identification of fuzzy variable(s), and the design of the distribution function [2].

Prior to the design of a fuzzy variable the domain of interest (the problem) must be identified along with the degree of uncertainty inherent in the problem. This requires identification of the fuzzy measures including: *imprecision, inaccuracy, ambiguity, randomness, vagueness,* and other measures including domain specific measures of uncertainty [2, 12, 13]. The design of the fuzzy variable to address the demands of CP and CM with PM has been discussed under intelligent context processing and the CPA; this section considers the implementation of the fuzzy variable; the design of the DF with the essential process of defuzzification.

The DF and fuzzy variable design and not independent but are interdependent; this section considers the design of the DF with defuzzification which can be viewed as a process in which a fuzzy result (e.g., in CM where PM is the general case) is converted into a *crisp* result [21] (for example: a Boolean decision in CM in the CPA). It is important to ensure that the use of a fuzzy variable results in a more realistic representation of the problem than can be realized using a crisp variable; this is explored in [20, 21] where the CPA development is discussed.

The approach adopted in the CPA is the use of DB's (or threshold). A DB is a conceptual threshold in the solution space (see Figs. 1, 2, and 3 where the solution space is the CM(X) axis) that defines decision categories. As discussed in this section and in [4] a DB is a crisp concept however the actions resulting from decisions can be fuzzy based on some criteria [2]: for example, the distance (e.g., a ED) from a boundary or boundaries. This relates to the DBPI and the proposed solution to mitigate these effects as discussed in this chapter.

As discussed in the sections addressing intelligent context processing the CPA and the ECPA are designed to produce a Boolean decision (*true* or *false*). In the ECPA the result of step 6 is the (*rv*) metric which is a normalized measure of the degree to which a specific user is a *qualified* recipient for targets service provision. Step 7 applies an adjustment to the (*rv*) metric; a ED [5, 20, 21] is applied to accommodate the DBPI. Additionally, step 7 applies rules to redefine the (*rv*) metric to produce the (*q*) metric which is the semantic conversion of the (*rv*) context match in terms of a semantic representation of the degree of *"qualifiedness"* (*q*) when considered in terms of everyday natural language.

Context matching using a distribution function implemented in the ECPA provides an effective basis upon which the granularity of the result derived from the context matching process can be increased and a more realistic representation of the result (based on the fuzzy variable) achieved than is the case with a simple crisp counterpart.

The following sections consider the design of the distribution functions with an analysis and conclusions. A discussion relating to the design choices for the distribution function set with consideration of issues, challenges, and open research questions is presented in the discussion.

4.1 The Design of the Distribution Function

In considering fuzzy variable design there are essentially two approaches identified in [2]: (1) *Linguistic Design*, and (2) *Data-Driven Design*. As discussed in sect. 3.2 the distribution function employs semantic descriptions to classify the context match; the classifications are: "LQ", "GQ", and "HQ". These semantic metrics are used to identify the CM and apply CS with stopping criteria in context processing. Given the nature of the contextual data utilized in context processing and the semantic representation implemented in the ECPA the adopted approach combines elements of both approaches. The contextual data is used to arrive at the (*rv*) metric (the *Data-Driven Design* approach) and is then converted into a semantic representation in natural language: "LQ", "GQ", and "HQ" (the Linguistic Design approach).

The DB(s) concept is an essential element for both *fuzzy variable* development and *rule formation*; variable and rule design being closely interconnected thus one cannot be performed without consideration of the other [2]. DB(s) can be viewed in terms of a mapping (of a context match) process. Identifying and defining the DB(s) is achieved using a *partitioning* technique as discussed in [2]; in its simplest form, *partitioning* is an attempt to find a DB(s) between (at least two) data points on a solution space. Figure 3 graphically models the *partitioning* of the solution space; shown are the notional (*crisp*) DB(s) and the heuristically defined DB(s).

The use of an approach to context matching based on fuzzy rule-based systems (predicated on fuzzy set theory in which there are degrees of membership of a set) results in a solution space with a *lower bound* of [0.00] and an *upper bound* of [1.00]. The *lower* and *upper* bounds of the solution space represent *crisp* data points on the CM(X) axis (see Figs. 2) which defines the solution space and represent the only known *a priori* data points and information relating to the solution space.

The solution space is initially partitioned to into two partitions each with defined data points on the CM(X) axis; the DB(s). The result is a classification for a specific context match into: "LQ", "GQ", and "HQ" where "LQ" represents a *false* match, "GQ" represents a *true* (acceptable) match, and "HQ" represents a *true* (good) match. Because data can be either *crisp* or *fuzzy* there are two partitioning cases: (1) partitioning based on *crisp* data, and (2) the partitioning solution based on (at least two) *crisp* data points [2]. Given that there is no *a priori* knowledge relating to the probability of a context match being achieved the *partitioning* utilizes the available (normalized) data range as modeled in Figs. 2 and 3. The distribution design identifies:

- The *crisp* data points and the related DB(s). This requires merely *linear extension surfaces* [2], the result being *crisp* DB's at {0.00} and {1.00}.
- Heuristically defined (*fuzzy*) data points DB(s) identified using available knowledge; this approach employs heuristic interpolation predicated on: (a) the results published in the literature relating to the success of recommender systems in targeting service provision, and (b) the data points and DB(s) identified using the *partitioning* approach as discussed in this section.

The heuristic interpolation technique employs a number of factors:

- The literature addressing recommender systems as discussed in [16, 18] has identified success rates (generally predicated on the *precision, recall,* and *fall out* metrics) as identified in [18] in a normalized range: 0.48–0.710 (48–71 % respectively).
- Recall that the only known *a priori* crisp data points are the *upper* and *lower* bounds of the solution space, {0.00} and {1.00} respectively (see Figs. 2 and 3). There is no *a priori* knowledge relating to the identification of the crisp points upon which to base the intermediate partitioning of the solution space.
- As discussed in the analysis (see Sect. 4.3—the *partitioning* and (*rv*) *adjustment* analysis) and in this section the contribution of each context property makes to the overall context is variable and in direct proportion to the number of context properties in an overall context (e.g., for 5 context properties each property contributes 0.20 (5 × 0.20 = 1.00) and 9 context properties each property contributes 0.111(9 × 0.111 = 0.999). Both cases result in a normalized value of [1.00] (the upper bound of the solution space).

Based on: (1) the literature addressing *recommender systems* [16, 18], and (2) the contribution each context property makes to the set of context properties that define a context; the following analysis relating to the partitioning of the solution space (as shown in Figs. 2 and 3) has been used to develop the distribution function.

There are 3 decision boundaries identified for the context matching: (a) the lower bound (*t1* − 0.60), (b) the upper bound (*t3* 1 − 1.00), and (c) the intermediate bound (*t2* − 0.80). The initial partitioning coincidentally falls within an experimental error range of the identified crisp points for recommender systems.

Considering the contribution of each context property makes to the overall context as discussed above the analysis set out in Sect 4.3. as it relates to the contribution made for differing numbers of context properties in an overall context supports the following conclusions:

- It is unlikely that a context match in which 5 or less true matches (for 9 individual context properties) will be acceptable (5 × 0.11 = 0.55). The analysis of the contribution made for differing numbers of context properties in an overall context supports the conclusion that in general less than 5 properties may produce an unreliable result.
- A context match of less than 7 context properties is not a very good match (7 × 0.11 = 0.77).

Based on these conclusions the heuristics applied to set the upper and lower bounds (DB's) are that 6 context properties (maximum 6 × 0.11 = 0.66 degree of membership) will be used for the lower bound for an acceptable match. Clearly 1.0 will be the upper bound, and for the intermediate bound 7 context properties (7 × 0.11 = 0.77 degree of membership) will identify a suitable demarcation point. The degrees of membership for the numbers of true context (property) matches are: 5 true matches: = 0.55, 6 true matches: = 0.66, 7 true matches: = 0.77, 8 true matches: = 0.88, and 9 true matches: = 0.99. Considering these degrees of membership the median points are: (1) between 5 and 6 true context matches m = 0.605, and between 7 and 8 true context matches m = 0.825.

Based on these assumptions the heuristics applied to set the data points on the CM(X) axis are: 6 context properties (maximum 6 × 0.11 = 0.66 degree of membership) will be used for the *t1* DB, 7 context properties (7 × 0.11 = 0.77 degree of membership) will identify the *t2* DB, and clearly 1.0 will be used for the *t3* DB. Thus, the heuristically defined lower (*t1*) and intermediate (*t2*) bounds are: 0.60 and 0.80 respectively; these relate to the semantic metrics: "LQ", "GQ", the upper bound (*t3*) when taken with the intermediate (*t2*) bound defines the "HQ" classification as discussed in this section.

Figure 3 graphically models the distribution function design, shown are: (1) the *linear extension surfaces* used in the initial partitioning of the solution space, (2) the *crisp* DB's (shown as a solid line), and (3) the heuristically defined DB's (shown as a dotted line).

In the presence of fuzzy data between (at least two) crisp points on a solution space, equidistance partitioning (a result of the first approach) can be modified to accommodate the desired boundaries by asymmetric (as opposed to symmetric) functions [2]. The approach adopted for this research is the second approach in which *linear extension surfaces* are used to identify data points for the decision boundary identified by fuzzy data; then the interpolation technique is applied to arrive at a set of heuristically defined DB(s).

Identifying the crisp data points on the continuous solution space (the CM(X) axis shown on Fig. 2) is based on the assumptions set out above. In Fig. 3 the crisp data points obtained using the *partitioning* technique are 50 % (or 0.50 normalized) which is the lowed bound for the "GQ" context match and 75 % (or 0.75 normalized) for the upper bound for the "GQ" context match and the lower

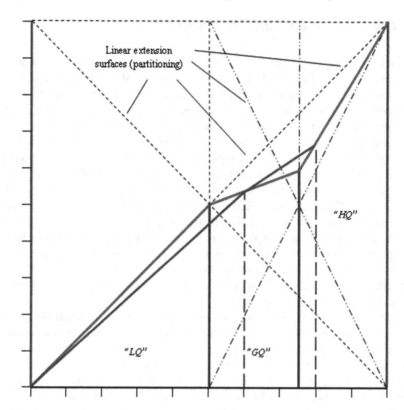

Fig. 3 The *Distribution Function* design with decision boundaries (thresholds): Shown are the linear extension surfaces used to identify *crisp* thresholds, the crisp thresholds, the heuristically defined thresholds, and the semantic representations of CM

bounds for the *"HQ"* context match. The heuristically defined DB(s) derived using the heuristic approach (as shown Figs. 2 and 3) are: (*t1*) 60 % (0.60 normalized) and (*t2*) 80 % (0.80 normalized) to represent the initial design decision parameters.

Thus the lower, and intermediate, and upper bounds (decision boundaries) set using the heuristics discussed above are: 0.60, 0.80, and 1.00 respectively. The (*t3*) DB is a special case as there can be no instance where (**rv**) can exceed [1.0]. The solution space classifications defined by these decision boundaries relate to the semantic metrics "LQ", "GQ", and "HQ" as discussed in Sect. 3.2. Therefore, in the presence of fuzzy data between (at least two) crisp points on a solution space, equidistance partitioning (a result of the first approach) can be modified using heuristics to accommodate the desired boundaries identified by asymmetric (as opposed to symmetric) functions.

4.2 The Decision Boundary Proximity Issue

Identifying the decision boundaries, while enabling defuzzification and providing a basis upon which predictable decision support can be realized, raises a significant issue. Consider for example a normalized context match of 0.595; this degree of membership when tested against the lower bound for the semantic measure "GQ" (*t1*—the 0.60 DB) fails. It is arguable that this is not logical given the very small difference (0.5 % of the overall solution space, and as discussed below 0.834 % of the "LQ" solution space). This example identifies the *Decision Boundary proximity Issue* (DBPI). To attempt to address this issue there is a need for an approach which at the very least attempts to mitigate such anomalies. To this end the concept of *Euclidean distance* (ED) (based on the triangle law) has been investigated in [20] based on the work documented in [5].

Considering the ED as it applies to a DB and the normalized degree of membership (*q*) computed in step 7 of the CPA (as discussed in Sect. 3.2) where ($q = rv * xc$) there are two data points: the context match (the *q*) and the decision boundaries *t1*, *t2*, and *t3* (which as noted is a special case as the *t3*, a context match will never exceed a normalized value of [1.0].

Additionally, adopting this approach provides a basis upon which the use of distance can be extended to other relationships between DB(s) and context match(s) (*q* values) including: (1) context matching achieved using other methodologies such as inference and reasoning and machine learning techniques, and (2) accommodating the relationship between the (*q*) value and the novel *uncertaintyboundary* located on the UC(Y) axis as shown in Fig. 4. A discussion on this topic is presented in the discussion later in this chapter where the location of the uncertainty decision boundary is discussed.

The CPA has been extended to incorporate a rule-based function which identifies IF a context match (the *rv* value) is within 1 % (i.e., 0.01) (on the negative side) of the DB. IF this is the case then, for example, the following rule is applied where: **t1** = the *lower* bound (the 0.60 DB, **xc** represents the adjustment applied to the *rv* value (resulting in *q*), **rv** = 0.595, and **q** represents the semantic context matching result:

IF ((**t1**) $\geq$ (**rv** +**xc**) <(**t2**)) THEN **q** = *"GQ"*)

It can be seen from the above that, within restricted parameters, this rule when implemented in the CPA provides a basis for mitigating the potential anomaly identified. The rule-based function is applied in step 7 of the CPA with the computation of the semantic context matching representation. In the example shown above the initial result "LQ" is promoted to "GQ".

In practice, as shown in the following analysis, the actual % adjustment reduces as the value for the (*rv*) metric increases. For the *t1* DB a 1 % adjustment (0.01) equates to 1.67 % and for the *t2* DB a 1 % adjustment (0.01) equates to 1.25 %

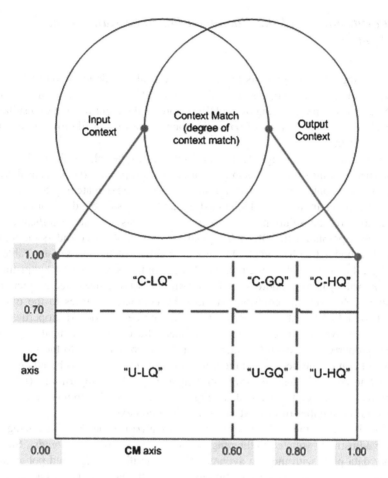

Fig. 4 The context match mapped onto the extended solution space. Shown is the CM(X) axis and UC(Y) axis with the DB's including the additional DB on the UC(Y) axis

(see Tables 3); thus with increasing degrees of confidence in the context matching the adjustment effectively reduces in direct proportion as shown in the analysis presented in the following section.

There are clearly issues in adopting this approach which are shared with the setting of decision support parameters in domain specific context-aware systems. Principal amongst the issues is how to set the distance metric. In practice this will be domain specific and will in actuality be based on expert knowledge and the domain of interest.

4.3 Partitioning and the Euclidean Distance Adjustment Analysis

The following analysis attempts to identify a logical foundation upon which: (1) the partitioning of a solution space—the identification of the data points on the CM(X) axis used in the setting of the DB(s), and (2) the identification of a suitable approach to the identification of (*rv*) adjustment values using ED for applications using CM in the CPA.

Tables 2 and 3 with Figs. 5, 6, and 7 set out: (a) the relationship(s) between the number of context properties in a set of context properties, (b) the contribution made by each context property to the normalized context matching total [1.00], and (c) the trends in the data which assist in identifying suitable data points on the CM(X) axis of the solution space to provide a basis for the identification of suitable data points for the DB(s). The analysis in Sect. 4.3.1 considers the trends in the data under two headings: (1) the identification of the DB(s) (see Table 3 with Figs. 4 and 5), and (2) the relative changes in the percentage (%) adjustment (to the *rv* value) used to address the DP proximity issue as discussed in Sect. 4.2.

Table 2 sets out the contribution made by context properties to the overall context for 1 to 12 context properties (i.e., the number of context properties that make up an overall context). Figure 5 models the contribution made by each context property to an overall context, the relative error related to the increasing number of context properties and the trends in the data (the error is based on 35 % error[1]). Shown are the data plots with moving average (MA), logarithmic (LT) and polynomial (PT) trends in the data. Figure 6 identifies the trends in the data relative to the number of context properties in a context.

Table 3 sets out the relative adjustments to the (*rv*) value; Fig. 7 showing the relative adjustments related to the range of context properties in a context. Shown are the 'data plot' with moving average (MA), logarithmic (LT), and polynomial (PT) trends in the data. The calculation to arrive at the percentage (%) adjustment value is e.g., for an adj(0.01) the adjustment to the (*rv*) is: {0.01/0.60 * 100 = 1.67 %} where the adj is divided by the rv(DB) value (e.g., 0.60) times 100.

4.3.1 An Analysis

Considering the tabulated results set out in Table 1 and Figs. 5 and 6 an analysis identifies a number of conclusions that can be supported:

---

[1] The term 'error' in the context of this analysis relates to the impact of the *weight* <W> property (in actuality the *Literal Value*). Based on an assumption that <W> *Literal Values* will be in the range [0.30, 1.00] the median *Literal Value* is [0.65]; this results in an 'error' of [0.35] or 35 %.

Table 2 Context property contributions for 1 to 12 context properties

Properties	1	2	3	4	5	6	7	8	9	10	11	12
	1.000	1.000	1.000	1.000	1.000	1.000	1.000	1.000	1.000	1.000	1.000	1.000
Contribution	1.000	0.500	0.333	0.250	0.200	0.167	0.143	0.125	0.111	0.100	0.091	0.083
1	1.00											
2	0.50	1.00										
3	0.33	0.67	1.00									
4	0.25	0.50	0.75	1.00								
5	0.20	0.40	0.60	0.80	1.00							
6	0.17	0.33	0.50	0.67	0.83	1.00						
7	0.14	0.29	0.43	0.57	0.71	0.86	1.00					
8	0.13	0.25	0.38	0.50	0.63	0.75	0.88	1.00				
9	0.11	0.22	0.33	0.44	0.56	0.67	0.78	0.89	1.00			
10	0.10	0.20	0.30	0.40	0.50	0.60	0.70	0.80	0.90	1.00		
11	0.09	0.18	0.27	0.36	0.45	0.55	0.64	0.73	0.82	0.91	1.00	
12	0.08	0.17	0.25	0.33	0.42	0.50	0.58	0.67	0.75	0.83	0.92	1.00

Table 3 Relative adjustments related to potential DB values

DB	rv (DB) value	Confidence level (%)	adj (0.01) (%)	adj (0.02) (%)	adj (0.03) (%)	adj (0.05) (%)
	0.00	0.00	0.00	0.00	0.00	0.00
	0.10	10.00	10.00	20.00	30.00	50.00
	0.20	20.00	5.00	10.00	15.00	25.00
	0.30	30.00	3.33	6.67	10.00	16.67
	0.40	40.00	2.50	5.00	7.50	12.50
	0.50	50.00	2.00	4.00	6.00	10.00
T1	0.60	60.00	1.67	3.33	5.00	8.33
	0.70	70.00	1.43	2.86	4.29	7.14
T2	0.80	80.00	1.25	2.50	3.75	6.25
	0.90	90.00	1.11	2.22	3.33	5.56
T3	1.00	100.00	1.00	2.00	3.00	5.00

- As shown in Table 1, the potential DB(s) values {0.00–1.00} the percentage (%) contribution made by each context property (in a context property set that combines to create a context) reduces as the *rv* (the context match) increases.
- Based on the assumption that a higher (*rv*) value provides a greater level of confidence in the context matching process it can be seen that the percentage (%) confidence level rises as the percentage (%) contribution reduces.
- While there is no clear statistical evidence relating to the identification of data points (on the CM(X) axis) and the DB(s) values (the partitioning of the solution space) a visual inspection identified similarities in the data plot, 'MA',' LT', and 'PT' trends in the data.

In considering Fig. 5 (see Sect. 4.3) it is arguable that in general a context made up of less than 5 context properties is a less reliable basis upon which context

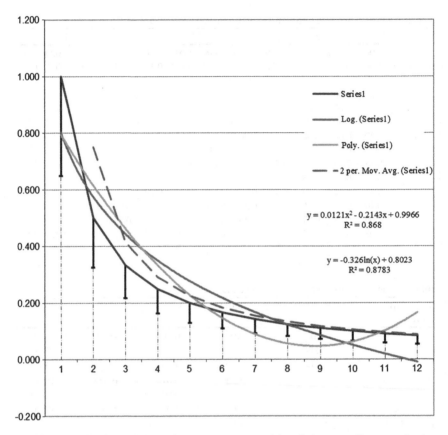

Fig. 5 The contribution made by each context property and the relative error. Shown are the data plots with MA, LT, and PT trends in the data. Fig. 6 identifies the trends in the data relative to the number of context properties in a context

matching can be achieved as the reliability (percentage (%) confidence) of a context match increases with increased numbers of context properties (in a context). This is a result of the variable contribution made to the overall context by individual context properties and a disproportionate contribution each context property makes to an overall context as the number of constituent properties reduces.

The trends in the data and the convergence (in the data plot and the trends in the data) provide support for a number of conclusions principal amongst which are:

- Supporting evidence that context property set's with less than 5 constituent properties provide low confidence levels relating to a context match; this provides evidence that based on the % contribution a *lower bound* (*t1*) of 0.60 is logically appropriate.

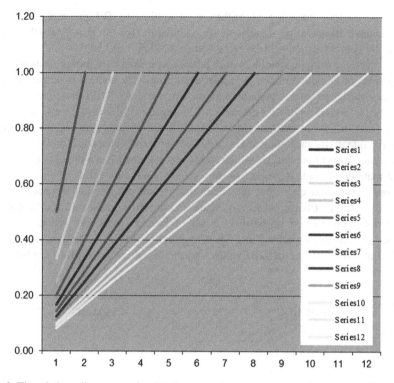

Fig. 6 The relative adjustments related to the range of context properties in a context. Shown are the tends in the data for 1 to 12 context properties in an overall context

- The setting of the *upper bound* (*t2*) at 0.80 is supported by the converging trends in the data around a point on the **X** axis which approximates to 80% of the solution space. This conclusion is also supported by the trends in the data when taken with the increasing level of confidence in the context matching.

In considering Fig. 5 it can be further seen that:

- The error is directly proportional the contribution made by each context property (in a set of context properties); the error reducing as the number of context properties that make up an overall context increases.
- As the number of context properties (that combine to create a context) increase the effect of the *weight* <w> applied to each property (as it applies to the overall context match) reduces. However, as the confidence levels of the context match increases the <w> factor becomes increasingly relevant as the granularity of the context matching (the *rv* value) and the number of context properties (and therefore the number and range of the <w> values) increases number and range of the <w> values) increases.

Additionally, in considering the error as shown on Fig. 5, the error is always negative. This is due to the promotion of a context match (e.g., "LQ" to "GQ") when the context matching algorithm as it applies to DB proximity issue is implemented. In actuality, there is never a situation (for the prototype PMLS) in which the context match will be relegated (e.g., "GQ" to "LQ").

However, the domain specific nature of context makes both the 'promotion' and 'relegation' of a context match probable, accordingly the context match algorithm has been designed to be generic and the (rule-based) DBPI function has the capability to address both 'promotion' and 'relegation' of a context match around a data point (a DB) in both single and multiple dimensions (see Fig. 4) achieved using Euclidean distance adjustments as discussed in this chapter. Notwithstanding these observations the analysis which states that the error has a direct relationship to the number of context properties (in a context) remains valid.

From Fig. 6 it is clear that the relationships and progression for each set of context properties (that combine to create a context) is linear and the % contribution for each context property in a set of context properties reduces as the number of context properties (in a set of context properties) increases. Shown in Table 3 are: the DB, the **rv**(DB) value, and the % values for adj(0.01), adj(0.02),

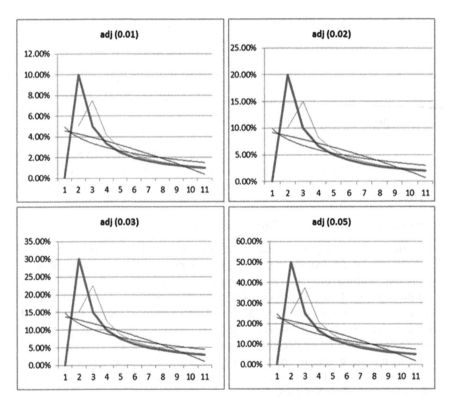

Fig. 7 The 'MA', 'LT', and 'PT' trends in the data as they relate to the logical identification of a partition (DB) there are convergences in the trends around potential DB values that relate to 5 and 8 context properties

adj(0.03), adj(0.05). The adjustment values are modeled in Fig. 7. An analysis of Table 2 and Figs. 5 and 6 supports a number of further conclusions:

- It can be seen that there are clear similarities in the trends in the data and visual analysis identifies: (a) for the CM(X) axis (as shown in Fig. 5) there is a high similarity in the trends in the data irrespective of the % values used for the adjustment, and (b) for the UC (Y) axis (as shown in Fig. 4) while the values clearly change in direct proportion to the % value used for the adjustment there is a clear similarity in the trends in the data.
- Considering the data plot in and the 'MA', 'LT', and 'PT' trends in the data (as shown in Figs. 5 and 7 as they relate to the logical identification of a partition (DB) there are convergences in the trends around potential DB values that relate to 5 and 8 context properties, this supports the observation that 0.60 (DB) and 0.80 (DB) represent suitable approximations for data points on the CM(X) axis and the related DB(s).

Considering the conclusions drawn relating to converging trends in the data shown in Figs. 5 and the converging trends in the data as shown in Fig. 7 provide supporting evidence that the partition (DB) values of 0.60 for DB (*t1*) and 0.80 for DB (*t2*) are based on a logical foundation.

4.4 Analysis Conclusions and Observations

The overall conclusions to be drawn from the analysis set out above [as it relates to the identification of suitable data points on the CM (X) axis and the related DB(s)] as discussed in Sect. 4 (The Distribution Function) are as follows.

Context is inherently domain specific; this attribute is shared by a distribution function which requires domain specific design using domain specific knowledge relating to the domain of interest. Identifying *crisp* data points upon which the location of DB(s) to enable defuzzification requires *a priori* domain specific knowledge however the *upper* (*t3*) and *lower* bounds of the solution space (see Figs. 4 and 5) provides the only *a priori* knowledge relating to the solution space. These data points do however they provide the initial (two) *crisp* data points for the partitioning as discussed in this chapter.

As identified in this analysis and as discussed in [13] the results obtained in research addressing recommender systems (generally based on the *recall* and *precision* metrics) fall in to the range: 48–71 %, these results provide supporting evidence for the initial partitioning of the solution space (as shown in Figs. 4 and 5 using the *crisp* data points (0.50 and 0.75). The two crisp data points provide a basis upon which the related heuristically defined DB(s) *t1* and *t2* (0.50 and 0.75 respectively) are identified.

While the percentage (%) contribution made by a context property to a context (the set of context properties) provides a logical foundation for the identification of the *crisp* and *heuristic* identification of the data points and therefore the related

DB(s) there is little discernable statistical evidence upon which the location of the DB's can be determined. Notwithstanding this observation a visual analysis of the trends in the data supports the overall conclusion that DB's set at 0.60 and 0.80 represent realistic options.

In considering the percentage (%) contributions as shown in Figs. 5 and 6 a context constructed using set of context properties with less than 5 properties arguably fails to provide a high degree of confidence in the overall context match as the percentage (%) contribution made by each context property is arguably too large. Conversely, if we consider a context created using 9 properties the percentage contribution made by each property is too low. The impact of these observations when taken with an analysis of Figs. 5 and 7 is that a context made up of between 6 and 9 context properties may represent an optimal solution.

From a design perspective the impact of the *prioritizing bias* (the *weight* <w> parameter) is arguably greater than the location (on the CM(X) axis of the solution space) of the data points and the related DB(s). This is demonstrated by considering the effects in context matching of high priority (<w> = 1.0) properties and low priority (<w> = 0.3) when implemented in the context matching algorithm as shown in the implementations and proof-of concept set out in [18, 20, 21].

Space restricts a the presentation of the implementations however in summary the example implementations show that in the implementing the CM in the ECPA: (a) the (*rv*) value for a context match with a *false* value for a high priority context property(s) results in a value for *rv* of 0.56, and (b) with a *false* value for a low priority context property(s) the results in a value for *rv* of 0.80. Thus, for a set of 5 context properties the range for the (*rv*) value may be in the range: [0.56, 0.80]; this demonstrates the scope for a context match to be from "LQ" to "HQ" classifications and supports the observations that:

- When designing a membership function and identifying suitable DB(s) the <w> property has a significant effect on the performance of an intelligent context processing
- The designation of the <w> *Literal Values* along with the DB(s) are closely connected and interconnected
- The error as discussed in the foregoing section forms an important factor in the identification of data points and the related DB(s).

In considering context processing the design of a distribution function must include both the *partitioning* technique (used to locate the data points and the related DB(s)) and the setting of the *weight*(s) (*w*) when applied to the context properties as considered in the discussion where issues and challenges identified in the research are considered along with extensions to the solution space to include the Y and Z axes.

Additionally, in the discussion and analysis relating to the decision boundary proximity issue there is little supporting evidence in the data to justify the selection of a specific percentage (%) adjustment value to be applied to the (*rv*) metric. In practice this will be derived based on *a priori* knowledge obtained from domain specific research in which *tacit* and *explicit* expert knowledge derived from knowledge engineering using surveys and investigations which will form a central factor in the setting of DB(s). This conclusion applies to both the data points (DB(s)) and the *weight* <w> value applied to each context property.

The posited approach to the design of the membership function when implemented using domain specific knowledge when used in combination with the known crisp data points as discussed above provides a basis upon which effective partitioning of the solution space and therefore defuzzification for fuzzy rule-based systems can be achieved.

5 Context Processing under Uncertainty

The diverse range of potential contextual information, the general lack of *a-priori* knowledge relating to the domain of interest, and the highly dynamic nature of context with its inherent complexity [15] imposes issues related to decision support under uncertainty as discussed in Sect. 4. To address this issue the use of an additional *uncertainty boundary* has been investigated in [21]. Figure 4 identifies 2 axes (the CM(X) axis and the UC(Y) axis; the initial investigations have used an additional uncertainty boundary located on the UC axis.

The context matching process initially produces 3 semantic results {"LQ", "GQ", "HQ"} in step 7. Step 8 extends the semantic metrics using the (*t4*) DB on the UC(Y) axis. The results are: ("U-LQ", "U-GQ", "U-HQ", "C-LQ", "C-GQ", "C-HQ" where "U" and "C" represent *uncertainty* and *certainty* respectively. These extensions to the range of context matching representation and the semantic descriptions of the context match increase markedly the granularity of the results obtained. The additional step 8 is as follows:

Step 8: Adjust uncertain environmental conditions to adapt with context matching and compute the relationship to the uncertainty boundary where $\{f_1, f_2 \ldots f_n\}$ is a be a set of ECA rules which represent domain specific uncertainty in environmental conditions. Note: **uc** : = the relationship of **rv** to the *uncertainty boundary*; i.e., (*positive*) (**+**) or (*negative*) (**−**) is defined using (**rv + xc**) where ((**t4**) $>$ = **uc(+)**) and ((**t4**) $>$ **uc(-)**).

IF (**t4** $>=$ **rv**) AND <*event*>:

Rule f_1: {<*condition$_1$*> AND <*condition$_2$*> AND <*condition$_n$*> THEN <*action*> **uc(+)**}
  ELSE IF (**t4** $<$ **rv**) AND <*event*>:

Rule f_2: {<condition$_1$> AND (<condition$_2$> OR <condition$_n$>) THEN <action> **uc(-)**}
  ELSE (0.70 $>=$ **rv**(t4) <*event*>):

Rule f_n: {<condition$_1$> AND <condition$_2$> AND <condition$_n$>
THEN <action> **uc**(+)}

An analysis of the ECPA is set out in Sect. 3 where the ECPA is presented. The addition of the new step 8 results in a number of conclusions:

- *Step 6* computes the resultant value (*rv*) for testing against DB values (*t1, t2, t3*).
- The result of *step 7* is a semantic conversion of the *rv* CM metric to one of: {"LQ", "GQ", and "HQ"}. As discussed in Sect. 3.2 there is a direct relationship between this and an expressed user preference relating to the degree of membership derived from context matching (*q*) the user is happy with.
- *Step 8* identifies the context match as a positive (+) or negative (-) condition relative to the uncertainty boundary (*t4*). Note that the posited approach enables multiple uncertainty boundaries thus introducing increased granularity in the classification of uncertainty. Figure 4 expands on Figs. 1 and 2 and shows the DB's and the UB with the semantic classifications ("U-LQ", "U-GQ", "U-HQ", "C-LQ", "CGQ", and "C-HQ") where "U" and "C" represent *uncertainty* and *certainty* respectively.
- The use of ED implemented in step 7 will be retained in step 8 where the application of ED metrics will apply to, for example, both the (*t1*) and the (*t4*) DB's.

The *partitioning* approach (for the UC axis) shares with the previous discussion the lack of *a priori* knowledge therefore a heuristic approach is required based on domain specific expert knowledge. In this research *t4* has been set at {0.70} based on the analysis set out for the CM axis partitioning.

6 Results and Discussion

This chapter has considered context and its complexity with intelligent context processing achieved using the CPA and the ECPA. The results presented demonstrate the ECPA in operation in an actual program run. An overview of fuzzy systems (as they relate to the CM problem and the CPA) has been presented. The design of the distribution function (a core function where fuzzy systems are applied to enable defuzzification which is an essential process where predictable decision support is a system requirement) has been discussed with consideration of the DBPI and an overview of CM under uncertainty. It has been shown that CM using a semantic membership function provides a basis upon which the granularity of the CM result can be improved. In considering context processing the design of a distribution function must include both the *partitioning* technique [used to locate the data points and the related DB(s)] and the setting of the *weight*(s) as applied to the context properties.

The ECPA lies at the core of the generic rule-based intelligent context-aware decision-support system [20, 21]. The posited approach has been shown to provide an effective solution to personalization in information systems in a broad range of domains, systems and technologies where decision support under uncertainty forms a systemic requirement. The analysis of the CPA and the design of the distribution function supports the conclusion that the posited approach using CM implemented in the ECPA using the semantic classifications has the potential to improve the targeting of service provision based and improve decision support under uncertainty in a broad range of domains and systems.

As discussed in this chapter the distribution function currently uses a decision boundary (or threshold) solution implemented in the ECPA. Current research is investigating:

1. In relation to the ECPA: (1) the use enhanced semantic representation, and (2) the extending of the solution space to include the UC(Y) axis and also the (Z) axis to create a two dimensional and ultimately a three dimensional solution space in which a context match can be classified.
2. In considering the design of the distribution function: (a) using enhanced semantic} representation, and (b) how heuristics are applied using the available *a priori* knowledge related to the domain of interest and the problem being addressed.

Points 2(a, b) represent important design parameters for a distribution function. These parameters when taken with the relative contribution each context property (in a set of context properties that combine to create a context) makes to the overall context definition [20, 21] are being used in ongoing research to attempt clarify the approach to the identification for the optimal values for the decision boundaries and derive a methodology to enable the identification of the optimal DB's generalisable across a broad range of domains, applications and systems where predictable decision-support forms a systemic requirement. The results derived from the research will be analyzed to attempt identify the optimal solution for the membership function design generalizable to a range of domains and applications. The identification of the optimal solution represents an open research question.

6.1 Situational Awareness

The developments in 'Cloud-Based' systems represents an interesting concept in relation to situational awareness where location and the use of *"people-as-sensors"* represents an interesting approach to the application of context where *Wide Area Networks* (WAN), *Local Area Networks* (LAN), and *Personal Area Networks* (PAN) form an integrated networked system. This may be viewed in terms of *"crowd-sourcing"* where individuals' locations are shared with possibly other contextual information (profiles) in collaborative activities. SA has been explored in [8, 9, 24] where social media has been investigated in relation to local community SA during emergency situations such as a major incident. In a medical setting and social media (*Twitter, Facebook*, etc), may provide a means by which advance notification of an impending or actual emergency situations may be identified.

The sharing of services and resources may be facilitated by knowledge of others based on their current context. This has been considered in Sect. 2 in relation to a hospital setting where shared contexts for individual's and resources offers the potential for multiple users to have knowledge of patients, staff, and resource availability. Additionally, in assisted living scenarios remote e-health monitoring using mobile and cloud-based solutions may enable significant improvements in the quality of life for patients with financial and resource efficiencies for all.

There are issues and challenges in the sharing of context as discussed in [18, 19, 20], the issues generally relating to data description and *persistent / in-memory* storage. A commonly adopted approach to address these requirements is *Ontology-Based Context Modeling* [14] however this approach involves a number of problems including ambiguous semantic terminology and language difficulties. Resolving these difficulties remains an open research question.

7 Concluding Observations

While the research has resolved many the issues in the processing of contextual information a number of challenges have been identified including: (1) persistent storage of dynamic contextual information, (2) the approach to prioritization of context properties in the CPA (the setting of the *weights* (w) parameter), and (3) alternative approaches to context processing including hybrid systems.

A discussion on the challenges identified is beyond the scope of this chapter however consideration of the nature and scope of the challenges identified, the design choices that influenced the context processing strategy adopted, and the use of ontology-based context modeling can be found in [15–21]. The challenges identified have grown out of the research and are open research questions which form the basis for future lines of research.

References

1. W. Beer, V. Christian, A. Ferscha, L. Mehrmann, *Modeling context-aware behavior by interpreted ECA rules, in Proceedings of the International Conference on Parallel and Distributed Computing (EUROPAR'03), LNCS 2790* (Springer, Klagenfurt, Austria, 2003), pp. 1064–1073
2. R.C. Berkan, S.L. Trubatch, *Fuzzy Systems Design Principles: Building Fuzzy IFTHEN Rule Bases* (IEEE Press, USA, 1997)
3. B. Berndtsson. Lings, logical events and ECA rules. (Tech Rep HS-IDA-TR-95-004, University of Skovde, 1995), http://citeseerx.ist.psu.edu/legacymapper?did=77063
4. A. K. Dey, G. D. Abowd, Towards a better understanding of context and context-awareness. (GVU Tech. Rep. GIT-GVU-99-22, College of Computing, Georgia Institute of Technology, Georgia, 1999).
5. J.C. Gower, Euclidean distance geometry. Math. Sci. **7**, 7–14 (1982)

6. I. Graham, P.L. Jones, *Expert Systems Knowledge* (Uncertainty and Decision, Chapman and hall, New York , 1988)
7. F. Hayes-Roth, Rule-based systems. Commun. ACM **28**(9), 921–932 (1985)
8. P. Heim, D. Thom, SemSor: Combining social and semantic web to support the analysis of emergency situations, in Proceedings of the 2nd Workshop on Semantic Models for Adaptive Interactive Systems SEMAIS (2011).
9. N. Ireson, Local community situational awareness during an emergency, in Proceedings of the IEEE International Conference on Digital Ecosystems and Technologies (DEST 2009), (2009), pp. 49–54.
10. P. Moore, B. Hu, X. Zhu, W. Campbell, M. Ratcliffe, A Survey of context modelling for pervasive cooperative learning, in IEEE Proceedings of the First International Symposium on Information Technologies and Applications in Education (ISITAE '07), Kunming, Yuan, 23–25 Nov 2007, (IEEE, USA 2007), pp. K5-1-K5-6.
11. G.J. Klir, T. Folger, *Fuzzy Sets, Uncertainty, and Information* (Prentice Hall, Englewood Cliifs, NJ, 1988)
12. G.J. Klir, B. Yuan, *Fuzzy sets and Fuzzy Logic: Theory and Applications* (Prentice Hall, NJ, 1995)
13. P. Lonsdale, C. Barber, M. Sharples, T.N. Arvanitis, *A context awareness architecture for facilitating mobile learning, in Proceedings of MLEARN 2003* (UK, London, 2003)
14. P. Moore, *The complexity of context in mobile information systems, in Proceedings of the 12th International Conference on Network-Based Information Systems (NbiS 2009), 1st International Workshop on Heterogeneous Environments and Technologies for Grid and P2P Systems (HETGP'2009), 19–21 Aug 2009* (Indianapolis, Indiana, USA, 2009), pp. 91–96
15. P. Moore, B. Hu, J. Wan. J, 'Intelligent Context' for Personalised Mobile Learning. In Architectures for Distributed and Complex M-Learning Systems: Applying Intelligent Technologies, (Eds) S. Caballe; F. Xhafa; T. Daradoumis; A. A. Juan, Hershey, PA, USA, IGI Global, 236–270 (2010).
16. P. Moore, M. Jackson, B. Hu, *Constraint Satisfaction in Intelligent Context-Aware Systems, in Proceedings of the Fourth International Conference on Complex, Intelligent and Software Intensive Systems (CISIS 2010), 15–18 Feb, 2010* (Krakow, Poland, 2010), pp. 75–80
17. P. Moore, B. Hu, M. B., Jackson, in Computational Intelligence for Technology Enhanced Learning, vol. 273, ed. by F. Xhafa, S. Caballe, A. Abraham, T. Dardoumis, A. Juan, Fuzzy ECA rules for pervasive decision-centric personalized mobile learning, (Springer, Heidelberg, 2010) ISBN: 978-3-642-11223-2, http://dx.doi.org/10.1007/978-3-642-11224-9. (Studies in Computational Intelligence)
18. P. Moore, B. Hu, M. Jackson, Rule strategies for intelligent context-aware systems: The application of conditional relationships in decision-support, in *Proceedings of the International Conference on Complex, Intelligent, and Software Intensive Systems (CISIS-2011), June 30–July 2, 2011, (Korean Bible University (KBU)* (Seoul, Korea, 2011), pp. 9–16
19. P. Moore, H.V. Pham, *Personalization and Rule Strategies in Human-Centric Data Intensive Intelligent Context-Aware Systems* (The Knowledge Engineering Review (Cambridge University Press, UK, 2012). (to appear)
20. P. Moore, H. V. Pham. Hai, Intelligent context with decision support under uncertainty. Second international workshop on intelligent context-aware systems (ICAS 2012), in Proceedings of the 6th International Conference on Complex, Intelligent, and Software Intensive Systems (CISIS 2012), 4–6 July 2012, (Palermo, Italy, 2012), pp. 977–982.
21. G.L.S. Shackle, *Decision Order and Time in Human Affairs* (Cambridge University Press. Cambridge, UK, 1961)
22. M. Sharples, D. Corlett, O. Wesmancott, The design and implementation of a mobile learning resource. Pers. Ubiquitous Comput. **3**, 220–234 (2002)
23. L. Winerman, Crisis communication. Nature **457**(7228), 376–8 (2009)
24. L.A. Zadeh, L. A., Fuzzy sets. Inf. Control **8**(3), 338–353 (1965)
25. L.A. Zadeh, Fuzzy logic. IEEE Comput. **2**, 83–93 (1988)

Autonomous Evolution of Access to Information in Institutional Decision-Support Systems Using Agent and Semantic Web Technologies

Desanka Polajnar, Jernej Polajnar and Mohammad Zubayer

Abstract This chapter addresses the question of how an institutional decision-support system built on legacy relational databases (RDB) can evolve from a traditional database access model to a modern system that provides its decision-making users with agent-assisted direct semantic query access. We introduce a novel approach in which the system ontologies, developed autonomously within the institution, gradually co-evolve with the related ontologies accessible on the Web. This is achieved through cooperative developement of system ontologies by human domain specialists and software agents. The agents assist with ontology-building expertise, discovery of relevant knowledge on the Web, and ontology mediation. The underlying RDB need not be modified, which allows seamless transition and coexistence between access models. The approach is concretized as Semantic Query Access System (SQAS), a distributed system architecture based on agent-oriented middleware, in which database servers develop reference ontologies, while the application-oriented clients import and overlay them with user-specific custom ontologies.

1 Introduction

The impact of computer-based information systems on the progress of human society is well acknowledged in all disciplines. Individuals and organizations increasingly rely on them for problem solving, decision making, and forecasting. As

D. Polajnar (✉) · J. Polajnar · M. Zubayer
University of Northern British Columbia, Prince George, BC V2N 4Z9, Canada
e-mail: desanka.polajnar@unbc.ca

J. Polajnar
e-mail: jernej.polajnar@unbc.ca

M. Zubayer
e-mail: zubayer@unbc.ca

F. Xhafa and N. Bessis (eds.), *Inter-cooperative Collective Intelligence:*   141
Techniques and Applications, Studies in Computational Intelligence 495,
DOI: 10.1007/978-3-642-35016-0_6, © Springer-Verlag Berlin Heidelberg 2014

a consequence of global integration, changing market dynamics, and deployment of new technologies, institutional and corporate decision-support systems face increasingly steep requirements in regard to flexible, fast, and intelligent access to their underlying information repositories. The requests for information are increasing in complexity and sophistication, while the time to produce the results is tightening. These trends compel researchers to look beyond traditional access techniques in order to meet modern requirements. A major constraint in their efforts is the fact that vast amounts of relevant information, accumulated over time, reside in legacy systems that do not adequately support modern access techniques.

In legacy information systems, the relational database model has been dominant for more than three decades. In order to extract the necessary information from relational databases (RDB) with traditional methods, non-technical users require technical assistance of database programmers, report writers, and application software developers, which involves delays, costs, and semantic gaps in human communication. In order to speed up access and give users more control, decision-support systems often rely on data warehousing techniques. Those techniques require information to be selected and extracted from operational databases, reorganized in terms of facts and dimensions, and stored in data warehouses [10]. Operational databases are designed to support typical day-to-day operations, whereas data warehouses are designed for analytical processing of large volumes of information accumulated over time. That approach still requires human mediation, time to restructure large amounts of data, and accurate foresight as to what information might be needed.

In this chapter, we explore an alternative approach, aimed at overcoming those limitations. It combines two fast-developing technologies—the Semantic Web (SW) [4] and multiagent systems (MAS) [16]—to provide the users of enterprise decision-support systems with direct, flexible, and customized access to information, through high-level semantic queries. Since it does not require modification of underlying databases, our proposed form of access can coexist with the more traditional ones. It allows continuous use of the legacy system. An important aspect of the new approach is that the transition from the traditional to the new access model can be effected gradually and autonomously within the host institution or company. This autonomy is significant because the resident domain expertise has an essential role in the transition, as well as because some of the relevant knowledge is often proprietary. A key premise underlying the approach is that the knowledge of the generic ontology-building process may be easier to standardize and formalize than the domain-specific knowledge particular to an organization and accumulated through work experience of its personnel. Accordingly, in the human-agent interactive and cooperative development of the institutional systems ontologies, the human partner should adopt the role of domain specialist and the agent the role of ontology-building specialist.

We present the approach in the form of an intelligent distributed system, called the Semantic Query Access System (SQAS) [11], with servers containing databases and clients providing access to users. Its basic functionality is provided by agent-oriented middleware. Many of the issues arising in SQAS are closely related to Semantic Web research. The SW project envisions a world-wide infrastructure providing universal integrated semantic access to a multitude of distributed knowledge sources. This requires a hierarchy of standard ontologies that correspond to various knowledge domains at different levels of abstraction, as well as languages, design techniques, software components, and tools. As SW technology matures, many of the SQAS development needs should be satisfiable from its repository. Differences stemming from the "closed world" [14] nature of enterprise systems (vs. "open world" SW) are also being studied (e.g., [12]). However, ontologies representing the meaning of database structures in SQAS must be specifically developed. In our approach that is done within the system itself.

An innovative feature of SQAS is the role of agents in ontology building. The system ontologies are built gradually. In a server, the meaning of the database structure is captured in the *reference ontology*. This includes automatic generation of the basic structures from RDB schemas and their incremental enhancement to full ontology through human-agent cooperative design. Reference ontologies are exported to clients that need them. In a client, a layer of *custom ontology* is constructed for each user, as an overlay that relies on the imported reference ontologies, again through human-agent cooperative design. The approach relies on agents endowed with the technical knowledge of ontology-building procedures that assist human actors in the design process. A part of the agents' role is to find, identify, reference, import, display, and apply relevant knowledge available on the Semantic Web. The development of ontologies in turn permits further delegation of operational tasks to agents. The approach is expected to become increasingly effective with the advancement of the SW infrastructure.

The rest of the chapter describes: the principles of our approach to agent-oriented semantic access in institutional decision-support systems (Sect. 2); the basic distributed architecture of SQAS (Sect. 3); the components constituting the intelligent middleware, their roles and interactions on the server (Sect. 4) and client (Sect. 5) subsystems, including agent-assisted development of reference and custom ontologies; a few closing remarks (Sect. 6); and the conclusions (Sect. 7).

2 Agent-Oriented Semantic Access

In this section we examine the basic requirements of user access to information stored in an existing relational database (RDB) in the context of an institutional or corporate decision support system. The user is a decision-making executive who formulates requests for information and receives reports from the system. Our model focuses on access and does not explicitly represent the various analytical processing that may be involved in report generation. The user is aware that the structure of the database and its contents may evolve over time. Our model assumes that such changes are introduced by the database administrator and does not represent the mechanisms by which they may be prompted or influenced by the user. The user is familiar with the knowledge domain of the information in the database, but may differ in specific expertise and interests from other users of the same system. The user is not a database management specialist.

The requirements are developed in three steps. We first describe the requirements for a generic system that represents user access to information in an RDB in a way that is common to its many possible implementations. We then focus on the user-system interaction in legacy RDB systems. Finally, we examine user-system interaction through high-level semantic queries as represented in SQAS, and discuss its perceived practical advantages.

2.1 The generic system

The requirements for a generic system are shown in Fig. 1. They are described in terms of high-level use cases and actors. A use case is a coherent unit of functionality expressed as a transaction among actors and the system. An actor may be a person, organization, or other external entity that interacts with the system [13].

The actors of primary interest for us are User and Database Administrator (DBA). The use cases are largely self-explanatory. The top four use cases of Fig. 1 capture the generic system functions performed on behalf of the user, regardless of how these functions are implemented. In particular, in the *Process Request* use case, the system accepts a request formulated by User, queries the database, and returns a report with the results formatted as requested. The *Manage Ontology* use case is concerned with bridging the semantic gap between User's domain-oriented terminology, often shaped by personal expertise and preferences, and the vocabulary of the database, whose meaning is specified in the database documentation, possibly with clarification of finer points provided by DBA. The last two use cases enable DBA and Data Entry Operator (DEO) to maintain the RDB structure and content respectively. In order to highlight the differences between legacy systems and SQAS, we next focus on two use cases, *Process Request* and *Manage Ontology*.

2.2 The Legacy RDB System

In a legacy RDB system, some of the generic system functions shown in Fig. 1 are performed on behalf of User by intermediary technical personnel, represented here by a human role called Report Writer; the rest are performed by the computer system. The actors and the basic high-level use cases of a legacy RDB system are shown in Fig. 2. Let us elaborate the two key generic use cases.

In the *Process Request* use case, User explains to Report Writer what information should be retrieved and how it should be presented; Report Writer then

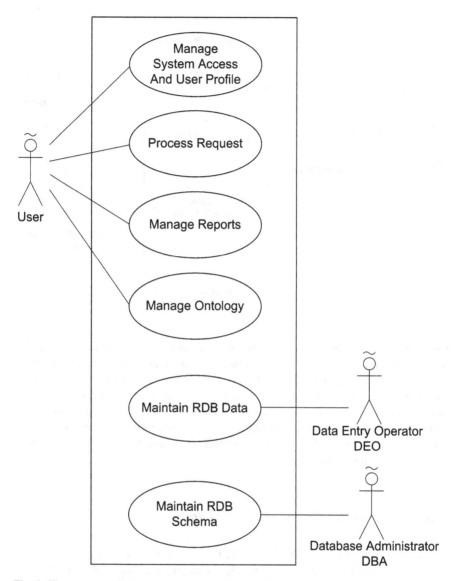

Fig. 1 The actors and high-level use cases of the generic system

queries the RDB to retrieve the information and presents it to User in the requested format. If the meaning of the request is not clear, Report Writer interacts with User in natural language in order to clarify it.

The *Manage Ontology* use case is concerned with the correspondence and translation between the database structures and their meaning, and the domain-oriented concepts and associated custom terminology employed by User. Apart from the basic relationships captured within the RDB schema, the meaning

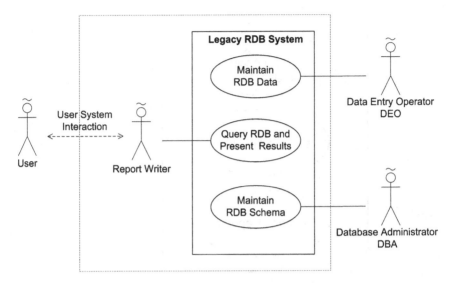

Fig. 2 User access to information in a legacy RDB system

of database structures is typically captured informally in natural-language documentation (and sometimes human knowledge) maintained by DBA. User's terminology and conceptual framework may differ from the ones presented by DBA.
In order to learn both, Report Writer typically depends on informal documentation,
and on natural-language interactions with both User and DBA. This process of
consultation, negotiation, and delegation between User, Report Writer, and DBA is
often time consuming and sometimes ambiguous, resulting in delays, costs, and
occasional misunderstandings.

2.3 The Semantic Query Access System

In SQAS, the user directly interacts with the system that performs the functionalities in the top four use cases of Fig. 1, eliminating the Report Writer role. We
briefly describe the key generic use cases.

The *Process Request* use case allows User to directly communicate requests to
the system, in a simplified natural language. In the request, User specifies what
information should be retrieved and how it should be presented. If User's request is
not clear, the system asks User for clarification of the request. This clarification
process is an interactive one in which the system ensures that it understands User's
request, similar to Report Writer in a legacy RDB system. It then retrieves the
information and presents it in the requested format.

In the *Manage Ontology* use case, the meaning of database structures is formally captured in the *reference ontology*. The reference ontology represents the combined knowledge originating from the underlying RDB structure, the human actors in the system, and external ontologies available on the Semantic Web. DBA interacts with the system in building and maintaining the reference ontology. Thus, the DBA's actor profile now includes the new role of managing the reference ontology in addition to the traditional role of managing the RDB system. Similarly, User's conceptual framework and associated terminology are formally captured in the *custom ontology*. The custom ontology is a layer on top of the client's imported reference ontology; it is custom-built for each specific user. It is also built within the system itself, in interaction with User, with access to the reference ontology and external ontologies available on the Semantic Web. The User actor now has the additional role of managing the custom ontology.

2.4 Decomposition of SQAS Use Cases

The functions of each high-level use case can be further specified through decomposition into more elementary use cases. In presenting the decompositions of key generic use cases in SQAS, we also decompose the functionality into its client part, related to User, and its server part, related to the RDB. The client and server subsystems can reside on different machines and communicate through a network. In general, a client can interact with multiple servers, and a server with multiple clients; this is discussed in more detail in Sects. 3 and 6. For the moment, we consider the case of one client and one server.

The decomposition of the generic *Process Request* use case is shown in Fig. 3. In the client, the *Process SNL Request* use case allows User to formulate a request for information in Simplified Natural Language (SNL). The request contains domain-specific terms that describe the information to be retrieved, and keywords that describe the format in which it should be presented. Once the request is accepted, the *Parse SNL Request* use case produces an intermediate representation of the request, and the *Verify Request Semantics* use case checks that each statement as a whole in the request is semantically correct, including its use of custom ontology terms. If the SNL request is valid, the *Generate SPARQL Script* use case creates a SPARQL script from the intermediate representation of the request. The ontologies in SQAS are represented as Resource Description Framework (RDF) structures [6], and SPARQL [15] is the standard query language for RDF. The client then sends the SPARQL script to the server, and receives the SPARQL results from it. Finally, the SPARQL results are formatted and presented by the *Format and Display Report* use case.

In the server, the *Process SPARQL Request* use case receives the SPARQL script and has it translated to equivalent SQL queries by the *Convert SPARQL Script to SQL Queries* use case. The *Query RDB and Present Results* use case then executes the SQL queries on the RDB system and passes the SQL results to

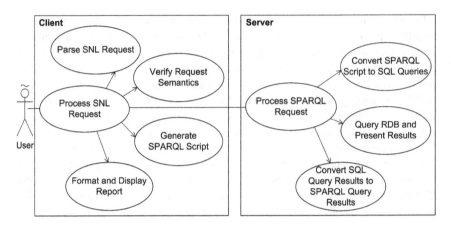

Fig. 3 The SQAS decomposition of the generic use case *Process request*

Convert SQL Query Results to SPARQL Query Results for translation. Finally, the *Process SPARQL Request* use case sends the results to the client. Note that the data remain permanently stored only in the RDB, and that RDF representations are created on demand as a request is processed. This approach does not require any modification in the RDB structure, and allows SQAS to coexist with other methods of accessing the legacy database.

The decomposition of the generic *Manage Ontology* use case is shown in Fig. 4. The use cases that primarily interact with User are assigned to the client subsystem, whereas the ones that primarily interact with DBA and RDB are assigned to the server subsystem.

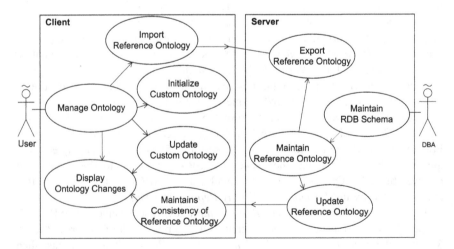

Fig. 4 Use case: *Manage ontology*

The client invokes the *Import Reference Ontology* use case when it connects to the server, relying on the functions of the *Export Reference Ontology* use case in the server. The *Initialize Custom Ontology* use case allows the user to create a conceptual framework specific to the user. The *Update Custom Ontology* use case lets the user modify definitions of user-specific concepts in the custom ontology. When the reference ontology is updated in the server, the *Maintain Consistency of Reference Ontology* use case ensures that the updates are also applied to the reference ontology in the client. Thus the reference ontology is pulled by the client subsystem when it connects to the server initially, or reconnects following a period of disconnected operation. When changes occur in reference ontology while the client is connected, the updates are pushed to the client by the server. The reference ontology updates are displayed to the user by the *Display Ontology Changes* use case.

In the server subsystem, the *Export Reference Ontology* use case sends a copy of the reference ontology as requested by the newly attached client. The *Maintain RDB Schema* use case allows DBA to modify the structure of the RDB. When DBA changes the RDB schema, the *Maintain Reference Ontology* use case incorporates the schema changes into the reference ontology with the help of the *Update Reference Ontology* use case, which also pushes the updates to the attached client.

3 The SQAS Architecture

3.1 The High-Level Architecture of SQAS

At the high level, SQAS consists of any number of clients of the type *User Subsystem (US)* and any number of servers of the type *Database Subsystem (DBS)*. A single US can support multiple users. These subsystems can reside on different machines and communicate through a wide area network using a standard transport protocol. Figure 5 depicts a simple configuration, consisting of one single-user client subsystem and one server subsystem, that is used in most of the current presentation to explain the principles of system operation.

A US consists of an agent, called the *User Interface Agent (UIA)*, attached to each user, and a collection of interacting software components called the *User Interface Environment (UIE)*. Similarly, DBS consists of the *Database Interface Agent (DBIA)*, the *Database Interface Environment (DBIE)*, and the RDB system. The rest of this section outlines the agent roles of UIA and DBIA. Given the diversity of roles, each of these agents should preferably be internally designed as an agent team. Those internal designs and the associated issues of agent teamwork are beyond the scope of the current presentation.

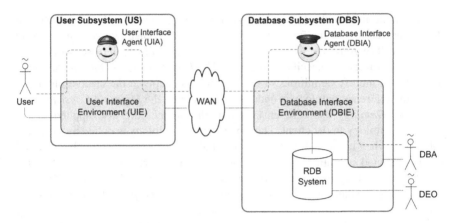

Fig. 5 A simple SQAS configuration with one single-user US and one DBS

3.2 The Roles of User Interface Agent

Assistance in SNL dialogue. User addresses the system in Simplified Natural Language (SNL). If the SNL processor generates a warning, UIA tries to autonomously resolve the issue in interaction with the subsystem components. If the SNL processor reports an error, UIA engages with User to correct it.

Searching the Semantic Web. The agent can search the Semantic Web for relevant external knowledge. For instance, it can look up synonyms and hypernyms of terms in natural language knowledge sources such as WordNet [9]. It can also look for relevant domain ontologies to standardize the usage of terms or complement the locally developed custom ontology.

Development of custom ontology. UIA helps User create and maintain a custom ontology, a user-specific conceptual framework that is translatable to reference ontology. In order to maintain consistency, UIA ensures that any updates to the reference ontology are reflected in the custom ontology. UIA has the technical knowledge of the required ontology development and mediation mechanisms.

Customizing the behavior of User Interface Environment. While assisting User with SNL dialogue, UIA may learn from observations of User's preferential choices and customize the behavior of the user interface and possibly other UIE components, assuming that User elects to enable such options.

Coordination of reference ontologies. When US interacts with multiple DBSs, UIA acts to resolve any conflicts between imported reference ontologies.

3.3 The Roles of Database Interface Agent

Assistance in SNL dialogue. Apart from the conventional RDB operations on RDB, DBA addresses the system in SNL, with the agent's assistance as in US.

Searching the Semantic Web. The agent's actions are similar as in US, but with primary emphasis on sources needed in the building of reference ontology.

Development of reference ontology. DBIA interacts with DBA in developing and maintaining the reference ontology. A DBIE component called the *Schema to Base Ontology Mapper* analyzes the RDB schema and generates a Mapping File, which contains Resource Description Framework (RDF) models of the RDB schema. The Mapping File then serves as the *base ontology* from which DBA incrementally builds a full reference ontology with the assistance of the agent. DBIA provides technical guidance in the ontology development process, and accesses ontologies on the Semantic Web.

Customizing the behavior of Database Interface Environment. Primarily customizes the behavior of the user interface component, as in US.

4 Agent-Oriented Middleware for Server Subsystems

The software of SQAS is distributed between its client and server subsystems. At each end, it consists of an agent and its environment that contains a set of interacting software components. The environment components can be designed and implemented using the conventional object-oriented software engineering (OOSE) methodology, with an emphasis on efficient performance. The agent can observe every component and interact with it, and it also interacts with the primary human actor. While these interactions are central to our present discussion, it is important to note that the agents along with core environment components constitute a layer of *intelligent middleware* that can offer support to other applications. An enterprise system normally includes a variety of business intelligence applications performing various types of analysis. In the context of SQAS, such applications would be realized as agent-oriented software, running on top of core SQAS. In general, agents in such applications would interact with SQAS agents, with the users, and with the Semantic Web.

This section focuses on the internal structure of DBS, shown in Fig. 6. In Sect. 4.1 we discuss the DBIE components that support the main subsystem functions, in Sect. 4.2 a strategy for SQAS middleware implementation, and in Sect. 4.3 an example of an ontology-building scenario executed within DBS.

4.1 The Database Interface Environment

This environment comprises all software components that communicate with DBIA, DBA, DEO, and RDB. The solid lines represent direct communication between components. The dashed line represents communication between DBA and DBIA. A dashed envelope groups the components directly interacting with RDB that we revisit in Sect. 4.2. DBIE includes the following main components:

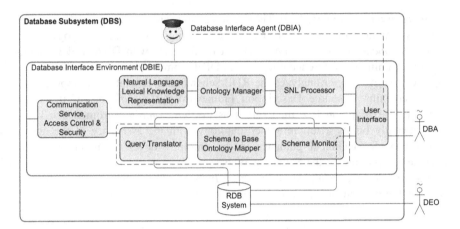

Fig. 6 The Database Subsystem

The *User Interface* provides an access point at which DBA interacts with DBS. Through it, DBA maintains the reference ontology with the assistance of DBIA and manages the RDB system. Based on its observation of user behavior and its learning abilities, the agent can intelligently adapt the interface to meet the user preferences.

The *SNL Processor* enables DBA to interact with the system using Simplified Natural Language (SNL), in addition to the more conventional user interface options. Given a user statement in SNL, the SNL processor first performs the lexical analysis and syntax analysis, using the SNL language definition, the vocabulary information from the reference ontology, and lexical information from the Natural Language Lexical Knowledge Representation component to generate an intermediate representation of the statement. After that, it performs semantic analysis, including ontology checking in interaction with the Ontology Manager, to verify that each statement as a whole is meaningful. Once the intermediate representation is generated and verified, the SNL Processor invokes the relevant components that execute the DBA's request. For instance, it can activate the Ontology Manager to update the reference ontology, or the agent to initiate a Web search.

The *Ontology Manager* is responsible for storage and maintenance of the reference ontology. DBS exports a copy of the reference ontology to the attached US. Thus the reference ontology is replicated in both subsystems. The Ontology Manager ensures that any modifications to the reference ontology in DBS are propagated to the instances of reference ontology in all participating USs. It also provides the SNL Processor the ontology information needed for semantic analysis.

The *Natural Language Lexical Knowledge Representation* component provides the meaning and semantic relations between natural-language concepts, as well as vocabulary knowledge, in both machine processable and human readable format. It provides a language ontology which can be enhanced by acess to external language ontologies. The agent and the SNL Processor communicate with this component to look up meanings and relationships between natural language terms.

The *Query Translator* generates SPARQL query results from RDB data in three steps. First, it converts the SPARQL script to SQL queries; second, it executes the SQL queries on the RDB system and retrieves the SQL query results; finally, it converts the SQL query results to SPARQL query results. The Query Translator then sends the SPARQL results to the US.

The *Schema to Base Ontology Mapper* automatically generates a base ontology in the RDF format from the underlying RDB schema. The base ontology represents an RDB table name as a class and the column names of the corresponding table as properties of the class. It also captures the relationships between RDB tables. The base ontology serves as a rudimentary ontology from which the reference ontology is incrementally developed. The Mapper re-generates the base ontology whenever it is alerted to a change in the RDB schema.

The *Schema Monitor* always listens for changes in the RDB schema made by DBA. When it detects a schema change it notifies Schema to Base Ontology Mapper to reflect the modifications in the base ontology, and then prompt the adjustments in the reference ontology.

The *Communication Service, Access Control, and Security* component facilitates all communications between US and DBS. By enforcing security features it ensures that no unauthorized access occurs.

The *RDB System* contains relational data which the user of SQAS is interested in. The Data Entry Operator (DEO) may insert, delete, or modify data in the RDB system. SQAS is not affected by such modifications. The structural changes to RDB are introduced by DBA as modifications to the RDB schema, which are intercepted by the Schema Monitor and further result in modifications to the reference ontology.

4.2 A Note on Implementation Strategy

The intelligent midldleware of SQAS includes a variety of components, some of which would require both research and implementation efforts. The most innovative and research-oriented aspect of SQAS is the role of agents in ontology building. Many of the other components could be adapted from existing or future solutions in the development of Semantic Web, natural language processing, and some other areas. A plausible implementation strategy would be to adapt the architecture of SQAS as necessary in order to take full advantage of independently developed solutions and software components. In this section we briefly illustrate this approach with the three components of Database Interface Environment presented in the dashed envelope in Fig. 6.

The three compnents, namely the *Query Translator*, *Schema to Base Ontology Mapper*, and *Schema Monitor*, jointly provide the necessary conversions between the RDB schema and the base level of the reference ontology in RDF format, as well as the actual translation of queries and results between the two formats. The functionalities of the first two components are provided by a number of existing tools, and most closely matched by the D2RQ platform [5]. The D2RQ Engine is the core of the platform which provides the conversion service. It analyzes the structure of the RDB and generates a Mapping File, which corresponds to an RDF representation of the RDB schema. In SQAS, the Mapping File represents the base ontology. The D2RQ Engine thus performs the role of Schema to Base Ontology Mapper. The Engine then uses the Mapping File to translate SPARQL queries to SQL queries, invokes the RDB, and translates the SQL results back to SPARQL results. The front end of the platform, the D2R Server, accepts SPARQL queries, passes them to the Engine, and presents the returned SPARQL results (RDF triples). The D2R Server and the D2RQ Engine thus match the role of the Translator component. The third component, the Schema Monitor, is a custom designed extension introduced in [17]. It is an interceptor component placed between DBA and RDB that recognizes the SQL commands which modify the RDB schema and prompts the D2RQ Engine to re-generate the Mapping File, i.e., the base ontology.

D2RQ can work with the Jade [3] agent platform with the assistance of Jena [8]. A Jade agent uses Jena's SPARQL capabilities for executing a SPARQL query on the D2RQ Platform.

4.3 A Scenario for Agent-Assisted Ontology Development

We will now have a closer look at the behavior of agents as ontology builders. In SQAS, the agents interact with human actors throughout the entire ontology development process. The agents perform some of the technical tasks and make suggestions, while the human actors make decisions. This human actor role in ontology development adds a new dimension to the traditional User and DBA profiles. However, this does not require them to become technical experts fully specialized in the ontology development process because the agents are responsible for executing some of the technical tasks.

The SQAS agents must have the requisite knowledge of how to build an ontology in order to fulfill their roles. This includes the ability to understand the semantics of general ontological notions, such as class, subclass, property, and relationship. Such conceptual knowledge itself represents an ontology, to which we refer as *meta-ontology* (noting that this use of the term differs from its established meaning in philosophy). The agents must also have the procedural expertise in the development of knowledge representations. They provide technical guidance and assist their human partners in the construction of concrete ontologies for the knowledge domains specific to the given databases.

We illustrate the process with a few examples from a scenario in which a human designer (DBA) interacts with an agent (DBIA) to construct a reference ontology from a university relational database. The elements of the reference ontology in these examples are constructed in Web Ontology Language (OWL). As a well-known ontology language, OWL is a convenient choice for presentation purposes. The choice is not intended to suggest that OWL representations are well suited for reasoning in SQAS, whose agents must deal both with the open world of the Semantic Web and with the closed world of the institutional information system. The questions related to the optimal choice of ontology language for SQAS are beyond the scope of this chapter.

As a first step in the development of reference ontology, its domain name is chosen, and a new name space is established, with a suitable prefix that allows one to differentiate between the names coming from different ontologies. Several types of names, each identified with a distinct standard prefix, may appear in the reference ontology. A concept may have several names, but those synonyms have different roles that are indicated by their standard prefixes.

A *base* name is introduced by the mapping of the RDB schema into base ontology. It is automatically derived from a term used in the RDB schema. For instance, from the RDB table name `Department` the mapper produces a base ontology entry `map:Department a d2rq:ClassMap`, which results in the base ontology class name entry `<owl:Class rdf:ID=''bn.Department''/>`, where the `bn` prefix identifies a base name. Base names of properties are constructed similarly. For instance, from the column name `FirstName` in `Student` table in the RDB, the mapper constructs `map:Student_First-Name a d2rq:PropertyBridge;` which results in the property base name `FirstName` of the base class `Student` in the reference ontology. The base names cannot be changed independently, because their role is to maintain the correspondence between the reference ontology and the underlying RDB schema.

When the mapping is completed, the agent presents each base name to the designer for the decision on the primary name of the same concept. The *primary* name of a concept is its unique official identifier within the reference ontology, as distinguished from other synonyms. The agents use the primary names (with prefix `pn`) in reasoning and interactions with environment components; the system allows user-specific synonyms (prefix `un`) in communication with human actors. The designer may adopt the base name as the primary name, or consider other choices. In the latter case, the agent may assist by offering natural language synonyms of the base name from a lexical knowledge source such as WordNet [9].

Once the base classes are defined, the designer and agent can define the more general classes. The superclasses can be defined in several ways. The designer may identify several existing classes that can be generalized into a new superclass, provide the primary name for the superclass, and let the agent create it. The agent may offer natural language synonyms before the name choice is finalized, or may look for natural language hypernyms that are common to the primary class names of all sunbclasses of the new class and offer them as candidate names for the new class. For example, the base classes `Student` and `FacultyMember` could be used to abstract a new superclass `Person`, for which there is no corresponding table in the RDB.

It should be noted that, since the entire reference ontology is ultimately derived from the RDB schema, the only concrete classes are the base classes; they are instantiated in the RDB, where all data reside. All other classes are abstract. This distinction influences the handling of property names in superclasses. If a superclass property is designated by identical primary names in all of its subclasses, and all the subclasses are abstract, the entries for that property can be removed from the subclass descriptions, as the property will be inherited from the superclass once it has been defined. However, base classes retain their properties with their associated base names in order to maintain the translatability to the RDB schema. Another observation is that when introducing new superclasses in the case where the primary names of the corresponding properties in subclasses do not match, the designer may be tempted to revisit the subclass definitions and rename the properties in order to remove the name conflicts. This may be simple in a very early design stage when the reference ontology has not been exported to client systems and the dependencies in the existing local software are few and easily traceable; later on, a change of primary property name in an existing class may require a lot of maintainance in derived ontologies and applications, and reliance on synonym management may be preferable to a primary name change. Again, agent's assistance and ability to track the implications of a potential change may be highly valuable to the designer.

The Schema to Base Ontology Mapper can recognize relations between RDB tables that result in class relations within the reference ontology. For instance, the following map entry identifes the ontology relation in which a university department offers a course:

```
map:Course_DepartmentName a d2rq:PropertyBridge;
    d2rq:belongsToClassMap map:Course;
    d2rq:property vocab:Course_DepartmentName;
    d2rq:refersToClassMap map:Department;
    d2rq:join "Course.DepartmentName
    => Department.DepartmentName";
```

When the building of reference ontology is completed, the agent dispays it as an editable graph to the designer for modification and approval (Fig. 7).

5 Agent-Oriented Middleware for Client Subsystems

The architectural structure of the US is shown in Fig. 8. The User Interface Environment (UIE) comprises the components that provide the main subsystem functions. The primary purpose of the UIE is to execute the routine user requests efficiently, without the need to engage in reasoning in the sense of artificial intelligence techniques. The User Interface Agent (UIA) can observe the events in the environment, including the behavior of individual components, and act on the environment to influence the behavior of its components. The agent provides the

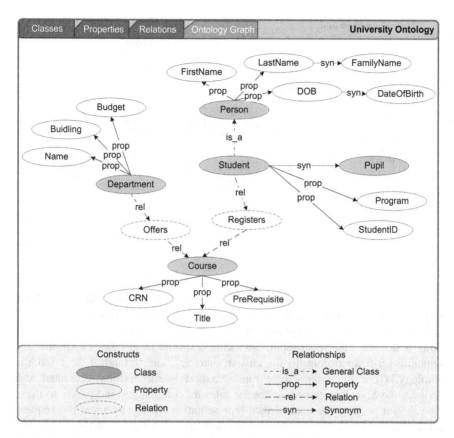

Fig. 7 The reference ontology graph for a small university RDB

practical reasoning (i. e., deliberation and planning) capabilities to the subsystem, enabling it to autonomously resolve arising problems without intervention of human experts. Its presence introduces the qualities of flexibility, adaptability, tolerance to variations in user preferences and practices, and evolution of the subsystem behavior according to changing user requirements. Those qualities are necessary in order for the system to meet its objectives without additional human assistance.

5.1 The User Interface Environment

All the components that communicate with the user and the UIA are grouped into the UIE. The solid lines represent direct communication between the user, the components, and the UIA. The dashed line represents communication between the user and the UIA. The UIE consists of the following main components:

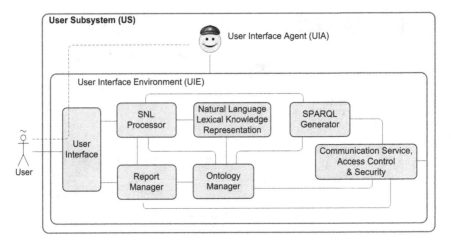

Fig. 8 The User Subsystem

The *User Interface (UI)* enables all communications between the user and the system. It provides the system access functionality as formulated in Sect. 2.3.

The *SNL Processor* component enables the user to interact with the system using SNL. The analysis of input statements is similar as in DBS, except that the semantic verification consults the custom ontology, and through it the reference ontology. Once the intermediate representation is successfully generated and verified, the SNL Processor invokes the relevant components and passes to them the relevant parts of the intermediate representation. If the statement is a request for information, the SNL Processor invokes the SPARQL Generator with the query-related information, and the Report Manager with the formatting instructions. Otherwise, it forwards the statements to the Ontology Manager.

The *SPARQL Generator* constructs a SPARQL script from the intermediate representation of user requests for information received from the SNL Processor. While constructing a script, it refers to the Ontology Manager for the RDB-specific names of terms used in the requests. Once a SPARQL script is generated, the UIA sends it to the DBS for further processing.

The *Report Manager* presents requested information in the form of reports. It receives SPARQL query results from the DBS and formats the results according to the user's formatting preferences. It communicates with the Ontology Manager to replace any database-specific name in the report with its primary name. The Report Manager allows the user to view, reformat, save, and delete reports.

The *Ontology Manager* is responsible for maintaining the custom ontology and providing ontological services to the SNL Processor and SPARQL Generator. The custom ontology defines user-specific concepts and their relationships using constructs from the reference ontology. Updates to the reference ontology may require updates to the custom ontology in order to maintain consistency. The updating process may require the involvement of UIA, and possibly User.

The *Natural Language Lexical Knowledge Representation* component has an identical role as in DBS. The UIA and the SNL Processor communicate with this component to look up meanings and relationships between natural language terms.

The *Communication Service, Access Control and Security* component facilitates communication between the US and the DBS. It provides user authentication, privileges, security, and the interactions with lower-layer communication services.

5.2 A Scenario for Agent-Assisted Semantic Access

We can now illustrate how a user accesses information in SQAS with a simple scenario in which a request is entered and a report is generated. This scenario assumes that the user requests information stored in the RDB through the User Subsystem (US), which has a copy of the reference ontology consistent with the original in the DBS. The US also has a custom ontology for the current user, which defines user-specific concepts using constructs in the reference ontology.

The user formulates a request for report in the Simplified Natural Language (SNL) and submits it through the User Interface (UI):

```
Generate list of students that registered
    for Fall 2012.
Include in the list student ID, first name,
    last name, date of birth, CGPA.
Format report using format-k with
    title: Registered Students;
    subtitle: Date: today's date;
    sort list alphabetically by last name.
```

Some of the words in an SNL request express the control structure and other relationships in simple English (e.g., that, with, using); others directly relate to system actions (generate, format); and some have a defined meaning in the custom or reference ontologies (student, last name). In all three categories the user has the flexibility of defining custom terms. In this scenario, we only consider the translation of custom ontology terms.

The current request has three related statements. The first statement tells what information is to be retrieved; the second statement gives additional details as to what specific information is to be included in the report; and the third statement describes how the information is to be formatted. The first two statements constitute a query, and the last specifies the report generation.

The SNL Processor analyzes the request online, allowing the agent and the user to deal with any arising problems. The Processor first performs lexical analysis in which it breaks the SNL text into a sequence of tokens, such as words and symbols. This is followed by syntax analysis that checks whether the text is grammatically correct and generates an initial intermediate representation. Next, the Processor performs semantic analysis to determine if the statements in the request are meaningful. In this step it may consult the language definition, as well as the custom and reference ontologies. In particular, this implies that each ontological term is ultimately translatable to the base ontology level (which implies that it meaningfully relates to the RDB schema). During this step, the Processor recognizes the actions that need to be performed and verifies whether all parameters that are needed for these actions are present. The fully verified intermediate representation is further transformed so that it can be executed by invoking the appropriate components in the environment.

In each of the described steps, the SNL processor may produce a warning or an error. For example, during the semantic analysis, the SNL processor searches both the natural language vocabulary in the lexical knowledge component and the custom and reference ontologies, to determine whether the word or phrase has a purely natural language meaning or a technical meaning. If both searches are successful, there may be an ambiguity to resolve. For example, the words *First* and *Name* both have general meaning in English, but the expression *First Name* is found as a property of the class *Student* in the reference ontology and hence has a technical meaning. The warning raised by the SNL Processor in situations of this type would be intercepted by the agent that would usually resolve it in favor of the technical interpretation autonomously, without invoking the option of consulting the user.

Another situation that prompts agent's intervention arises when a term, such as registered in our current example, appears from the context to have a technical meaning, but cannot be found in that exact form. The agent may look for lexically similar terms and find that there is a class called registration. The user may be consulted for clarification and also prompted to add the definition of the term registered to the custom ontology.

An important aspect of SNL processing is that all custom ontology terms in the query must be eliminated and replaced by reference ontology terms. The intermediate representation of the query contained in the request will be passed to SPARQL Generator, and the SPARQL script that it produces will proceed for further processing to DBS, where the custom ontology is not known. Since custom ontology terms are more likely to be involved in ambiguity resolution that requires linguistic analysis and occasional interventions by the agent and user, the SNL Processor is the most suitable component to effect the custom term elimination with the assistance of Ontology Manager. The immediate representation links the reference terms to their ontology definitions. It is left to SPARQL Generator, as its preprocessing step, to translate the reference ontology terms into base ontology terms, which are then used to generate the SPARQL script. In our current example, the primary name date of birth is replaced by the base name DOB. As a part of the preprocessing, SPARQL Generator removes the leading tag (bn.) from each base term.

The SNL Processor executes the commmands, such as *Generate*, *Include*, and *Format*, by structuring the intermediate representation and passing its relevant parts to specific components in the environment. When executing the request in our current example, the SNL Processor forwards the intermediate representation of the query derived from the first two statements to SPARQL Generator, and the formatting information derived from all three statements to Report Manager. The SPARQL Generator then translates the query information to SPARQL script and sends it on to DBS, while Report Manager uses the information to produce the formatted report when SPARQL query results come back from DBS. Note that it is not necessary to purge the custom terms from the information passed by the SNL Processor to Report Manager, as the report may indeed employ some user-specific terminology involved in the formulation of the request.

The SPARQL Generator constructs a SPARQL script from the commands and parameters received from the SNL Processor. In general, the script may contain multiple SPARQL queries; in the current example there is only one. The SPARQL Generator divides the technical terms in the first statement into two categories: the 'basic terms' set, consisting of terms appearing before the keyword *that*, and 'conditional terms' set consisting of the terms that follow. This distinction later helps the generator in constructing the script. A SPARQL query is made up of three components: the PREFIX declaration, the SELECT clause, and the WHERE clause. The generator gets the base URI `base=''http://local-host:2020/vocab/resource/''` from the reference ontology header and includes it in the query as a PREFIX. In the prefix declaration, it replaces the equals symbol (=) with a colon (:), and the quotes with opening (<) and closing (>) tags. It then constructs the body of the query consisting of a SELECT clause and a WHERE clause. The SELECT clause identifies the variables to appear in the query results. Those variables are taken from the technical terms appearing in the second statement of the SNL request. The generator appends a leading "? " symbol to each base name to make it a variable. In our current example, the variables are `?StudentID`, `?FirstName`, `?LastName`, `?DOB`, and `?CGPA`.

In the WHERE clause, a number of triples are constructed. A triple consists of a subject, a predicate, and an object. The subject is a variable created by appending the "? " symbol to the class name from the 'basic terms' set (`student`). The predicate is a technical word in the URI format (`PREFIX:Class_Property`), constructed in two steps. First, SPARQL Generator concatenates a class name from the 'basic terms' set and a property name from the SELECT clause with an underscore symbol (`_`). Second, it concatenates the prefix (`base`) and the previously created segment (`Class_Property`) with a colon symbol (`:`). The object variable is constructed using the property name. Following this method the generator constructs a triple for each variable appearing in the SELECT clause. Finally, the generator constructs a triple for each property name from the 'conditional terms' set (`Semester`, `Year`), using the class name variable from the 'conditional terms' set (`?registration`) as subject, and the specified property value as object. These two groups of triples are then linked with a third triple whose predicate has the property `StudentID`, which is a common property between the class in the 'basic terms' set and the class in the 'conditional terms' set.

The complete SPARQL script for our current example is:

```
PREFIX base: <http://localhost:2020/base/resource/>
SELECT ?StudentID ?FirstName ?LastName ?DOB? ?CGPA
WHERE {
    ?student a vocab:Student.
    ?registration a vocab:Registration.
    ?student base:Student_StudentID ?studentID.
    ?student base:Student_FirstName ?FirstName.
    ?student base:Student_LastName ?LastName.
    ?student base:Student_DOB ?DOB.
    ?student base:Student_CGPA ?CGPA.
    ?registration base:Registration_StudentID ?student.
    ?registration base:Registration_Semester "Fall".
    ?registration base:Registration_Year "2012".
}
```

Once the SPARQL script is constructed, the Communication Service, Access Control and Security component sends it to the destination DBS. The corrresponding component in DBS receives the SPARQL script. By verifying credentials of the sender, it ensures that no unauthorized access occurs to the RDB system. It then passes the SPARQL script to the Translator component, which decomposes the script into one or more SPARQL queries. The D2RQ Engine within Translator generates the equivalent SQL queries. The SQL query generated in our example is:

```
SELECT Student.StudentID, Student.FirstName,
       Student.LastName, Student.DOB, Student.CGPA
FROM Student, Registration
WHERE Student.StudentID = Registration.StudentID
      AND Registration.Semester = 'Fall'
      AND Registration.Year = '2012'
```

The D2RQ Engine executes the SQL queries on the RDB system and retrieves SQL results. Query Translator then converts them from SQL format to SPARQL format and Communication sends them to the US. A subset of the generated SPARQL results is shown in Fig. 9.

The Report Manager component in US receives the SPARQL results from DBS. It then formats the results according to the instructions provided in the request by the user. A user selected template (format-k) is used for displaying the report. It also sorts the SPARQL results alphabetically by LastName. Report Manager refers to Ontology Manager to replace any base name with its primary name or user-specific name. It then displays the formatted report to the user. The report generated from the SPARQL results is shown in Fig. 10.

Fig. 9 The SPARQL results

StudentID	FirstName	LastName	DOB	CGPA
98988	Shen	Ming	1988-12-22	3.25
44553	Phill	Cody	1990-05-10	3.7
98765	Emily	Brandt	1978-10-29	2.85
70665	Jie	Zhang	1990-08-26	3.4
76543	Lisa	Brown	1992-06-01	3.7
19991	Shankar	Patel	1986-02-17	3.65
70557	Amanda	Snow	1989-01-17	3.1
76653	Tom	Anderson	1984-03-20	3.5

. . .

6 Closing Remarks

A few points remain to be made on issues that become apparent when envisioning a full-scale architecture of a system such as SQAS. These issues have been indicated but not elaborated while we presented the main principles in a simplified setting.

In a large and complex information system, a client subsystem would typically not import the complete reference ontology that a server has developed, but only its part, or view, that is relevant to the users of the particular client subsystem. This requires methods for specifying views and for their coordinated maintenance by the Ontology Manager components of the client and server.

A client subsystem would in general connect to multiple servers and form its reference ontology by composing the views imported from them. Therefore each entry in the client subsystem's reference ontology must carry a tag indicating its source, which enables SPARQL Generator to produce separate scripts for queries directed to different servers, and Report Generator to correctly integrate the retrieved information. While the basic mechanisms require little adjustment, the objective of smooth and user-friendly integration of views increases the complexity of ontology management and involves various research questions.

When a group of users naturally share a common composite view of reference ontology, the designer may choose to give each user a separate replica or let them share access to the same copy. In the latter case, individual users may still need distinct custom ontologies, which increases the complexity of ontology management.

Even a brief enumeration of developmental issues indicates the centrality of ontology management tasks to the semantic access framework formulated above. Many of the arising questions appear to belong to mainstream research topics motivated by a wide variety of potential applications. There are strong indications that the resulting ontology management techniques will require careful balancing of conflicting objectives, reasoning, autonomous judgment, learning from experience, and intelligent interaction with other entities. The development of such techniques along the lines proposed in this chapter could find immediate

Registered Students
Date: January 10, 2012

Student ID	First Name	Last Name	Date Of Birth	CGPA
76653	Tom	Anderson	1984-03-20	3.5
98765	Emily	Brandt	1978-10-29	2.85
76543	Lisa	Brown	1992-06-01	3.7
44553	Phill	Cody	1990-05-10	3.7
98988	Shen	Ming	1988-12-22	3.25
19991	Shankar	Patel	1986-02-17	3.65
70557	Amanda	Snow	1989-01-17	3.1
70665	Jie	Zhang	1990-08-26	3.4
...				

Fig. 10 The formatted report

application in the processing layer of context aware systems [2], in acquisitional query processing systems [7], and in a number of other rapidly advancing areas. This reinforces our view that agents will play a significant role in ontology building and management in complex intelligent systems of the future.

7 Conclusions

This chapter makes the case for the use of intelligent agents in ontology-building tasks as a means for autonomous evolution of conventional decision-support systems in institutional or corporate environments towards modern systems that provide flexible and direct access to information through high-level semantic queries. This novel approach outlines an incremental evolutionary path that permits continuous operation of the system, requires no modification of the legacy databases and allows conventional access to them, preserves organizational autonomy, and supports direct semantic-query interactions between the decision-making user and the software system, without intervening personnel.

The approach is based on an innovative combination of multiagent systems and Semantic Web technologies, in which agents assist human partners in the development and maintenance of system ontologies, which in turn permits further delegation of operational tasks to agents. The envisioned system has a distributed architecture with any number of client and server subsystems connected by a wide area network and functionally integrated through a layer of agent-oriented middleware. A server contans a legacy relational database along with a reference ontology, autonomously developed within the system through human-agent cooperation, that represents the semantics of the database schema. This allows online translation of SPARQL queries into SQL queries and conversion of retrieved SQL results back into SPARQL format. A client subsystem allows its users to formulate semantic queries in a simplified natural language, using high-level terms from the composite reference ontology whose component views are imported from servers, as well as user-specific terms from a custom ontology that is co-developed by the user and an agent as a layer on top of the reference ontology. A semantic query is translated into SPARQL scripts for distributed execution on appropriate severs. The client integrates and delivers the retrieved results. The co-development of reference ontology and the execution of a semantic query are ilustrated by typical scenarios.

The approach suggests that significant practical benefits could result from endowing agents and agent teams with ontology management capabilities. Many of the necessary preconditions for this, both in terms of formal understanding and modeling of ontology management processes and also in terms of available software tools for development, mediation, and maintenance of ontologies, are either already appearing or likely to to be brought about by research in Semantic Web technologies and applications. The motivation, feasibility, and potential benefits of agent-oriented ontology management applications are likely to be enhanced by increased availability of public knowledge resources that agents could access. These considerations motivate further studies in meta-ontologies and techniques for agent-oriented ontology management. Possible directions of research into further applications of the current approach include context-aware systems, acquisitional query processing systems, and other rapidly advancing areas.

References

1. G. Antoniou, F. Harmelen, Web ontology language: Owl, in: Handbook on Ontologies, International Handbooks on Information Systems, edited by S. Staab, R. Studer (Springer, Berlin Heidelberg 2009), pp. 91–110
2. M. Baldauf, S. Dustdar, F. Rosenberg, A survey on context-aware systems. Int. J. Ad Hoc Ubiquitous Comput. **2**(4), 263–277 (2007)
3. F. Bellifemine, G. Caire, D. Greenwood, *Developing Multi-Agent Systems with JADE* (Wiley, Wiltshire, 2007)
4. T. Berners-Lee, Semantic Web Road Map. W3C Design Issues Architectural and Philosophical Points (1998). Retrieved May 03, 2010 from http://www.w3.org/DesignIssues/Semantic.html
5. Bizer, C., Seaborne, A.: D2RQ-Treating non-RDF Databases as Virtual RDF Graphs. In: Proceedings of the 3rd International Semantic Web Conference (ISWC2004). Hiroshima (2004).
6. D. Brickley, R. Guha, RDF Vocabulary Description Language 1.0: RDF Schema. Tech. rep., W3C (2004). Retrieved October 06, 2010 from http://www.w3.org/TR/2004/REC-rdf-schema-20040210/
7. S.R. Madden, M.J. Franklin, J.M. Hellerstein, W. Hong, TinyDB: An acquisitional query processing system for sensor networks. ACM Trans. Database Syst. **30**(1), 122–173 (2005)

8. B. McBride, D. Boothby, C. Dollin, An Introduction to RDF and the Jena RDF API (2010), Retrieved June 20, 2011 from http://openjena.org/tutorial/RDF_API/index.html
9. G. Miller, WordNet: a lexical database for English. Comm. ACM **38**, 39–41 (1995)
10. C. Olszak, E. Ziemba, Approach to building and implementing business intelligence systems. Interdisc. J. Inf., Knowl., Manage. **2**,134–148 (2007)
11. D. Polajnar, M. Zubayer, J. Polajnar, A multiagent architecture for semantic access to legacy relational databases. In: 2012 IEEE International Systems Conference (SysCon), pp. 1–8 (2012). doi 10.1109/SysCon.2012.6189521.
12. F. Ricca, L. Gallucci, R. Schindlauer, T. Dell'Armi, G. Grasso, N. Leone, OntoDLV: an ASP-based system for enterprise ontologies. J. Logic Comput. **19**, 643–670 (2009)
13. J. Rumbaugh, I. Jacobson, G. Booch, *Unified Modeling Language Reference Manual*, 2nd edn. (Pearson, Higher Education, 2004)
14. S. Russell, P. Norvig, *Artificial Intelligence: A Modern Approach*, 2nd edn. (Prentice Hall, New Jersey, 2003)
15. W3C: SPARQL Query Language for RDF (2008). Retrieved January 18, 2013 from http://www.w3.org/TR/rdf-sparql-query/
16. M. Wooldridge, *An Introduction to Multiagent Systems*, 2nd edn. (Wiley, Glasgow, 2009)
17. M. Zubayer, *A Multiagent Architecture for Semantic Query Access to Legacy Relational Databases* (University of Northern British Columbia, Canada, 2011). Master's thesis

A Research Survey on Large XML Data: Streaming, Selectivity Estimation and Parallelism

Muath Alrammal and Gaétan Hains

Abstract Semi-structured data sets in the form of XML documents have many practical uses and they have motivated a very large amount of work in theoretical, applied and industrial computing. Their efficient exploitation requires specific methods for filtering and querying them, using techniques that are neither keyword searches nor relational methods. In this chapter we survey a large body of recent research on efficient querying methods for XML data. Our analysis of the literature follows the three dimensions of stream-processing, parallel processing and performance variability.

1 Introduction

Extensible markup language (XML) [23] is a simple, very flexible text format derived from SGML, the standard generalized markup language. It has gone from the latest buzzword to an entrenched e-business technology in record time. XML is currently being heavily motivated by the industry and community as one of the most popular language for data representation and exchange on the Internet. The

M. Alrammal (✉)
Laboratoire d'Informatique Fondamentale d'Orléans (LIFO), Université d'Orléans,
Rue Léonard de Vinci, B.P. 6759 45067 Orléans, France
e-mail: muath.alrammal@univ-orleans.fr; muath.alrammal@khawarizmi.com
URL: www.kic-uae.com

M. Alrammal
Al-Khawarizmi International University-College, P.O. Box 25669 Abu Dhabi,
United Arab of Emirates (UAE)

G. Hains
Laboratoire d'Algorithmique, Complexité et Logique (LACL), Faculté des Sciences et
Technologie, Université Paris-Est, 61 avenue du Général de Gaulle,
94010 Créteil Cedex, France
e-mail: gaetan.hains@u-pec.fr

F. Xhafa and N. Bessis (eds.), *Inter-cooperative Collective Intelligence:*
Techniques and Applications, Studies in Computational Intelligence 495,
DOI: 10.1007/978-3-642-35016-0_7, © Springer-Verlag Berlin Heidelberg 2014

popularity of XML has created several important applications like information dissemination, processing of the scientific data, and real time news.

Data Stream Mining is the process of extracting knowledge structures from continuous, rapid data records. A data stream is an ordered sequence of instances that in many applications of data stream mining can be read only once or a small number of times using limited computing and storage capabilities. Examples of data streams include computer network traffic, phone conversations, ATM transactions, web searches, and sensor data. Data stream mining can be considered a subfield of data mining, machine learning, and knowledge discovery. In [5] we showed how to use XML data stream mining for extracting performance meta-data.

Many modern applications require processing massive amounts of XML data, creating difficult technical challenges. Among these, there is the design and implementation of applications to optimize the processing of XPath queries [14] and to provide an accurate cost estimation for these queries processed on massive amounts of XML data.

As a result, operations based on XML data are extremely demanding of query processing performance, either because a user must have instantaneous results on very large data sets or because an advanced algorithm for data-mining or machine learning must repeatedly issue queries on its reference data set.

When processing very large data sets one is either motivated towards solutions that read data in streaming mode [99] or parallel algorithms [48] that take advantage of the very large internal memory and processing speed that multiprocessing systems can provide. Selectivity estimation techniques are used to estimate the number of matches for queries evaluated on data sets. It is desirable in interactive and internet applications. With it, the system could warn the end user for example that his/her query is of low selectivity so that the amount of results will be insufficient. Or it could warn that the expected output size is unreasonable leading to a refined query.

This chapter surveys the existing work on selectivity estimation techniques, stream processing, and on parallel processing of XPath queries. The remainder of the chapter is structured as follows: Sect. 1.1 introduces some preliminaries. Section 2 presents the important properties needed for the selectivity estimation techniques and it explains them. Section 3 presents the different techniques used to evaluate XPath queries on streams of XML data. Section 4 surveys the different techniques on parallel processing of XPath queries. Finally, Sect. 5 concludes this survey.

1.1 Preliminaries

In this section, we present and define some terminology. Then, we illustrate through examples the efficiency of using the different processing approaches for large XML data sets: streaming, selectivity estimation and parallelism.

An XML data set (document) is modeled as a rooted, ordered, labelled tree, where each node corresponds to an element, attribute or a value, and the edges represent (direct) element-subelement or element-value relationships. In Fig. 1, D is an example of XML data set.

XPath [14] is a language that describes how to locate specific elements (and attributes, processing instructions, etc.) in a document. It operates on the abstract, logical structure of an XML document, rather than its surface syntax. In Fig. 1, Q: $//A[./F]/C$ is an example of XPath query, where the single line edges represents child ('/'), the double line edges represents descendant ('//'), single dashed line represents $[./node()]$ (double dashed line represents $[.//node()]$) and the result node which is in this example the black colored node C.

A stream of XML data is the depth-first, left-to-right traversal of an XML data set [23]. In the streaming model queries must be known before any data arrives, so queries can be preprocessed by building automata for query evaluation. Query evaluation of a stream of XML data can be done by a lazy approach or an eager one. The lazy approach processes an element when it encounters its closing tag in the incoming stream. The eager approach processes an element when it encounters its opening tag in the incoming stream.

In Fig. 1, by evaluating Q on D, the buffer size B of the so-called lazy approach is $B = n$ or in other words $B = |D|$ since the predicate of A is not evaluated until $</A_i>$ arrives. In this case all nodes starting from C_1 to C_n have to be buffered, which will increase the buffering size remarkably. In the so-called eager approach $B = 0$ because the predicates of A is evaluated to be true the moment element $<F>$ arrives. Thus, each $<C_i>$ can be flushed as a query result the moment it arrives and does not need to be buffered at all. Obviously this will improve the buffering space performance.

In our example, it is possible to optimize the buffer size of the lazy approach by using selectivity estimation techniques. A preprocessing for D is required to create a summary of D (see the figure). The number in the bracket to the right of the each element (in the summary of D) represents its frequency. Therefore, by applying a selectivity estimation technique (evaluation of Q on the summary of D using the lazy approach) we buffer only one element C with its frequency (n) and we get the number of matches that is the value of n. Furthermore, this evaluation informs the end user (query's sender) in the number of matches which allows him to refine the query if needed.

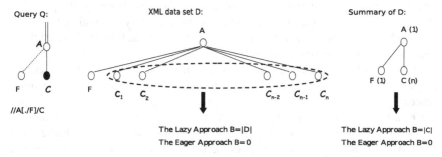

Fig. 1 XML data set D, XPath query Q, and a summary of D

Parallel processing for accelerating a single query on a, potentially large, dataset is dual to stream processing in the following manner. Streaming techniques minimize memory usage and require in principle a single processor, but allow unlimited processing time and dataset size. On the other hand (pure) parallel processing assumes that the dataset is loaded into main memory which must therefore scale with the documents being processed. This large amount of memory is provided by multi-processor machines and their very fast internal interconnexions. The goal is then to minimize querying time at a large expense in memory and processors. For example in Fig. 1 the complete set of C_i nodes children of root A can be processed in parallel i.e. in potentially constant time, given the appropriate data structures and algorithm. But there are several obstacles to it being generally and easily possible: namely query complexity, load balancing and communication costs. In our last bibliographic section we survey the current state of knowledge and experiments on this problem of parallel processing for XML queries.

2 Selectivity Estimation

In this section, we start by introducing some of the important properties of selectivity estimation techniques. After that, we give an overview of the literature related to this domain, as of 2012.

2.1 Properties of Selectivity Estimation Techniques

The design and the choice of a particular selectivity estimation technique depends on the problem being solved with it. Therefore, the technique needs to be constructed in a way related to the needs of the particular problem being solved [3].

In general, we would like to construct the synopsis structure (a summary of the structure of an XML data set) in such a way that it has wide applicability across broad classes of problems. The applicability to streams of XML data makes the space and time efficiency issue of construction critical. When looking for an efficient, capable (general enough) and accurate selectivity estimation technique for XPath queries, there are several issues that need to be addressed. Some of these issues can be summarized as follows:

(1) It must be practical: in general, one of the main usages of the selectivity estimation techniques is to accelerate the performance of the query evaluation process. The selectivity estimation process of any query or sub-query must be much faster than the real evaluation process. In other words, the cost savings on the query evaluation process using the selectivity information must be higher than the cost of performing the selectivity estimation process. In addition, the required summary structure(s) for achieving the selectivity estimation process must be

efficient in terms of memory consumption. (2) It should support structural and data value queries: in principle, all XML query languages can involve structural conditions in addition to the value-based conditions. Therefore, any complete selectivity estimation system for the XML queries requires maintaining statistical summary information about both the structure and the data values of the underlying XML documents. (3) One pass constraint: for streaming applications or techniques, the streams of XML data typically contain a large number of points, the contents of the stream cannot be examined more than once during the course of computation. Therefore, all summary structure/data values construction algorithms should be designed under a one pass constraint. (4) It should cover a larger enough query language fragment: the standard query languages for XML [23] namely XPath [14] and XQuery [17] are very rich languages. They provide rich sets of functions and features. These features include structure and content-based search, join, and aggregation operations. (5) It must be accurate: providing an accurate estimation for the query optimizer can effectively accelerate the evaluation process of any query. However, on the other hand, providing the query optimizer with incorrect selectivity information will lead it to incorrect decisions and consequently to inefficient execution plans. (6) It must evolve and be incremental: when the underlying XML document is updated, i.e. some elements are added or deleted, the selectivity estimation technique should be updated (without the need of rebuilding it) as well to provide an accurate selectivity estimation for a given query. (6) It should be independent: it is recommended that the selectivity estimation process be independent of the actual evaluation process which facilitates its use with different query engines that apply different evaluation mechanisms. This property is an advantage for software engineering of the corresponding module(s). (7) Time and Space Efficiency: In many traditional synopsis methods on static data sets (such as histograms), the underlying dynamic programming methodologies require super-linear space and time. This is not acceptable for a data stream [3]. For the case of space efficiency, it is not desirable to have a complexity which is more than linear in the size of the stream.

2.2 Path/Twig Selectivity Estimation Techniques

In this section, we give an overview of the literature related to the selectivity estimation approaches in the XML domain. Estimation techniques can be classified in terms of the structure used for collecting the summary information into two main classes: (1) Synopsis-based estimation techniques: this class of estimation techniques uses tree or graph structures for representing the summary information of the source XML documents. (2) Histogram-based estimation techniques: this class of estimation techniques uses the statistical histograms for capturing the summary information of the source XML documents.

2.2.1 Synopsis-Based Estimation Techniques

Aboulnaga et al. [1] have presented two different techniques for capturing the structure of the XML documents and for providing accurate cardinality estimations for the path expressions. The presented techniques only support the cardinality estimations of simple path expressions without predicates and so-called *recursive* axes (repeated node-labels in the expression). Moreover, the models cannot be applied to twigs.

The first technique presented in this chapter is a summarizing tree structure called a path tree. A path tree is a tree containing each distinct rooted path in the database (or data set) where the nodes are labeled by the tag name of the nodes.

To estimate the selectivity of a given path expression p in the form of $s1/s2/.../sn$, the path tree is scanned by looking for all nodes with tags that match the first tag of the path expression. From every such node, downward navigation is done over the tree following child pointers and matching tags in the path expression with tags in the path tree. This will lead to a set of path tree nodes which all correspond to the query path expression. The selectivity of the query path expression is the total frequency of these nodes.

The problem is the size of the path tree constructed from a large XML document is larger than the available memory size for processing. To solve this problem, the authors described different summarization techniques based on the deletion of low frequency nodes, and on their replacement by means of $* - nodes$ (*star nodes*). Each $* - node$, denoted by a special tag name "$*$", denotes a set of deleted nodes, and inherits their structural properties as well as their frequencies. The second technique presented in this chapter is a statistical structure called Markov table (MT). This table, implemented as an ordinary hash table, contains any distinct path of a length up to m and its selectivity. Thus, the frequency of a path of length n can be directly retrieved from the table if $n \leq m$, or it can be computed by using a formula that correlates the frequency of a tag to the frequencies of its $m - 1$ predecessors if $n > m$. Since the size of a Markov table may exceed the total amount of available main memory, the authors present different summarization techniques which work as in the case of a path tree and delete low frequency paths and replace them with $* - paths$.

Figure 2a illustrates an example of XML document D and the representation of its corresponding path tree Fig. 2b and Markov table Fig. 2c.

XPATHLEARNER [66] is an online learning method for estimating the selectivity of XML path expressions. In this method, statistical summaries are used to build a Markov histogram on path selectivities gathered from the target XML data set. XPATHLEARNER employs the same summarization and estimation techniques as presented in [1]. The novelty of XPATHLEARNER is represented by the fact that it collects the required statistics from the query feedback in an online manner, without accessing and scanning the original XML data, which is in general resource-consuming. These statistics are used to learn both tag and value distributions of input queries, and, when needed, to change the actual configuration of the underlying Markov histogram in order to improve the accuracy of

(a)

```
<A>
    <B> </B>
    <B>
        <D> </D>
    </B>
    <C>
        <D> </D>
        <E> </E>
        <D> </D>
        <E> </E>
        <E> </E>
    </C>
</A>
```

(b)

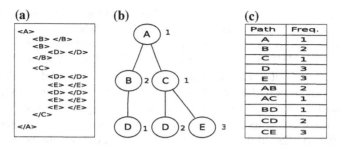

(c)

Path	Freq.
A	1
B	2
C	1
D	3
E	3
AB	2
AC	1
BD	1
CD	2
CE	3

Fig. 2 An XML document D and its both path tree and markov table [1]. **a** XML document D. **b** Path tree of D **c** Markov table of D

approximate answers. From this point of view, XPATHLEARNER can be intended as a workload-aware method. An important difference between XPATHLEARNER and the MT of [1] is that the XPATHLEARNER supports the handling of predicates (to further refine the selected node-set) by storing statistical information for each distinct tag-value pair in the source XML document.

The authors of [108] have presented a similar technique called Comet (Cost Modeling Evolution by Training) for cost modeling of complex XML operators. It exploits a set of system catalogue statistics that summarizes the XML data, the set of simple path statistics and a statistical learning technique called transform regression instead of detailed analytical models to estimate the selectivity of path expressions. The technique used to store the statistics is the path tree of [1]. COMET is more oriented toward XML repositories consisting of a large corpus of relatively small XML documents. In COMET, initial focus is only on the CPU cost model. To do that, they developed a CPU cost model for XNAV operator which is an adaptation of TurboXPATH [59]. Their idea to build from previous work in which statistical learning method are used to develop cost models of complex user-defined functions [55, 62]. COMET can automatically adapt to changes over time in the query workload and in the system environment. The optimizer estimates the cost of each operator in the query plan (navigation operator, join operator) and then combines their costs using an appropriate formula. The statistical model can be updated either at periodic intervals or when the cost-estimation error exceeds a specified threshold. Updating a statistical model involves either re-computing the model from scratch or using an incremental update method.

The authors of [35] proposed a correlated sub-path tree (CST), which is a pruned suffix tree (PST) with set hashing signatures that helps determine the correlation between branching paths when estimating the selectivity of twig queries. The CST method is off-line, handles twig queries, and supports substring queries on the leaf values. The CST is usually large in size and has been outperformed by [1] for simple path expressions.

The Twig-Xsketch described in [87] is a complex synopsis data structure based on XSketch synopsis [83] augmented with edge distribution information. It was shown in [87] that Twig-Xsketch yields selectivity estimates with significantly smaller errors than correlated sub-path tree (CST). For the data set XMark [94] the ratio of error for CST is 26 versus 3 % for Twig-Xsketch.

TreeSketch [86] is based on a partitioned representation of nodes of the input graph-structured XML database. It extends the capabilities of XSketch [83] and Twig-Xsketch [87]. It introduces a novel concept of count-stability (C-stability) which is a refinement of the previous F-stability of [83]. This refinement leads to a better performance in the compression of the input graph-structured XML database.TreeSketch builds its synopsis in two steps. First, it creates an intermediate count-stability (C-stability) synopsis that preserves all the information of the original XML data set in a compact format. After that, the Tree-Sketch synopsis is built on top of the C-stability synopsis by merging similar structures. The construction time of TreeSketch for the complex data set TreeBank 86MiB (depth 36) took more than 4 days, this result was confirmed in [69]. Moreover, the TreeSketch synopsis does not support the recursion in the data set as it is explained in [109].

The authors of [109] have addressed the problem of deriving cardinality estimation (selectivity) of XPath expressions. In this work, the authors are mainly focusing on the handling of XPath expressions that involve only structural conditions. Authors define a summary structure for summarizing the source XML documents into a compact graph structure called XSeed. The XSeed structure is constructed by starting with a very small kernel which captures the basic structural information (the uniform information) as well as the recursion information of the source XML document. The kernel information is then incrementally updated through the feedback of queries. The XSeed kernel is represented in the form of a label- split graph summary structure proposed by Polyzotis and Garofalakis [84]. In this graph, each edge $e = (u; v)$ is labeled with a vector of integer pairs $(p0 : c0, p1 : c1, \ldots, pn : cn)$. The ith integer pair $(pi : ci)$ indicates that at recursion level i, there are a total of pi elements mapped to the synopsis vertex u and ci elements mapped to the synopsis vertex v.

Figure 3 illustrates an example of XML document D and its corresponding XSeed kernel. The high compression ratio of the kernel can lead to a situation where information is lost. This loss of information results in the occurrence of significant errors in the estimation of some cases. To solve this problem, the authors introduce another layer of information, called hyper-edge table (HET), on top of the kernel. This HET captures the special cases that are not addressed by

Fig. 3 An XML document D and its XSeed kernel [109]

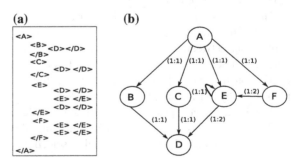

original assumptions made by the kernel (irregular information). For example, it may store the actual cardinalities of specific path expressions when there are large errors in their estimations. Relying on the defined statistics graph structure and its supporting layer, the authors propose an algorithm for the cardinality estimation of the structural XPath expressions. The main contribution of this work is the novel and accurate way of dealing with recursive documents and recursive queries. By treating the structural information in a multi-layer manner, the XSeed synopsis is simpler and more accurate than the TreeSketch synopsis [86]. However, although the construction of XSeed is generally faster than that of TreeSketch, it is still time-consuming for complex data sets.

Paper [85] introduced XCLUSTER, which computes a synopsis for a given XML document by summarizing both the structure and the content of document. XCLUSTER is considered to be a generalized form of the XSketch tree synopses which is a previous work of the authors presented in [84]. On the structure content side, an XCLUSTER tree synopsis is a node-labeled graph where each node represents a sub-set of elements with the same tag, and an edge connects two nodes if an element of the source node is the parent of elements of the target node. Each node in the graph records the count of elements that it represents while each edge records the average child count between source and target elements. On the value content side, XCLUSTER has borrowed the idea of the XMill XML compressor [65] which is based on forming structure-value clusters which groups together data values into homogeneous and semantically related containers according to their path and data type. Then, it employs the well-known histogram techniques for numeric and string values [30, 88] and introduces the class of end-biased term histograms for summarizing the distribution of unique terms within textual XML content. The XCLUSTER estimation algorithm relies on the key concept of a query embedding, that is, a mapping from query nodes to synopsis nodes that satisfies the structural and value-based constraints specified in the query. To estimate the selectivity of an embedding, the XCluster algorithm employs the stored statistical information coupled with a generalized path-value independence assumption that essentially de-correlates path distribution from the distribution of value-content. This approach can support twig queries with predicates on numeric content, string content, and textual content.

Figure 4 illustrates an example of XML document D and its corresponding XCLUSTER. However, XCLUSTER address the summarization problem for structured XML content, but its construction time is unknown. Furthermore, as it is mentioned in [93] it does not process a nested expressions (nested predicates).

The authors of [43] have proposed the SLT (Straight line tree) XML tree synopsis. The idea of this work is based on the fact that the repetitive nature of tags in the XML documents makes tag mark-ups re-appears many times in a document. Hence, the authors use the well-known idea of removing repeated patterns in a tree by removing multiple occurrences of equal subtrees and replacing them by pointers to a single occurrence of the subtree. The synopsis is constructed by using a tree compression algorithm to generate the minimal unique directed acyclic graph (DAG) of the XML tree and then representing the resulting DAG structures

(a)

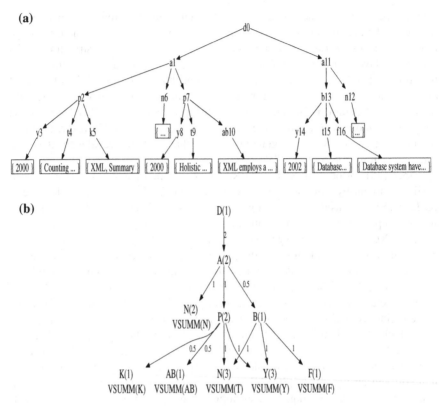

(b)

Fig. 4 An XML document *D* and its XCLUSTER synopsis [85]

using a special form of grammars called an straight line tree grammar (SLT grammar).

Additionally, the size of this grammar is further reduced by removing and replacing certain parts of it, according to a statistical measure of multiplicity of tree patterns. This results in a new grammar which contains size and height information about the removed patterns. The authors have described an algorithm for a tree automaton which is designed to run over the SLT grammars to estimate the selectivity of queries containing all XPath axes. This algorithm converts each XPath query into its equivalent tree automaton and describes how to evaluate this tree automaton over a document to test whether the query has at least one match in the document and returns the size of the result of the query on a document. The proposed synopsis of this work can deal only with structural XPath queries. Furthermore, as it is mentioned in [93], this approach does not support any form of predicate queries.

The design and implementation of a relational algebraic framework for estimating the selectivity of XQuery expressions was described in [91, 92, 98]. In this approach (Relational algebraic), XML queries are translated into relational algebraic plans [52]. Analysis of these relational plans is performed to annotate each operator with a set of special properties [51]. These annotations are produced during a single pass over the relational plan and use a set of light-weight inference rules which are local in the sense that they only depend on the operator's immediate plan inputs. Summary information about the structure and the data values of the underlying XML documents are kept separately. Then by using all these pieces together with a set of inference rules, the relational estimation approach is able to provide accurate selectivity estimations in the context of XML and XQuery domains. The estimation procedure is defined in terms of a set of inference rules for each operator which uses all of the available information to estimate the selectivity of not only the whole XQuery expression but also of each sub-expression (operator) as well as the selectivity of each iteration in the context of FLWOR expressions. The framework enjoys the flexibility of integrating any XPath or predicate selectivity estimation technique and supports the selectivity estimation of a large subset of the XML query language XQuery.

In [69], authors proposed a sampling method named subtree sampling to build a representative sample of XML which preserves the tree structure and relationships of nodes. They examine the number of data nodes for each tag name starting from the root level. If the number of data nodes for a tag is large enough, a desired fraction of the data nodes are randomly selected using simple random sampling without replacement and then the entire subtrees rooted at these selected data nodes are included, as sampling units, in the sample. They call each such set of subtrees to which random sampling is applied a subtree group. If a tag has too few data nodes at the level under study, then all the data nodes for that tag at that level are kept and they move down to check the next level in the tree. The paths from the root to the selected subtrees are also included in the sample to preserve the relationships among the sample subtrees. This sampling scheme assumes that the sizes of the subtrees in the same subtree group are similar. This is because the root nodes of these subtrees have the same tag name, i.e. they are nodes of the same type. These root nodes reside in the same level. Consequently, subtrees in the same subtree group tend to have similar structures, thus similar sizes. Based on this observation, the sampling fraction of the subtree groups f'_i, where f_i is the sampling fraction of the ith subtree group, can be simply set to f_t, which is the sampling fraction of the whole data set.

If the number of nodes n for a tag satisfies the minimum requirements $n * ft > = 1$, they consider it large enough. The sample trees are just a portion of the original XML data tree. These sample trees differ only in magnitude from the original XML data tree. Therefore, ordinary twig query evaluation methods, such as TwigStack [24] can be applied directly to the sample trees synopsis to derive approximate answers. Though a subtree sampling synopsis can be applied to aggregations functions such as SUM, AVG, etc., this approach is based on an essential assumption that nodes of the same type and at the same level have similar subtrees. Moreover, it is shown in [69] that XSeed [109] outperforms subtree sampling for queries with Parent/Child on simple data set e.g. XMark [94], while it is the inverse for *recursive* data sets.

2.2.2 Histogram-Based Estimation Techniques

As we mentioned before, this class of estimation techniques uses the statistical histograms for capturing the summary information of the source XML documents. Below, we give a survey on the existing work.

Authors of [45] have presented an XML Schema-based statistics collection technique called StatiX. This technique leverages the available information in the XML Schema to capture both structural and value statistics about the source XML documents. These structural and value statistics are collected in the form of histograms. The StatiX system is employed in LegoDB [18]. LegoDB is a cost-based XML-to-relational storage mapping engine, which tries to generate efficient relational configurations for thecXML documents. The StatiX system consists of two main components. The first component is the XML schema validator which simultaneously validates the document against its associated schema and gathers the associated statistics. It assigns globally unique identifiers (IDs) to all instances of the types defined in the schema. Using these assigned IDs, structural histograms are constructed to summarize information about the connected edges. Value histograms are constructed for types that are defined in terms of base types such as integers. The storage of the gathered statistics is done using equi-depth histograms (wherein the frequency assigned to each bucket of the histogram is the same). The second component is the XML schema transformer which enables statistics collection at different levels of granularity. Although, StatiX is used in the context of the LegoDB system and the presented experimental results indicate highlyaccurate query estimates, the technique can only be applied to documents described by XML schemas with no clear view as to how it can be extended to deal with schema-less documents. Moreover, the paper [45] does not show a clear algorithm for estimating the cardinality of the XQuery expression and there is no clear definition of the supported features and expressions of the language.

In [105], the authors have presented an approach for mapping XML data into 2D space and maintaining certain statistics for data which fall into each predefined grid over the workspace. In this approach, each node x in an XML document D is associated with a pair of numbers, $start(x)$ and $end(x)$, numeric labels representing the pre-order and post-order ranks of the node in the XML document D. Each descendant node has an interval that is strictly included in its ancestors interval. For each basic predicate P a two-dimensional histogram summary data structure is built and collectively named as position histograms. In the position histograms data structure, the *start* values are represented by the $x - axis$ while the *end* values are represented the $y - axis$. Each grid cell in the histogram represents a range of start position values and a range of end position values. The histogram maintains a count of the number of nodes satisfying the conditions of predicate P and has start and end positions within the specified ranges of the grid cell. Since the start position and end position of a node always satisfies the formula $start < = end$, none of the nodes can fall into the area below the diagonal of the matrix. So, only the grid cells that reside on the upper left of the diagonal can have a count of more than zero. Given a predicate P_1 associated with the position histograms H_1 and a predicate P_2 associated with the position histograms H_2, estimating the number of pair of nodes u, v where u satisfies P_1 and v satisfies P_2 and u is an ancestor of v is done either in an *ancestor–based* fashion or in a *descendant–based* fashion. The *ancestor–based* estimation is done by finding the number of descendants that joins with each ancestor grid cell. The *descendant–based* estimation is done by finding the number of ancestors that are joined with each descendant grid cell.

Follow-up work has improved on the ideas of interval histograms by leveraging adaptive sampling techniques [101]. In this work, the proposed technique treats every element in a node set as an interval, when the node set acts as the ancestor set in the join or a point or when the node set acts as the descendant set. Two auxiliary tables are then constructed for each element set. One table records the coverage information when the element set acts as the ancestor set, while the other captures the start position information of each element when the element set acts as the descendant set. To improve the accuracy of the estimated results, sampling-based algorithms are used instead of the two-dimensional uniform distribution assumption as used in [105].

In [102], the authors have proposed a framework for XML path selectivity estimation in a dynamic context using a special histogram structure named *bloom histogram* (BH). *BH* keeps a count of the statistics for paths in XML data. Given an XML Document D, the path-count table $T(path, count)$ is constructed such that for each $path_i$ in D, there is a tuple t_i in T with $t_i.path = path_i$ and $t_i.count = count_i$ where $count_i$ is the number of occurrences of $path_i$. Using T, a bloom histogram H is constructed by sorting the frequency values and then grouping the paths with similar frequency values into buckets. Bloom filters are used to represent the set of paths in each bucket so that queried paths can be quickly located.

To deal with XML data updates and the dynamic context, the authors proposed a dynamic summary component which is an intermediate data structure from which the bloom histogram can be recomputed periodically. When data updates arrive, not only the XML data is updated but the updated paths are also extracted, grouped and propagated to the dynamic summaries. Although, the bloom histogram is designed to deal with data updates and the estimation error is theoretically bounded by its size, it is very limited as it deals only with simple path expressions of the form $/p1/p2/\ldots/pn$ and $//p1/p2/\ldots/pn$.

Authors of [63] have described a framework for estimating the selectivity of XPath expressions with a main focus on the order-based axes (following, preceding, following-sibling, and preceding-sibling). They used a path encoding scheme to aggregate the path and order information of XML data. The proposed encoding scheme uses an integer to encode each distinct root-to-leaf path in the source XML document and stores them in an encoding table. Each node in the source XML document is then associated with a path id that indicates the type of path where the node occurs. Additionally, they designed a PathId-Frequency table where each tuple represents a distinct element tag and aggregates all of its associated element tags with path ids and their frequency. To capture the order information, they used the Path-order table associated to each distinct element tag name to capture the sibling-order information based on the path ids.

For estimating the cardinality of XPath expressions, the authors introduced the Path Join algorithm. Given an XPath query Q, it retrieves a set of path ids and the corresponding frequencies for each element tag in Q from the PathId-Frequency table.

For each pair of adjacent element tags in Q, they use a nested loop to determine the containment of the path ids in their sets. Path IDs that clearly do not contribute to the query result will be removed. The frequency values of the remaining path ids will be used to estimate the query size. The algorithm uses the information of the path-order table to compute the selectivity of the (following-sibling, preceding-sibling) axes that may occur in Q. The authors introduced two compact summary structures called p-histogram and o-histogram, to summarize the path and order information of XML data respectively. Though the proposed model is the first work to address the problem of cardinality estimation of XPath expression with order-based axes, it is unfortunately not clear how an extension can be introduced to support predicates.

In [5] we proposed a stream-based selectivity estimation technique. It is an innovative technique, which consists of (1) the path tree synopsis data structure, a succinct description of the original document with low computational overhead and high accuracy for processing tasks like selectivity estimation, (2) a streaming selectivity estimation algorithm which is efficient for path tree traversal. To enable the selectivity estimation process, authors inspired their selectivity estimation algorithm from LQ (the extended lazy stream- querying algorithm of Gou and Chirkova work [49]). Therefore, the advantages of this algorithm are the same as for the lazy stream-querying algorithm. Detailed explanations about LQ and its advantages are in [4]. The current version of our estimation algorithm processes queries which belong to the fragment of Forward XPath. The estimation algorithm takes two input parameters. The first one is the XPath query that will be transformed to a query table statically using Forward XPath Parser. After that, the main function is called. It reads the second parameter (the path tree) line by line repeatedly, each time generating a tag. Based on that tag a corresponding *startBlock* or *endBlock* function is called to process it. Finally, the main function generates as output the estimates needed for the given query. Experiments demonstrated that this technique is accurate and outperforms the existing approaches in generality and preprocessing time.

However, none of the selectivity estimation techniques process XML data and XPath queries in confidential manner.

In Sect. 3, we present several stream-processing approaches.

3 Stream-Processing Approaches

Much research has been conducted to study the processing of XML documents in streaming fashion. The different approaches to evaluate XPath queries on streams of XML data can be categorized as follows (1) *stream-filtering*: determining whether there exists at least one match of the query Q in the XML document D, yielding a boolean output, for example XTrie [28]. (2) *Stream-querying*: finding which parts of D match the query Q. This implies outputting all answer nodes in a XML document D i.e. nodes that satisfy a query Q. An example of stream-querying research is XSQ [81].

Below, we present some existing algorithms for each category.

3.1 Stream-Filtering Algorithms

Various stream-filtering systems have been proposed. Below we explain some of them.

XFilter [7] is the first filtering system that addresses the processing of streaming XML data. It was proposed for for selective dissemination of information (SDI). For structure matching, XFilter adopts some form of Finite State Machine (FSM) to represent path expressions in which location steps of path expressions are mapped to machine states. Arriving XML documents are then parsed with an event-based parser, the events raised during parsing are used to drive the FSMs through their various transitions. A query is said to match a document if during parsing, an accepting state for that query is reached. In the filtering context, large numbers of queries representing the interests of the user community are stored and must be checked upon the arrival of a new document. In order to process these queries efficiently, XFilter employs a dynamic index over the states of the query FSMs and includes optimizations that reduce the number of path expressions that must be checked for a given document. In large-scale systems there is likely to be significant commonality among user interests, which could result in redundant processing in XFilter.

YFilter [39] is an XML filtering system aimed at providing efficient filtering for large numbers (e.g., 10's or 100's of thousands) of queries. The key innovation in YFilter is an Nondeterministic Finite Automaton (NFA)-based representation of path expressions which combines all queries into a single machine. Figure 5 illustrates an examples of this NFA, where all common prefixes of the paths are represented only once in the NFA. A basic path matching engine of YFilter is designed to handle query that are written in a subset of XPath. YFilter focuses on two common axes: the parent-child axis '/', and the ancestor-descendant axis '//'. It supports node tests that are specified by either an element name or the wildcard '*' (which matches any element name). Predicates can be applied to address contents of elements or to reference other elements in the document.

In [22] a SAX Based approach is introduced to evaluate the XPath queries that support all axes of *Core XPath*. Each input query is translated into an automaton that consists of four different types of transitions. The small size of the generated automata allows for a fast evaluation of the input stream of XML data within a small amount of memory. The authors implemented a prototype called XPA. The query processor decomposes and normalizes each XPath query, such that the

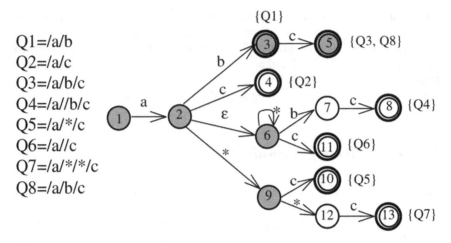

Q1=/a/b
Q2=/a/c
Q3=/a/b/c
Q4=/a//b/c
Q5=/a/*/c
Q6=/a//c
Q7=/a/*/*/c
Q8=/a/b/c

Fig. 5 XPath queries and a corresponding NFA [7]

resulting path queries contain only three different types of axes, and then converts them into lean XPath automata for which a stack of active states is stored. The input SAX event stream is converted into a binary SAX event stream that serves as input of the XPath automata. In [22], it is shown that XPA consumes far less main memory than YFilter [39]. XPA consumes from 20 % of the document size on average for simples XPath queries without predicate filters up to 50 % of the document size on average for paths with predicate filters.

The XTrie [28] technique is built on top of the XFilter approach and claims 2–4 times improvement in speed over the XFilter [7] system. Its authors proposed a trie-based index structure, which decomposes the XPath expressions (XPEs) to substrings that only contain parent-child axis. As a result, the processing of these common substrings among XPath expressions (XPEs) can be shared. The three key prominent features of XTrie can be summarized as follows: (1) it can filter based on complex and multiple path expressions, (2) it supports both ordered and un-ordered matching of XML documents, (3) since XTrie uses substrings, instead of elements names to index, the authors claim that XTrie can reduce both the number of unnecessary index probes and avoid redundant matching.

XTrie is designed to support online filtering of streaming XML data and is based on the SAX event-based interface that reports parsing events, the search procedure for the XTrie, which accepts as input an XML document D and an XTrie index (ST, T), processes the parsing events generated by D, and returns the identifiers of all the matching XPEs in the index. XTrie outperformed XFilter [7]. But YFilter [39] has demonstrated a better performance than XTrie on certain workloads.

In [10] the authors initialized a systematic and theoretical study of lower bounds on the amount of memory required to evaluate XPath queries over streams of XML data. They present a general lower bound technique, which given a query, specifies the minimum amount of memory that any algorithm evaluating the query on a stream would need to incur. The first memory lower bound is the *query frontier size*. When a query Q is represented as a tree, the frontier size at a node of this tree is the number of siblings of this nodes, and its ancestors' siblings. The query frontier size of Q is the largest frontier over all nodes of Q. The second lower bound is the document *recursion depth*. The recursion depth of a tree t with respect to a query Q is the maximal number of nested nodes matching a same node in Q. The third lower bound is the logarithmic value $log(d)$, where d is the depth of the document t. Based on these bounds a stream-filtering algorithm was proposed to optimize the space complexity. The algorithm transforms the query into NFA and uses different arrays for matching the stream of XML data. For queries in the fragment of *univariate XPath*, the space complexity of the algorithm is $O(|Q|.r.(log|Q| + logd + logr))$, where $|Q|$ is the query size, r is the document recursion depth, and d is the document depth. The time complexity is $O(|D|.|Q|.r)$, where $|D|$ is the document size.

The XPush machine [53] was proposed to improve the performance of stream-filtering. It processes a large number of XPath expressions, each with many predicates, on a stream of XML data. It is constructed lazily by creating an AFA (Alternating Finite Automaton) for each expression, and then transforming the set of AFAs into a single DPDA (Deterministic Pushdown Automaton). This is similar to the algorithm for converting an NFA to a DFA as described in the standard textbook on automata where stack automata are defined [56]. Existing systems (e.g. YFilter [39]) can identify and eliminate common subexpressions in the structure navigation part of XPath queries. This technique focuses on eliminating redundant work in predicate evaluation part. For examples, given the following two path expression $P1 = //a[./b/text() = 1$ *and* $.//a[@c > 2]]$ *and* $P2 = //a[@c > 2$ *and* $./b/text() = 1]$, previous techniques cannot exploit the fact that the predicate $[./b/text() = 1]$ is common. Since inherently the XPush machine cannot be partially updated, addition of a single expression necessitates recalculation (i.e., reconstruction) of the XPush machine as a whole. In other words, the cost of updating an automaton depends on the total number of AFAs (or expression). To solve this problem [96] proposed an integrated XPush machine, which enables incremental update by constructing the whole machine from a set of sub-XPush machines. The evaluation result positively demonstrates that efficient partial change of the AFAs is possible without significantly affecting all of the state transition tables.

SFilter [74] indexes the queries compactly using a query guide and uses simple integer stacks to efficiently process the stream of XML data. A *query guide G* is an ordered tree representation of all the path expressions that exploits the prefix commonality between the path expressions such that (1) the root of G is the same as the dummy root 'r' of the path path expressions and (2) the root-to-result node path of each path expression appears in G as a path that starts at node 'r' and ends at a descendant node and the path has the same node labels and edge constraints (i.e., P-C or A-D edge) of the path expression. The basic idea of this approach is to process the streaming XML data one tag at a time using the query guide representing the given path expressions. At any time during execution, the algorithm maintains a sequence of elements S in the stream whose open-tags have been seen but close-tags are yet to arrive. It maintains an integer stack at every query guide node to keep track of the current sequence of tags S in the stream. Each value in the stack represents the depth of an element in the stream that matches with the query guide node to which the stack is associated. Note that this number can uniquely identify a node in the stream as there will be exactly one node at a given depth in the current (or active) path in the document tree, represented by S.

Stream-filtering approaches deliver whole XML documents which satisfy the filtering condition to the interested users. Thus, the burden of selecting the interesting parts from the delivered XML documents is left upon the users. We therefore concentrate on stream-querying as more general and useful approach for our performance prediction (cost) model.

3.2 Stream-Querying Algorithms

Holistic XML matching algorithms are prevalent for matching pattern queries over stored XML data. They demonstrate good performance due to their ability to minimize unnecessary intermediate results. In particular,[24] proposed the first merge-based algorithm, which scans input data lists sequentially to match twig patterns. Such merge-based algorithms can be further improved by structure indexes that can reduce sizes of input lists [32]. Index-based holistic joins [58] were also proposed to speedup the matching of selective queries, as an improvement over merge-based algorithms.

In contrast, **streaming** algorithms assume that XML documents are not parsed in advance and they come in the form of SAX events. Sometimes even ad-hoc XML documents can be regarded as streams of XML data if using a SAX parser is the best way to access them. A large amount of work has been conducted to process XML documents in streaming fashion. The different stream-querying approaches to evaluate XPath queries on XML data streams can be categorized by the processing approach they use. Most of them are *Automata based*, for example: XPush [53] XSQ [81] SPEX [75] or *Parse tree based*, for example [12, 34, 49]. We highlight below some of the existing work.

In [81], authors proposed XSQ a method for evaluating XPath queries over streams of XML data to handle closures, aggregation and multiple predicates. Their method is designed based on hierarchical arrangement of pushdown transducers augmented with buffers. Automata is extended by actions attached to states, extended by a buffer to evaluate XPath queries.

The basic idea of XSQ is to use a pushdown transducer (PDT) to process the events that are generated by a SAX parser when it parses XML streams. A PDT is a pushdown automaton (PDA) with actions defined along with the transition arcs of the automaton. A PDT is initialized in the start state. At each step, based on the next input symbol and the symbols in the stack, it changes state and operates the stack according to the transition functions. The PDT also defines an output operation which could generate output during the transition. In the XSQ system, the PDT is augmented with a buffer so that the output operation could also be the buffer operation. Notice that the PDT generated for each location step of an XPath expression is called a basic pushdown automaton (BPDT. The BPDTs are combined into one Hierarchical PDT (HPDT). As it is shown in [49] XSQ does not support the AND operator. Furthermore, XSQ does not support same node-labels in a query, and requires that each axis node have at most one ('/') predicate node child.

The authors of [34] proposed a lazy stream-querying algorithm, TwigM, to avoid the exponential time and space complexity incurred by XSQ. TwigM extends the multi-stack framework of the TwigStack algorithm [24]. It uses a compact data structure to encode patterns matches rather than storing them explicitly which is a memory advantage. After that, it computes query solution by probing the compact data structure in Lazy fashion without computing pattern matches. The output consists of XML fragments. In [34], it is shown that TwigM can evaluate univariate XPath in polynomial time and space in the streaming environment. Specifically, TwigM works in $O(|D|.|Q|(|Q| + d_D.B))$ time and uses $O(|Q|.r)$ caching space. Where r is the recursion in D and d_D is the maximum depth of D. However, like XSQ, TwigM might have to buffer multiple physical copies of a potential answer node at a time, which is a space problem for recursive documents or data sets.

The SPEX [25, 75] system processes XPath expressions with forward axes by mapping it to a network of transducers. Query re-writing methods [76] are used to transform expressions with backward axes to ones containing only forward axes. Most transducers used are single-state pushdown automata with output tape. For path expressions without predicates, the transducer network is a linear path; otherwise, it is a directed acyclic graph. Each transducer in the network processes, in stepwise fashion, the stream of XML data it receives and transmits it unchanged or annotated with conditions to its successor transducers. The transducer for the result node holds potential answers, to be output when conditions specified by the query are found to be true by the corresponding transducers. Due to the absence of built-in order information, the system processes and caches large number of stream elements which will be found useless later.

TurboXPath [59] is an XML stream processor evaluating XPath expressions with downward and upward axis, together with a restricted form of for-let-where (FLOWR in XQuery) expressions. Hence, TurboXPath returns tuples of nodes instead of nodes. In TurboXPath [59] the input query is translated into a set of parse trees. Whenever a matching of a parse tree is found within the stream of XML, the relevant data is stored in form of a tuple that is afterward evaluated to check whether predicate and join conditions are fulfilled. The output is constructed out of those tuples of which have been evaluated to true.

In [54] the authors studied the problem of extracting flattened tuple data from streaming, hierarchical XML data. For this goal, they proposed StreamTX, in this approach they adapt the holistic twig joins for tuple-extraction queries on streaming XML with two novel features: first, they use the block-and-trigger technique to consume streaming XML data in a best-effort fashion without compromising the optimality of holistic matching; second, to reduce peak buffer sizes and overall running times, they apply query-path pruning and existential-match pruning techniques to aggressively filter irrelevant incoming data.According to their experiments, StreamTX has demonstrated superior performance advantage over TurboXPath [59] for positive queries.

Some stream-querying systems for evaluating XQuery queries have been developed, such as BEA/XQRL [44], Flux [61], and XSM [68].

Authors in [107] introduced a streaming XPath algorithm (QuickXScan). It is based on the principles similar to that of attribute grammars. There is a nice solution of using compact stacks to represent a possibly combinatorial explosive number of matching path instantiations with linear complexity like [58], therefore, QuickXScan extends the idea of compact stacks in a technique called matching grid, which is used also in [90]. QuickXScan represents queries using a query tree, together with a set of variables and evaluation rules associated with each query node. In this approach, there is a set of interrelated stacks, one for each query node to keep XML data nodes that match with the query node. Active query nodes can be precisely tracked with maximum up to the query size. Though, this approach handles queries containing child and descendant axes with complex predicates, it is not clear whether it supports queries with wildcard. The time complexity of this QuickXScan $O(|Q|.r.|D|)$ while the space complexity is $O(|Q|.r)$, where $|Q|$ is the query size, $|D|$ is the document size, and r is the recursion in the document.

The authors of [33] presented a model of data processing for information system exchange environment. It consists of a simple and general encoding scheme for servers, and algorithms of streaming query processing on encoded stream of XML data for data receivers with constrained computing abilities. The EXPedite query processor takes an encoded stream of XML data and an encoded XPath query as input, and outputs the encoded fragment in the stream of XML data that matches the query. The idea of the query processing algorithm is taken from different proposed techniques [38, 50] for efficient query evaluation based on XML node labels for XML data stored in the database.

In [49] authors proposed two algorithms to evaluate XPath over streams of XML data, they are (1) Lazy streaming algorithm (LQ). (2) Eager streaming algorithm (EQ). Algorithms accept XML document as a stream of SAX events. The fragment of XPath used is called *univariate XPath*. The goal of both algorithms is to prove that univariate XPath can be efficiently evaluated in $O(|D|.|Q|)$ time in the streaming environment and to show that algorithms are not only time-efficient but also space-efficient.

These algorithms take two input parameters. The first one is the XPath expression (which respects univariate XPath to allow stream-processing) that will be transformed to a query table throughout stream processing and statically stored on the memory. After that, the main function is called. It reads the second parameter (XML in SAX events syntax) line by line repeatedly, each time generating a tag. Based on that tag a corresponding *startBlock* or *endBlock* function is called to process it. Finally, the main function generates as output the result of the XPath query.

Both algorithms were proposed to handle two challenges of stream-querying that were not solved by XSQ [81] and TwigM [34]. These challenges are: recursion in the XML document and the existence of same node-labels in the XPath expression. Based on their experiments, both LQ and EQ algorithms show very similar time performance in practice. In non-recursive (there are no nodes of a certain type can be nested in another nodes of the same type) cases, LQ and TwigM [34] has the same buffering space costs, as well as, EQ and XSQ [81] has the same cost.

In [72] the authors proposed an approach for encoding and matching XPath queries with forward (child, descendant, following, following-sibling) axes against streaming XML data. For this purpose, they propose an *Order-aware Twig* (OaT) that is a tree structure rooted at a node labeled 'r' known as the root of the OaT. There are three types of relationship edges P-C edge, A-D edge and closure edge (it is used to handle XPath expressions containing an axis step with *following – sibling*). Moreover, OaT has two types of constraint edges *LR* edge and *SLR* edge. The match of an OaT against an XML document is a mapping from nodes in the OaT to nodes in the document satisfying the node labels and relationships and constraints between the nodes of the OaT. The algorithm processes branches of the twig in left-to-right order. A branch is never processed unless constraints specified in the preceding branches are satisfied by the stream. Also, the algorithm avoids repeated processing of branches whose constraints have already been satisfied by the stream. The complexity of this algorithm is not given, and only experimentally studied. Recently, the authors also investigate the streaming evaluation of backward axes [73].

In [77] authors proposed an XPath processing algorithm which evaluates XPath queries in XP $\{\downarrow,\rightarrow,\star,[]\}$ over XML streams. An XPath query is expressed with axes, which are binary relations between nodes in XML streams: ? identifies the child/descendant axes and $\rightarrow$ indicates the following/following-sibling axes. The proposed algorithm evaluates XPath queries within one XML parsing pass and outputs the fragments found in XML streams as the query results. The algorithm uses double-layered non-deterministic finite automata (NFA) to resolve the following/following-sibling axes. First layer NFA is compiled from XPath queries and is able to evaluate sub-queries in XP $\{\downarrow,\rightarrow,\star\}$. Second layer NFA handles predicate parts. It is dynamically maintained during XML parsing: a state is constructed from a pair of the corresponding state in the first layer automaton and the currently parsed node in the XML stream. Layered NFA achieves $O(|D|.|Q|)$ time complexity by introducing a state sharing technique, which avoids the exponential growth in the state size of Layered NFA by eliminating redundant transitions. The proposed technique outperforms XSQ [81] and SPEX [25]. However, though this approach processes following/following-sibling axes, it does not process predicates with boolean operators ('*and*', '*or*', '*not*'). It will be interesting to check how to update the XPath fragment covered by this approach to process predicates with boolean operators, then, to check whether it will have the same linear time complexity or not.

In [26] authors presented through an empirical study that this lower bound proposed by Bar-Yossef et al. [11] can be broken by taking semantic information into account for buffer reduction. To demonstrate this, They built SwinXSS, a SAX-based XML stream query evaluation system and designed an algorithm that consumes buffers in line with the concurrency lower bound. They designed several semantic rules for the purpose of breaking the lower bound and incorporated these rules in the lower bound algorithm. However, using semantic information requires having DTD/Schema for the XML data set which means that this system functions only with XML data sets which have DTD/Schema. Furthermore, the fragment of XPath covered by this system does not consider the descendant axe '//' not the wildcard node ☆.

In [9] authors proposed Chimera, a stream-oriented XML filtering/querying engine. Chimera is an XPath querying engine that works for massive XML data sets stored and multiple queries input with light-weight pre-processing of data, instead of requiring heavy pre-processing and human operations like traditional stored data processing engines.

Chimera first decomposes XPath patterns into a set of primitive patterns. Then, they use XSIGMA, which is a matching engine for primitive patterns. XSIGMA generates occurrence event streams for all primitive patterns They assume that the input XML data is transformed into its binary representation called BML data. So pre-processing is required to obtain this representation. Finally, Chimera applies Lazy Filter and Twig Evaluator to the event streams for evaluating all XPath patterns. The fragment covered by this system is Univariate XPath [11] which is smaller than the Forward one [4]. Unfortunately, the details of the evaluation algorithm are not explained.

For further information on stream-processing for XML data, we recommend to read [13, 104]. In [13] recommend to read it is a survey gives an overview of formal results on the XML query language XPath. Authors identify several important fragments of XPath, focusing on subsets of XPath 1.0. Then they present results on the expressiveness of XPath and its fragments compared to other formalisms for querying trees, algorithms and complexity bounds for evaluation of XPath queries, and static analysis of XPath queries. In [104], author survey the state of the art in XML streaming evaluation techniques. They classify the XPath streaming evaluation approaches according to the main data structure used for the evaluation into three categories: automaton-based approach, array-based approach, and stack-based approach. Then they compare the major techniques proposed for each approach.

In Sect. 4, we survey parallel methods for XPath query processing.

4 Parallel Query Processing

We now survey parallel methods for XPath query processing, or more precisely: the parallel acceleration of an individual query on a large XML document. Works on: parallel search for similarities between XML documents, parallel syntax analysis of XML or parallel keyword search in XML documents are not covered here if they do not relate to XPath query processing. Readers interested in high-performance XML parsing will learn about recent work in papers [27, 31, 71, 78–80, 103, 106]. We classify a first set of papers among: foundational studies on complexity, data-distribution research and general implementation issues and DBMS techniques applied to XPath. Then we cover papers that yield more specific concepts and techniques for parallel processing of XPath queries: XPath fragment and their parallel complexity, physical distribution and scalability.z

4.1 General Research on Decidability and Complexity

Berstel [15] presents an algebraic study of XML formal languages i.e. sets of documents generated by DTD grammars. Clean definitions are used to show that inclusion between XML grammars is decidable, that they are recognized by deterministic stack automata. XML languages are characterized as a subclass of context-free languages. Such foundational work delimits complexity issues and may be reused in the design of XML languages. *Disadvantage*: The decidability and complexity classes studied here are much too broad to yield positive information about parallel processing. The paper by Acciai et al. [2] describes a pi-calculus encoding of queries on XML documents, which is a theoretical basis for concurrent document processing. *Disadvantage*: The pi-calculus' very high computational complexity prevents any kind of performance model or high-performance parallel processing. More specifically: properties like scope extrusion tend to require centralized name management and lead to either shared-memory implementations or coherence problems in a distributed memory.

4.2 Data Distribution

Chand and Felber [29] study a subscriber-consumer system for XML documents following a scenario where many clients are served concurrently. An algorithm called XSEARCH combines queries into a tree structure before processing them. The algorithm is precisely described and detailed experimental data on processing speed is given. Numbers on bandwidth are also given which is interesting and not so common in such publications. *Disadvantage*: The graphs do not express parallel speed-ups and the XSEARCH concept is not meant to accelerate a single query on a large document.

4.3 Distribution and Parallel XPath Query Processing

All works surveyed here deal with intra-query parallelism i.e. speed-up methods applied to a single query. The problem of speeding up sets of multiple queries is not our main concern as its practical solutions are either independent from intra-query parallelism, or based on replications of its techniques.

Tang [97] presents a new data placement strategy called WIN. Nodes of the XML tree are split into three classes: "duplicates", "intermediates" and "trivial". The first ones are subtree roots that are copied on every processor. The trivial ones are placed on the same processor as their father and the intermediate ones are used to guide data placement. The WIN heuristic operates by expansion operations that spread data just enough to maximize performance. The cost model measures workload on individual processors. Performance tests show speed-up factors of 2 or 2.5 on 5 processors when applied to documents of 40 or 80 MB. *Disadvantage*: The cost model is blind to communication costs.

Gottlob conducted a detailed study of the parallel complexity of XPath query processing and this is published in [47]. It is shown that the fragment Core XPath is PTIME-complete which suggests that the whole query language is not efficiently parallelizable. But large fragments are shown to belong to complexity class NC2 (problems solvable in theoretical time $O(\log n)^2$ with a polynomial number of processors and a uniform "program" in SPDM style). DTD validation of an XML document is also shown to be theoretically parallelizable. Closer study of the interesting language fragments uncovers sufficient conditions for reasonable parallel complexity: either avoid nesting of boolean operations (in the XPath query) or limit the depth of negation operators. Following [47], Filliot et al. [42] propose a new restriction of XPath for which the query problem is theoretically parallelizable. *Disadvantage*: Parallel complexity classes are necessary but unsufficient for efficient parallel algorithms: they do not model communication or synchronization costs that can cancel theoretical speed-up factors.

Despite this theoretical generality, Gottlob's paper gives a clear outline for efficient sequential and parallel algorithms of very large data sets. His theorem 8.6 states that a query can be evaluated in sequential time $O(|Q|^2 * |t|^4)$ and space (related to the number of processors) $O(|Q|^2 * |t|^2)$ where Q is the XPath query and t the XML document. The basis for these upper bounds is the pre-processing of axis relations in t. The local relations like son, next-sibling etc, are computable in linear sequential time. The transitive closure relations like descendant can be computed by repeated doubling of the local relations. Once all relations are available complex XPath queries can be processed by following the query's structure.

The thesis by Li [64] contains a chapter on stream processing and another on parallel processing of queries. The latter is an application of automatic parallelization techniques for sequential programs. Loop fusion and loop parallelization techniques are applied to query-processing sequential programs. The loops come from an automaton in charge of processing the query and its efficiency is increased by transforming its source code. Tests are conducted up to 8 processors and speedups are on the order of 2 when taking communication into account. *Disadvantage*: This method is blind to document structure or query structure. In our opinion it is unlikely to scale-up well, unless coupled with more algorithmic or structural information.

Paper [46] presents a parallel algorithm called PAX3, and whose complexity properties appear to be excellent: communication proportional only to query size, only 2 or 3 times more total operations than the sequential version, minimal communcation of sub-trees. Previous methods combined the sub-trees at the end of processing but this has been improved. The algorithm's overall control is bottom-up, it first produces candidate nodes that could be part of a solution and then moves information upward to retain only a subset of those. Tests were performed on up to 10 processors with documents of 28 MB. Accelerations by a factor of 3 were measured. *Disadvantage*: There seems to be no study of PAX3's scalability or of the effect of XML tree imbalance.

In a short and concise paper [8] Amagasa et al. describe a parallel query processing method based on path analysis. The XML document is first compiled into a set of paths that constitute a relational table. Another table is created for document nodes. The tables are dispersed on a multiprocessor. From the paper's explanations this distribution is not horizontal. Finally, an XPath query is realized by a relational join computation which encodes the conjuction of all the constraints it expresses on the document paths. A theoretical cost model is given and covers data volumes but not communication costs. *Disadvantage*: The experimental measurements do not seem to be scalable but could become so if a finer analysis of communication overheads was conducted.

The authors of [19] propose a document fragmentation method to improve efficiency of stream- or parallel XML document processing. Existing methods are based on a token semantics and subject to imbalance. The proposed method uses histograms that encode the document's tree structure. The tree splitting problem is defined thus: to break the tree t into disjoint parts whose size, width and depth are bounded by a priori constants. The authors express the opinion that there exists no better splitting method and that the one presented here only needs adjustments. They thus propose a generic data structure with operations, but not a specific algorithm for query processing. Fragmentation experiments were conducted on files from Nasa.xml to measure the number of fragments, as well as the number of "hops" on a P2P network of 500 peers. *Disadvantage*: They show an improvement of 25 % on this last parameter but no idea is given of the effect on a given algorithm for absolute times and parallel speedup.

Waldvogel et al. paper [100] is a well-structured and lucid study of the problem of splitting and XML document into sub-documents for efficient, therefore balanced processing. The authors consider five heuristics: 1. Level Split 2. Fanout Split 3. Semantic Split 4. Postorder Split 5. PostorderSem Split. Methods 1 and 2 are naive and based on the tree structure. Method 3 uses token names and is based on an optimistic hypothesis. Methods 4 and 5 are novel and more sophisticated: they are based on the size of subtrees. Experiments confirm that Postorder and

PostorderSem split behave better. Tests shown a parallel efficiency between 60 and 100 % on 1–10 processors. The techniques shown here are similar to sampling-based (partial information) algorithms that are used by the best parallel sorting algorithms.

A paper by Lu [67] describes a relatively naive but dynamic technique for parallelizing operations on an XML document: work stealing. Static load balancing would lead to bad performance so load balancing is local and dynamic. A less busy process will steal a sub-tree to be processed by another, busy process. The authors show speedups of 4 on 4 threads or processes. *Disadvantage*: The absolute overhead of work stealing is not measured or estimated so scalability is at best unknown.

Bordawekar, Lim, and Shmueli published [21] and ingenious and detailed experiment to measure parallel speedup of XPath queries on a shared-memory architecture. Multi-threaded programs use version 1.10 of the C++ implementation of Xalan, an XPath processor presented as begin among the best of its kind. Two parallelization methods are studied: data-partitioning where data and query are split vertically, and query-partitioning where the tree is split horizontally so as to apply the same query many times. A hybrid algorithm is also shown. The hardware used for experimenting had up to 8 processing units and the documents being processed were: XMark, DBLP and Peen Treebank so as to test both the effect of depth and width. Parallel efficiency was mostly around 50 % except for a few super-linear accelerations that were probably due to a pre-emptive parallel search algorithm. In such a search one stops the query as soon as one thread has found an answer, which can happen before n/p operations where n is the number of operations required for a sequential algorithm and p the number of processing units in parallel. This phenomenon is probably inevitable but orthogonal to our main concern here: efficient processing of complex queries on large random documents. The experimental results of [21] are encouraging and should be extended to larger documents and larger machines, as the authors mention in their conclusion. *Disadvantage*: The shared-memory model is not supported by many large-scale parallel systems. It avoids the effects of non-uniform memory access typical of distributed or hierarchical architectures.

We recommend Bordawakar et al. paper [20] as introductory reading. It is one of the best-written papers to date on the problem of intra-query XPath parallelism. The problem's definition and context are well summarized, the XPath language fragment is defined clearly and a cost (performance) model is given covering selectivity and processing time. The data- and query-partitioning strategies are illustrated. Some problems are processed with pure data-partitioning, others with pure query-partitioning and some with a mix of both. The experimental results are disapointing because they never show speedup factors higher than two, which does not improve over many papers published since 2005. The target architectures are again limited to shared memory machines (SMP or multi-core) which leads to a uniform memory-access time hypothesis, and therefore a cost model which ignores communication costs. The XPath fragment that is processed in limited to for-wardXPath (only forward axes like son, descendant are allowed) and without recursive predicates. *Disadvantage*: The work described here is very well presented and is well designed but suffers from several incompleteness factors, without leading to high scalability.

Recent work presented in [31] concerns XML parsing and parallel XPath query processing on multicore systems. The XPath fragment covered contains forward axes, the * operator, and nested predicates. The method description is not very formal or technical. An adjacency matrix describes the XML tree once parsed. Modest speedups on the order of 30 % are observed. *Disadvantage*: Work restricted to small-scale parallelism, possible waste of space due to the document data structure.

Feng et al. [41] published one of the best publications so far on parallel processing of XPath queries. A parallel version of the PathStack algorithm is described. It operates in three phases, 1° the document is split vertically in a balanced manner, 2° queries are applied to the various sub-documents or "slices" 3° the results are recombined. The algorithm description is precise and detailed, as well as the presentation of experimentes with OpenMP on an eight-core architecture. The optimal number of threads is sought, and speedups of 4–6 are observed which amounts to 50–75 % of efficiency. *Disadvantage*: It will be interesting to extend this work to distributed-memory architectures and measure or model its communication overheads.

Recent work by Zinn et al. [111] concerns an experimental approach based on Map-Reduce operations. It targets simple-path queries with son and descendent axes only. The authors study three data-flow composition strategies as pipelines which process a document token. The first, called the naive strategy, splits the XML document by subtrees that correspond to the successive tokens in the path query. The second strategy pre-splits the document on the google distributed file system. The third strategy splits the document in memory before its parallel processing. Experiments run on Hadoop have shown speeup factors of up to 20 on 30 processors. This efficiency factor of 2/3, the number of processors and distributed memory architecture are innovative elements by comparison with almost all existing work. But the algorithm descriptions are very brief and it would not be easy to reuse the key elements of this work. *Disadvantage*: An apparently scalable technique but designed for, and applied to a very small query language fragment.

4.4 Low-Level Implementation Techniques

Article [40] is a performance study of XML document processing from the point of view of AON (application-oriented networking). It uncovers parameters that are familiar to parallel algorithm design and tuning: communication cost management relating high-level algorithms to hardware parameters. The study concentrates on low-level factors like the number of cores or processors (which is bound to 2 here) and bus bandwidth. This study aims at informing hardware designers of the key factors related to XML document processing.

An original technique is presented by Shao et al. in [95]. Bitmap filtering is an implementation method for efficient query processing on XML documents. It consists of a compact encoding of the document one node at a time: a description of the node by its position in the tree (prefix path) and its subtree (suffix subtree). For a given query, one pre-filters the document to keep only nodes that may satisfy part of the query. The rate of filtering (document reduction factor) is then proportional to the speedup for a complete query-processing algorithm. Recall that *filtering* only solves the decision problem of detecting a solution to the query. *Disadvantage*: Like most papers, large-scale efficiency and communication costs are not studied in depth.

Paper [16] presents an experimental work on reducing L2 cache misses so as to increase throughput when processing XML documents on 4–8 core processors. *Disadvantage*: Not specific to query processing or to large-scale parallelism.

Moussalli et al. [70] present an original experiment for *filtering* ForwardXPath queries (those with forward axes, ∗ and possibly recursive document structures) with FPGA hardware. The query is compiled into an automaton table with parallel FPGA implementation through which the XML document is then streamed. *Disadvantage*: An interesting approach but its adaptation to full *querying* is not clear due to the low-level style of programming and possibly limited local storage.

4.5 DBMS and Join-Based Implementations

The short paper [89] defines the problem of XML query processing by encoding the document in a database system. It recalls the two main approaches for doing so: to build table structures with DTD information, or to use general DTD-independent techniques. *Disadvantage*: This paper is only 5 pages long and too short for proposing a technical contribution.

Pleshachkov and Kuznetsov [82] study the speedup possibilities for queries in a mixed relational-and-XML database. They observe that existing DBMS do not process independent subtrees in parallel. It is implemented in MS SQL Server 2005. A novel idea is the use of an SXTM transaction manager, receiving the queries or transaction orders. Implementation and algorithms are described in detail but the performance tests are not detailed. The authors only mention that their implementation introduces a low overhead. Parallel speedup is therefore unknown.

Machdi et al. [57] propose a parallel execution model for a set of queries on a set of XML documents. They tackle a relatively general problem: a dynamic flow of queries is received by a machine similar to a PC cluster, and this system must process them on a whole set of documents. This implies inter-query parallelism as well as intra-query parallelism, in an unpredictable mix. The authors use two types of algorithms: one where each processor operates sequentially, and a more original algorithm using intra-query parallelism. Task splitting is either static or dynamic. An analytical performance model is used to estimate both computation and communication (or I/O) which is remarkably rare. Their experimental conclusion are encouraging but not necessarily definitive due to the vast space of possible input situations. *Disadvantage*: The paper is very dense and its very short treatment of the main originality, namely the parallel "holistic twig join" (recursive query processing) is too short to make it realisticly reusable by the research community.

To summarize this section, it appears that practical feasibility for parallel processing has been broadly outlined by theoretical work and initial experiments. But this line of research appears younger than stream processing which is not surprising. Stream processing is limited by memory consumption and the use of parallel systems makes memory more abundant. But load-balancing, communication overheads and general scalability are in general more daunting problems than predictable and reasonable use of memory. Our survey of parallel processing for XML query processing illustrates this fact. Unfortunately, among the papers surveyed, none considers all three necessary dimensions of parallel scalability namely: predictability, load-balancing and communication overheads.

One last remark on DBMS-based methods. If such an approach was to revolutionize *parallel* XML query processing, it would have already done so for sequential XML query processing. Since this has not been the case one is allowed to have serious doubts on the long-term value of methods that reduce query-processing on semi-structured data to table-structured methods.

4.6 Map-Reduce, Hadoop Implementations

Some of the most recent work in this very active area target massively-distributed platforms like Hadoop and the Map-Reduce paradigm. The advantages of cloud-based platform for massive document processing are the same as for all parallel hardware plus potentially unlimited scalability and fault tolerance through middleware support.

Khatchadourian, Consens and Siméon take a declarative programming approach with the ChuQL system, a Hadoop implementation of XQuery supporting Map-Reduce parallelism [60]. Declarative support for document statistics is shown with high-level semantics for the language extensions that include the map-reduce primitives. Performance analysis and measurement is not presented but high-level explicit parallelism is exposed in the source code so that significant speedups are possible if communication costs can be limited.

The HadoopXML of Choi et al. [36] is a system for inter-query parallelism i.e. processing multiple twig-queries on very large XML data over a Hadoop platform. Run-time balancing is supported by Hadoop and the innovative software architecture shares partial results among twig-queries, which saves much I/O time. The subset of the XPath language is a fragment of forward XPath and performance has been tested on datasets of hundreds of GB.

The most serious work to date in this direction is that of Cong, Fan, Kementsietsidis, Li and Liu [37] who present an original distributed-memory algorithm for evaluating a forward XPath query. The authors compare their algorithm with a centralized version to ensure that the total amount of computation is not excessive, use partial evaluation for distributing the operations and show that the amount of communication is proportional to the query's selectivity, which is theoretically optimal whenever the output must transit along the network. Scalability has been tested on datasets of hundreds of GB. The comparison with a MonetDB memory-based implementation shows that the new algorithm compensates a higher execution time by its economy of I/O time.

5 Conclusion

In this survey, we reviewed the state of the art from the viewpoint of three critical areas where we developed our research work, namely.

1. selectivity estimation techniques for XML queries: we categorized them into synopsis-based and histogram-based estimation techniques. In addition, we justified the need of a new selectivity estimation technique that is stream based. In analogy with the well-studied case of databases, the degree of query selectivity i.e. the size and quantity of query output(s) is the most variable and critical performance parameter.
2. stream-processing for XML data: we categorized existing works into stream-filtering and stream-querying algorithms. Furthermore, we explained the reason of proposing the lazy stream-querying algorithm LQ to be used for traversing the path tree synopsis structure. Both path tree and LQ compose a stream-based selectivity estimation technique [6] which outperforms the current existing selectivity estimation techniques.
3. Parallel Query Processing: we surveyed basic research on the parallel complexity of XML query processing which isolated some query-language fragments that may benefit from speed-ups. We also covered works that analyzed the effect of data distribution on efficiency, since XML data is strongly non-regular in shape and bad load-balancing can offset eny parallel speed-up. We then described and analyzed the existing parallel implementation techniques for accelerating a single XPath query, assuming that multi-query parallelism is either a separate issue or must itself be based on single-query techniques. A minority of works adapt relational parallel techniques to the XML querying context, and those were also covered in this section. More recently, cloud-based algorithms have emerged for inter- and intra-query parallel processing on the forward XPath language fragment.

The current state of techniques allows predictive selectivity estimation when very large datasets pre-exist the queries that are performed on them. This approach can lead to very high scalability but it is not clear how to adapt it to subscriber-producer type systems. One possibility that should be investigated is the general and transparent use of performance meta-data in very large datasets. Stream querying appears to be a mature technology than can now be applied in very diverse directions. On the contrary, parallel processing is still evolving and although scalable algorithms are emerging, they still require investigation for enlarging the XPath language fragment covered, pre-emptive load balancing and more precise communication cost estimations. There is no reason why such techniques cannot mature within the next decade and support query processing speeds unimagined today but for restrictive query types.

References

1. A. Aboulnaga, A.R. Alameldeen, J.F. Naughton, Estimating the selectivity of XML path expressions for internet scale applications, in *Proceedings of the 27th International Conference on Very Large Data Bases (VLDB)* (2001), pp. 591–600
2. L. Acciai, M. Boreale, S. Zilio, A typed calculus for querying distributed XML documents, in *Trustworthy Global Computing, volume 4661 of Lecture Notes in Computer Science*, ed. by U. Montanari, D. Sannella, R. Bruni (Springer, Berlin Heidelberg, 2007), pp. 167–182
3. C.C. Aggarwal, P.S. Yu, A survey of synopsis construction in Ddata streams, in *Data Streams Models and Algorithms* (2007), pp. 169–207
4. M. Alrammal, Algorithms for XML stream processing: massive data, external memory and scalable performance. Thesis, Université Paris-Est, 2011. http://lacl.univ-paris12.fr/Rapports/TR/muth_thesis.pdf
5. M. Alrammal, G. Hains, A stream-based selectivity estimation technique for forward XPath, in *Proceedings of 2012 International Conference on Innoviations in Information Technology (IIT)* (2012), pp. 309–214
6. M. Alrammal, G. Hains, M. Zergaoui, Path tree: document synopsis for XPath query selectivity estimation, in *2011 International Conference on Complex, Intelligent and Software Intensive Systems (CISIS)*, (2011). IEEE, 30 2011-july 2, pp. 321–328
7. M. Altinel, M.J. Franklin, Efficient filtering of XML documents for selective dissemination of information, in *Proceedings of 26th International Conference on Very Large Data Bases (VLDB)* (2002), pp. 53–64
8. T. Amagasa, K. Kido, H. Kitagawa, Querying XML data using PC cluster system, in *Database and Expert Systems Applications. 18th International Workshop on Computing and Processing (Hardware/Software)*. IEEE, Sept 2007, pp. 5–9
9. T. Asai, S. Tago, H. Inakoshi, S. Okamoto, M. Takeda, H. Kitagawa, Y. Ishikawa, Q. Li, C. Watanabe, Chimera: Stream-oriented XML filtering/querying engine, in *Proceeding of DASFAA*, vol. 5982 (Springer, Berlin, Heidelberg, 2010), pp. 380–391
10. Z. Bar-Yossef, M. Fontoura, V. Josifovski, On the memory requirements XPath evaluation over XML streams, in *Proceedings of the 23rd ACM SIGMOD Symposium on Principles of Database Systems*, June 2004, pp. 177–188
11. Z. Bar-Yossef, M. Fontoura, V. Josifovski, On the Memory Requirements XPath Evaluation over XML Streams. PODS., June 2004, pp. 177–188
12. C. Barton, P. Charles, D. Goyal, M. Raghavachari, M. Fontoura, V. Josifovski, Streaming XPath processing with forward and backward axes, in *Proceedings of the International Conference on Data Engineering (ICDE)* (2003), pp. 455–466
13. M. Benedikt, C. Koch, XPath Leashed. ACM Comput. Surv. **41**(1), 3:1–3:54 (Jan 2009)

14. A. Berglund, S. Boag, D. Chamberlin, M.F. Fernández, M. Kay, J. Robie, J. Siméon, XML path language (XPath) 2.0, 14 Dec 2010. http://www.w3.org/TR/2010/REC-xpath20-20101214/
15. J. Berstel, L. Boasson, Formal properties of XML grammars and languages. Acta Informatica **38**, 649–671 (2002)
16. R. Bhowmik, M. Govindaraju, Cache performance optimization for processing XML-based application data on multi-core processors, in *10th IEEE/ACM International Conference on Cluster, Cloud and Grid Computing* (2010). IEEE, pp. 455–463
17. S. Boag, D. Chamberlin, M.F. Fernández, D. Florescu, J. Robie, J. Siméon, XQuery 1.0: An XML Query Language (Second Edition), 14 Dec 2010. http://www.w3.org/TR/2010/REC-xquery-20101214/
18. P. Bohannon, J. Freire, P. Roy, J. Siméon, From XML schema to relations: a cost-based approach to XML storage, in *Proceedings of the 18th International Conference on Data Engineering (ICDE)* (2002), pp. 64–75
19. A. Bonifati, A. Cuzzocrea, Efficient fragmentation of large XML documents, in *DEXA 2007, volume 4653 of Lecture Notes in Computer Science*, ed. by R. Wagner, N. Revell, G. Pernul (Springer, New York, 2007), pp. 539–550
20. R. Bordawekar, L. Lim, A. Kementsietsidis, B.W.-L. Kok, Statistics-based parallelization of XPath queries in shared memory systems, in *Proceedings of the 13th International Conference on Extending Database Technology, EDBT '10*, New York, NY, USA, 2010. ACM, pp. 159–170
21. R. Bordawekar, L. Lim, O. Shmueli, Parallelization of XPath queries using multi-core processors: challenges and experiences, in *EDBT, volume 360 of ACM International Conference Proceeding Series*, ed. by M. L. Kersten, B. Novikov, J. Teubner, V. Polutin, S. Manegold. ACM, 2009, pp. 180–191
22. S. Böttcher, R. Steinmetz, Evaluating XPath queries on XML data streams, in *Proceedings of the 24th British Inational Conference on Databases (BNCOD)* (2007), pp. 101–113
23. T. Bray, J. Paoli, C.M. Sperberg-McQueen, F. Yergeau, in *Extensible Markup Language (XML) 1.0*, 5th edn., 26 Nov 2008. http://www.w3.org/TR/REC-xml/
24. N. Bruno, N. Koudas, D. Srivastava, Holistic twig joins: optimal XML pattern matching, in *Proceedings of the 2002 ACM SIGMOD International Conference on Management of Data* (2002), pp. 310–321
25. F. Bry, F. Coskun, S. Durmaz, T. Furche, D. Olteanu, M. Spannagel, The XML stream query processor SPEX, in *Proceedings of the International Conference on Data Engineering (ICDE)* (2005), pp. 1120–1121
26. C. Yang, C. Liu, J. Li, J.X. Yu, J. Wang, A query system for XML data stream and its semantics-based buffer reduction. J. Res. Pract. Inf. Technol. **42**, 111–128 (2010)
27. R.D. Cameron, K.S. Herdy, D. Lin, High performance XML parsing using parallel bit stream technology, in *CASCON* ed. by M. Chechik, M.R. Vigder, D.A. Stewart. IBM, (2008), p. 17
28. C. Chan, P. Felber, M. Garofalakis, R. Rastogi, Efficient filtering of XML documents with XPath expressions, in *Proceedings of the 18th International Conference on Data. Engineering*, vol. 11 (2002), pp. 235–244
29. R. Chand, P. Felber, Scalable distribution of XML content with XNet. IEEE Trans. Parallel Distrib. Syst **19**, 447–461 (2008)
30. S. Chaudhuri, V. Ganti, L. Gravano, Selectivity estimation for string predicates: overcoming the underestimation problem, in *Proceedings of the 20th International Conference on Data Engineering (ICDE)* (2004), pp. 227–238
31. R. Chen, W. Chen, A parallel solution to XML query application, in *2010 3rd IEEE International Conference on Computer Science and Information Technology (ICCSIT)*, vol. 6, july 2010, pp. 542–546
32. T. Chen, J. Lu, T.W. Ling, On boosting holism in XML twig pattern matching using structural indexing techniques, in *Proceedings of the 2005 ACM SIGMOD International Conference on Management of Data* (2005), pp. 455–466

33. Y. Chen, S.B. Davidson, G.A. Mihaila, S. Padmanabhan, Expedite: a system for encoded XML processing, in *Proceedings of the thirteenth ACM international conference on Information and knowledge management (CIKM)* (2004), pp. 108–117

34. Y. Chen, S.B. Davidson, Y. Zheng, An efficient XPath query processor for XML streams, in *Proceedings of the 22nd International Conference on Data Engineering (ICDE)* (2006), p. 79

35. Z. Chen, H.V. Jagadish, F. Korn, N. Koudas, S. Muthukrishnan, R.T. Ng, D. Srivastava, Counting twig matches in a tree, in *Proceedings of the 17th International Conference on Data Engineering (ICDE)* (2001), pp. 595–604

36. H. Choi, K.-H. Lee, S.-H. Kim, Y.-J. Lee, B. Moon, HadoopXML: a suite for parallel processing of massive XML data with multiple twig pattern queries, in *Proceedings of the 21st ACM International Conference on Information and Knowledge Management, CIKM '12*, New York, NY, USA, 2012. ACM, pp. 2737–2739

37. G. Cong, W. Fan, A. Kementsietsidis, J. Li, X. Liu, Partial evaluation for distributed XPath query processing and beyond. ACM Trans. Database Syst. 37(4), 32:1–32:43 (Dec 2012)

38. D. DeHaan, D. Toman, M.P. Consens, M.T. Ozsu, A comprehensive XQuery to SQL translation using dynamic interval encoding, in *Proceedings of the 2003 ACM SIGMOD International Conference on Management of Data* (2003), pp. 623–634

39. Y. Diao, M. Altinel, M. Franklin, H. Zhang, P. Fischer, Efficient and scalable filtering of XML documents, in *Proceedings of the 18th International Conference on Data, Engineering* (2002), pp. 341–342

40. J.J. Ding, A. Waheed, Dual processor performance characterization for XML application-oriented networking, in *Proceedings 2007 International Conference on Parallel Processing (36th ICPP'07)*, Xi-An, China, Sept 2007. IEEE Computer Society, p. 52

41. J. Feng, L. Liu, G. Li, J. Li, Y. Sun, An efficient parallel PathStack algorithm for processing XML twig queries on multi-core systems, in *Database Systems for Advanced Applications, volume 5981 of Lecture Notes in Computer Science* ed. by H. Kitagawa, Y. Ishikawa, Q. Li, C. Watanabe (Springer, Berlin, Heidelberg, 2010), pp. 277–291

42. E. Filiot, J. Niehren, J.-M. Talbot, S. Tison, Polynomial time fragments of XPath with variables, in *PODS '07: Proceedings of the twenty-sixth ACM SIGMOD-SIGACT-SIGART Symposium on Principles of Database Systems*, New York, NY, USA, 2007. ACM, pp. 205–214

43. D.K. Fisher, S. Maneth, Structural selectivity estimation for XML document, in *Proceedings of the 23rd International Conference on Data Engineering (ICDE)* (2007), pp. 626–635

44. D. Florescu, C. Hillery, D. Kossmann, P. Lucas, F. Riccardi, T. Westmann, M. J. Carey, A. Sundararajan, G. Agrawal, The BEA/XQRL streaming XQuery processor, in *Proceedings of the 2003 International Conference on Very Large Data Bases (VLDB)* (2003), pp. 997–1008

45. J. Freire, J. Haritsa, M. Ramanath, P. Roy, J. Siméon, StatiX: making XML count, in *Proceedings of the ACM SIGMOD International Conference on Management of Data* (2002), pp. 181–191

46. A.K. Gao Cong, W. Fan, Distributed query evaluation with performance guarantees, in *Proceedings of the 2007 ACM SIGMOD International Conference on Management of Data*

47. R.P. Georg Gottlob, C. Koch, L. Segoufin, The parallel complexity of XML typing and XPath query evaluation. J. ACM 52(2), 284–335 (2005)

48. A. Gibbons, W. Rytter, in *An Introduction to Parallel Algorithms* (Addison-Wesley, Massachusetts, 1992)

49. G. Gou, R. Chirkova, Efficient algorithms for evaluating XPath over streams, in *Proceedings of the 2007 ACM SIGMOD International Conference on Management of Data* (2007), pp. 269–280

50. T. Grust, Accelerating XPath location steps, in *Proceedings of the 2002 ACM SIGMOD International Conference on Management of Data* (2002), pp. 109–120

51. T. Grust, Purely relational FLWORs, in *Proceedings of the 2nd International Workshop on XQuery Implementation, Experience and Perspectives (XIME-P)*, in cooperation with ACM SIGMOD, 2005
52. T. Grust, S. Sakr, J. Teubner, XQuery on SQL hosts, in *Proceedings of the 29th International Conference on Very Large Data Bases (VLDB)* (2004), pp. 252–263
53. A. Gupta, D. Suciu, Stream processing of XPath queries with predicates, in *Proceedings of the 2003 ACM SIGMOD International Conference on Management of Ddata* (2003), pp. 219–430
54. W.-S. Han, H. Jiang, H. Ho, Q. Li, StreamTX: extracting tuples from streaming XML data, in *Proceedings of the VLDB Endowment* (2008), pp. 289–300
55. Z. He, B. Lee, R. Snapp, Self-tuning UDF cost modeling using the memory-limited quadtree, in *Proceedings of the 9th International Conference on Extending Database Technology (EDBT)* (2004)
56. J. Hopcroft, J. Ullman, *Introduction to Automata Theory, Language, and Computation* (Addison-Wesley, Massachusetts, 1979)
57. T.A. Imam Machdi, H. Kitagawa, GMX: an XML data partitioning scheme for holistic twig joins, in *Proceedings of the 10th International Conference on Information Integration and Web-based Applications & Services*
58. H. Jiang, W. Wang, H. Lu, J.X. Yu, Holistic twig joins on indexed XML documents, in *Proceedings of the 29th International Conference on Very Large Data Bases (VLDB)*, vol. 29 (2003), pp. 273–284
59. V. Josifovski, M. Fontoura, A. Barta, Querying XML streams. VLDB J. **14**, 197–210 (2005)
60. S. Khatchadourian, M. Consens, J. Siméon, Having a ChuQL at XML on the cloud, in *Proceedings of the 5th Alberto Mendelzon International Workshop on Foundations of Data Management, volume 749 of CEUR Workshop Proceedings* ed. by P. Barcelo and V. Tannen, Santiago, Chile, 2011. http://CEUR-WS.org
61. C. Koch, S. Scherzinger, N. Schweikardt, B. Stegmaier, Schema-based scheduling of event processors and buffer minimization for queries on structured data streams, in *Proceedings of the 2004 International Conference on Very Large Data Bases (VLDB)* (2004), pp. 228–239
62. B. Lee, L. Chen, J. Buzas, V. Kannoth, Regression-based self-tuning modeling of smooth user-defined function costs for an Object-Relational Database Management System query optimizer. Comput. J. 673–693 (2004)
63. H. Li, M. L. Lee, W. Hsu, G. Cong, An estimation system for XPath expressions, in *Proceedings of the 22nd International Conference on Data Engineering (ICDE)* (2006), p. 54
64. X. Li, *Thesis: Efficient and Parallel Evaluation of XQuery* (Ohio State University, OhioLINK, 2006)
65. H. Liefke, D. Suciu, Xmill: an efficient compressor for XML data, in *Proceedings of the ACM SIGMOD International Conference on Management of Data* (2000), pp. 153–164
66. L. Lim, M. Wang, S. Padmanabhan, J.S. Vitter, R. Parr, XPathLearner: an on-line self-tuning Markov histogram for XML path selectivity estimation, in *Proceedings of 28th International Conference on Very Large Data Bases (VLDB)* (2002), pp. 442–453
67. W. Lu, D. Gannon, Parallel XML processing by work stealing, in *Proceedings of the 2007 workshop on Service-oriented Computing Performance: Aspects, Issues, and Approaches, SOCP '07*, New York, NY, USA, 2007. ACM, pp. 31–38
68. B. Ludascher, P. Mukhopadhyay, Y. Papakonstantinou, A transducer-based XML query processor, in *Proceedings of the 2002 International Conference on Very Large Data Bases (VLDB)* (2002), pp. 227–238
69. C. Luo, Z. Jiang, W.-C. Hou, F. Yu, Q. Zhu, A sampling approach for XML query selectivity estimation, in *Proceedings of the International Conference on Extending Database Technology (EDBT)* (2009), pp. 335–344
70. R. Moussalli, M. Salloum, W. Najjar, V. Tsotras, Accelerating XML query matching through custom stack generation on FPGAs, in *High Performance Embedded Architectures*

and Compilers, volume 5952 of Lecture Notes in Computer Science (Springer, Berlin, Heidelberg, 2010), pp. 141–155

71. M. Nicola, J. John, XML Parsing: a threat to database performance, in Proceedings of the Twelfth International Conference on Information and Knowledge Management. ACM New York, NY, USA, 2003, pp. 175–178

72. A. Nizar, S. Kumar, Efficient evaluation of forward XPath axes over XML streams, in Proceedings of the 14th International Conference on Management of Data (COMAD) (2000), pp. 222–233

73. A. Nizar, S. Kumar, Ordered backward XPath axis processing against XML streams, in Proceedings of the 6th International XML Database Symposium (XSym) (2009), pp. 1–16

74. M.A. Nizar, G.S. Babu, P.S. Kumar, SFilter: a simple and scalable filter for XML streams, in Proceedings of the 15th International Conference on Management of Data (COMAD) (2009)

75. D. Olteanu, SPEX: streamed and progressive evaluation of XPath. IEEE Trans. Knowl. Data Eng. 934–949 (2007)

76. D. Olteanu, H. Meuss, T. Furche, F. Bry, XPath: looking forward, in Proceedings of the 2002 XML-Based Data Management and Multimedia Engineering (EDBT) Workshops (2002), pp. 109–127

77. M. Onizuka, Processing XPath queries with forward and downward axes over XML streams, in Proceedings of the 13th International Conference on Extending Database Technology (EDBT) (2010), pp. 27–38

78. Y. Pan, W. Lu, Y. Zhang, K. Chiu, A static load-balancing scheme for parallel XML parsing on multicore CPUs, in Proceedings of the Seventh IEEE International Symposium on Grid (2007). IEEE Computer Society, pp. 351–362

79. Y. Pan, Y. Zhang, K. Chiu, Parsing XML using parallel traversal of streaming trees, in HiPC, volume 5374 of Lecture Notes in Computer Science ed. by P. Sadayappan, M. Parashar, R. Badrinath, V. K. Prasanna (Springer, New York, 2008), pp. 142–156

80. Y. Pan, Y. Zhang, K. Chiu, Simultaneous transducers for data-parallel XML parsing, in IPDPS (2008). IEEE, pp. 1–12

81. F. Peng, S.S. Chawathe, XPath queries on streaming data, in Proceedings of the 2003 ACM SIGMOD International Conference on Management of Data (2003), pp. 431–442

82. P.O. Pleshachkov, S.D. Kuznetsov, Transaction management in RDBMSs with XML support. Program. Comput. Softw. 32, 243–254 (2006)

83. N. Polyzotis, M. Garofalakis, Statistical synopses for graph-structured XML databases, in Proceedings of the 2002 ACM SIGMOD International Conference on Management of Data (2002), pp. 358–369

84. N. Polyzotis, M. Garofalakis, Structure and value synopses for XML data graphs, in Proceedings of the 28th International Conference on Very Large Data Bases (VLDB) (2002), pp. 466–477

85. N. Polyzotis, M.N. Garofalakis, XCluster synopses for structured XML content, in Proceedings of the International Conference on Data Engineering (ICDE) (2006), p. 63

86. N. Polyzotis, M.N. Garofalakis, Y. Ioannidis, Approximate XML query answers, in Proceedings of the 2004 ACM SIGMOD International Conference on Management of Data (2004), pp. 263–274

87. N. Polyzotis, M.N. Garofalakis, Y. Ioannidis, Selectivity estimation for XML twigs, in Proceedings of the International Conference on Data Engineering (ICDE) (2004), pp. 264–275

88. V. Poosala, Y.E. Ioannidis, P.J. Haas, E.J. Shekita, Improved histograms for selectivity estimation of range predicates, in Proceedings of the ACM SIGMOD International Conference on Management of Data (1996), pp. 294–305

89. J. Qin, S. Yang, W. Dou, Parallel storing and querying XML documents using relational DBMS, in APPT, volume 2834 of Lecture Notes in Computer Science ed. by X. Zhou, S. Jähnichen, M. Xu, J. Cao (Springer, New York, 2003), pp. 629–633

90. P. Ramanan, Evaluating an XPath query on a streaming XML document, in *International Conference on Management of Data (COMAD)* (2005)
91. S. Saker, Cardinality-aware and purely relational implementation of an XQuery processor. PhD thesis-University of Konstanz (2007), pp. 58–82
92. S. Saker, Algebra-based XQuery cardinality estimation. Int. J. Web Inf. Syst. 7–46 (2008)
93. S. Sakr, Towards a comprehensive assessment for selectivity estimation approaches of XML queries. Proc. Int. J. Web Eng. Technol. **6**, 58–82 (2010)
94. A. Schmidt, R. Busse, M. Carey, M.K.D. Florescu, I. Manolescu, F. Waas, Xmark: An XML Benchmark Project, 2001. http://www.xml-benchmark.org/
95. F. Shao, G. Chen, L. Yu, Y. Bei, J. Dong, Bitmap filtering: an efficient speedup method for XML structural matching, in *Proceedings of the 8th ACIS International Conference on Software Engineering, Artificial Intelligence, Networking and Parallel/Distributed Computing, SNPD* ed. by W. Feng, F. Gao (IEEE Computer Society, 2007), pp. 756–761
96. H. Takekawa, H. Ishikawa, Incrementally-updatable stream processors for XPath queries based on merging automata via ordered hash-keys, in *Proceedings of the 18th International Conference on Database and Expert Systems Applications (DEXA)* (2007), pp. 40–44
97. N. Tang, G. Wang, J.X. Yu, K.-F. Wong, G. Yu, WIN: an effcient data placement strategy for parallel XML databases, in *ICPADS* (IEEE Computer Society, 2005), pp. 349–355
98. J. Teubner, T. Grust, S. Maneth, S. Sakr, Dependable cardinality forecasts for XQuery, in *Proceedings of the VLDB Endowment (PVLDB)* (2008), pp. 463–477
99. S. Viglas, Introduction to stream data management, in *Chapter 2—Stream Data Management*, vol. 30 (Springer, New York, 2005), pp. 16–33
100. M. Waldvogel, M. Kramis, S. Graf, Distributing XML with focus on parallel evaluation, in *DBISP2P* (2008), pp. 55–67
101. W. Wang, H. Jiang, H. Lu, J.X. Yu, Containment join size estimation: models and methods, in *Proceedings of the 2002 ACM SIGMOD International Conference on Management of Data* (2003), pp. 145–156
102. W. Wang, H. Jiang, H. Lu, J.X. Yu, Bloom histogram: path selectivity estimation for XML data with updates, in *Proceedings of the 13th International Conference on Very Large Data Bases (VLDB)* (2004), pp. 240–251
103. K.C. Wei Lu, Y. Pan, A parallel approach to XML parsing, in *7th IEEE/ACM International Conference on Grid Computing* (2006). IEEE, pp. 223–230
104. X. Wu, D. Theodoratos, A survey on XML streaming evaluation techniques. VLDB J. 1–26 (2012)
105. Y. Wu, J. Patel, H. Jagadish, Estimating answer sizes for XML queries, in *Proceedings of the 8th International Conference on Extending Database Technology (EDBT)* (2002), pp. 590–608
106. K.C. Yinfei Pan, Y. Zhang, W. Lu, Parallel XML parsing using Meta-DFAs, in *Proceedings of the Third IEEE International Conference on e-Science and Grid Computing, E-SCIENCE '07*, Washington, DC, USA (IEEE Computer Society, 2007), pp. 237–244
107. G. Zhang, Q. Zou, Quickxscan: efficient streaming XPath evaluation, in *Proceedings of the International Conference on Internet Computing* (2006), pp. 249–255
108. N. Zhang, P. Haas, V. Josifovski, G. Lohman, C. Zhang, Statistical learning techniques for costing XML queries, in *Proceedings of the 31st VLDB Conference on Very Large Data Bases (VLDB)* (2005), pp. 289–300
109. N. Zhang, M.T. Ozsu, A. Aboulnaga, I.F. Ilyas, XSeed: accurate and fast cardinality estimation for XPath queries, in *Proceedings of the 20th International Conference on Data Engineering* (2006), p. 61
110. Y. Zhang, Y. Pan, K. Chiu, A parallel XPath engine based on concurrent NFA execution, in *International Conference on Parallel and Distributed Systems*, vol. 0 (2010), pp. 314–321
111. D. Zinn, S. Bowers, S. Köhler, B. Ludäscher, Parallelizing XML data-streaming workflows via MapReduce. J. Comput. Syst. Sci. **76**(6), 447–463 (2010). Special Issue: Scientific Workflow 2009, The 2nd International Workshop on Workflow Management and Application in Grid Environments

Techniques and Applications to Analyze Mobility Data

Radu-Corneliu Marin, Radu-Ioan Ciobanu, Ciprian Dobre
and Fatos Xhafa

Abstract Mobility is intrinsic to human behavior and influences the dynamics of all social phenomena. As such, technology has not remained indifferent to the imprint of mobility. Today we are seeing a shift in tides as the focus is turning towards portability, as well as performance; mobile devices and wireless technologies have become ubiquitous in order to fulfil the needs of modern society. Today the need for mobility management is gradually becoming one of the most important and challenging problems in pervasive computing. In this chapter, we present an analysis of research activities targeting mobility. We present the challenges of analyzing and understanding the mobility (is mobility something that is inherently predictable? are humans socially inclined to follow certain paths?), to techniques that use mobility results to facilitate the interaction between peers in mobile networks, or detect the popularity of certain locations. Our studies are based on the analysis of real user traces extracted from volunteers. We emphasize the entire process of studying the dynamics of mobile users, from collecting the user data, to modelling mobility and interactions, and finally to exploring the predictability of human behavior. We point out the challenges and the limitations of such an endeavour. Furthermore, we propose techniques and methodologies to study the mobility and synergy of mobile users and we show their applicability on two case studies.

R.-C. Marin · R.-I. Ciobanu · C. Dobre (✉)
University Politehnica of Bucharest, Splaiul Independentei 313, Bucharest, Romania
e-mail: ciprian.dobre@cs.pub.ro

R.-C. Marin
e-mail: radu.marin@cti.pub.ro

R.-I. Ciobanu
e-mail: radu.ciobanu@cti.pub.ro

F. Xhafa
Universitat Politecnica de Catalunya, Girona Salgado 1–3 08034 Barcelona, Spain
e-mail: fatos@lsi.upc.edu

F. Xhafa and N. Bessis (eds.), *Inter-cooperative Collective Intelligence:*
Techniques and Applications, Studies in Computational Intelligence 495,
DOI: 10.1007/978-3-642-35016-0_8, © Springer-Verlag Berlin Heidelberg 2014

1 Introduction

In recent years, the ubiquitousness of mobile devices has led to the advent of various types of mobile networks. Such is the case of opportunistic networks (ONs), which are based on the store-carry-and-forward paradigm: a node stores a message, carries it until it encounters the destination or a node that is more suitable to bring the message closer to the destination, and then finally forwards it. Thus, an efficient routing algorithm for ONs should be able to decide if an encountered node is suitable for the transport of a given message with a high probability. It should also be able to decide whether the message should be copied to the encountered node, or moved altogether.

Since opportunistic networks are composed of human-carried mobile devices, routing and dissemination algorithms deployed in such networks should take advantage of the properties of human mobility, in order to be effective. Although for a time people have used mathematical models to simulate human mobility, it has been shown in recent years that the properties that were believed to be true regarding human mobility are actually incorrect. For example, contrary to what was believed, it was shown that human interactions follow a power law function, while the times between two successive contacts are described by a heavy-tailed distribution [1]. These results have shifted the focus from synthetic mobility models to real life traces that can offer a far better view of human interaction.

In this chapter, we present an analysis of current research activities on mobility. We highlight research efforts designed towards the collection and analysis of mobility traces. We present two case studies, on traces collected with the purpose of acquiring human interaction data in an academic environment. Both traces (entitled UPB 2011 and UPB 2012) were collected at the University POLITEHNICA of Bucharest, their participants being students and professors at the faculty.

Knowledge about the distributions of encounters in a trace, or the dependence on the time of day, is necessary in designing routing and forwarding algorithms. Therefore, our second contribution is to analyze the collected traces in terms of contact times distribution and highlight each trace's properties. Moreover, since we are dealing with academic environments where the participants' fixed schedules make them interact with a certain regularity, we attempt to prove the predictability of a node's future encounters. We propose doing this by approximating a node's behavior as a Poisson distribution and using the chi-squared test to demonstrate our assumption. Finally, we also analyze the predictability of an ON node's contacts with fixed wireless access points (APs), while also proposing a methodology for studying traces which involve the scanning of APs.

Preliminary versions of our work were previously published in [2, 3]. In this chapter we present more extensive results, describing the proposed traces in detail and analyzing them in regard to various ON-specific metrics (such as contact and inter-contact times) and the impact that these metrics have on the outcome of the trace. Furthermore, we describe the two tracing applications that were used to collect the data, and highlight a series of benefits and limitations of using mobility traces instead of mathematical models.

The chapter is organized as follows. Section 2 emphasizes on the process of collecting tracing data from real mobile users, discussing both the advantages and pitfalls of conducting such an endeavour. Sections 3 and 4 focus on the techniques for exploring mobility and interactional patterns applied on two case studies, as well as the resulted experimental data. Section 5 concludes our chapter with the implications of our work and thoughts for future development.

2 Collecting Inter-cooperative Mobility Data

The challenge in working with mobility arises from two difficult problems: formalizing mobility features and extracting mobility models. Currently, there are two types of mobility models in use: real mobile user traces and synthetic models [4]. Basically, traces are the results of experiments recording the mobility features of users (location, connectivity), while synthetic models are pure mathematical models which attempt to express the movement of devices.

Although they have been regarded as suspect models due to the limitations in mapping over reality [5], synthetic models have been largely used in the past, the two most popular models being random walks on graphs [7], which are similar to a Brownian motion, and the random waypoint mobility model [8], in which pauses are introduced between changes in direction or speed.

In 2005, Barabási [6] introduced a queueing model which disproved the claims of synthetic models based on random walks on graphs. Furthermore, Barabási's model showed that the distributions of inter-event times in human activity are far from being normal as they present bursts and heavy tails. This happens because people do not move randomly, but their behavior is activity-oriented [9–11]. This endeavour has paved the way for researchers in human dynamics, as the Barabási model [6] is continuously being developed [12–15] and experiments with it are using a variety of new interesting sources: web server logs [16–18], cell phone records [19–21] and wireless network user traces [2, 22].

The remainder of this section briefly describes two important sources for mobility analysis and modelling: the Haggle project and the CRAWDAD archives. Furthermore, it presents two mobility tracing applications developed and deployed at the University POLITEHNICA of Bucharest, and finally it highlights the pros and cons of such applications and the lessons learned from using them.

2.1 The Haggle Project

Haggle[1] is a European Commission-funded project that designs and develops solutions for opportunistic networks communication, by analyzing all aspects of the main networking functions, such as routing and forwarding, security, data dissemination and (most importantly for the work we present here) mobility traces and models [23]. The results proposed in Haggle were soon followed by a series of subsequent other research projects targeting similar interests: SCAMPI [24], SOCIALNETS [25], etc. Haggle is today seen by many as the project that created the premises for the advancements on human mobility for information and communications technology-related aspects.

In order to obtain mobility models, Haggle deals with the analysis and modelling of contact patterns between devices, introducing notions such as contact duration and inter-contact time. The contact duration, or contact time (CT), is the time when two devices are in range of each other, while the inter-contact time (ICT) is the period between two successive contacts of the same two devices. The contact duration influences the capacity of the network, while the inter-contact time affects the feasability and latency of the network.

Several mobility traces have been performed in the context of Haggle, mostly using Bluetooth-enabled devices such as iMotes. These are mobile devices created by Intel, based on the Zeevo TC2001P SoC, with an ARMv7 CPU and Bluetooth support. Two iMote traces, called Intel and Cambridge, have been presented and analyzed in [1]. The Intel trace was recorded for three days in the Intel Research Cambridge Laboratory, having 17 participants from among the researchers and students at the lab. The Cambridge trace was taken for five days, at the Computer Lab of the University of Cambridge, having as participants 18 doctoral students from the System Research Group. For both traces, the iMotes performed five-second scans at every two minutes, and searched for in-range Bluetooth devices. Each contact was represented by a tuple (MAC address, start time, end time). Both internal as well as external contacts were analyzed, where encounters between two devices participating in the experiment were considered internal contacts, while encounters with other devices were external contacts. The authors analyzed the distribution of CT and ICT, as well as the influence of the time of day on encounter opportunities. Regarding inter-contact time, the traces showed that it exhibits an approximate power law shape, which means that inter-contact distribution is heavy-tailed. The authors showed this observation to hold regardless of the time of day, by splitting a day into three-hour time intervals and noticing that the resulting distributions still maintained power law shapes. Contact durations were also noticed to follow power laws, but with much narrower value ranges and higher coefficients.

In addition to the Intel and Cambridge traces, another trace entitled Infocom was presented and analyzed in [26]. It was conducted during the IEEE INFOCOM

---

[1]   http://www.haggleproject.org/

conference in Miami in 2005, and had 41 conference attendees as participants, for a total duration of four days. The conclusions were similar to the ones above, namely that the distribution of the inter-contact times between two nodes in an opportunistic network is heavy-tailed over a large range of values, and that it can be approximated to a power law with a less than one coefficient. The authors showed that certain mobility models in effect at the time the paper was written (such as the random waypoint model) did not approximate the real life traces correctly. Similar to the Infocom trace, another trace was performed the following year at the same conference, but on a larger scale. There were 80 participants, chosen so that 34 of them formed four groups based on their academic affiliations. Apart from the 80 mobile devices, 20 other long-range iMotes were also deployed at strategic positions around the conference site [27]. Moreover, a trace was also performed in Hong Kong, where 37 people from a bar were given iMotes and were asked to return after five days [27].

2.2 CRAWDAD

The Community Resource for Archiving Wireless Data At Dartmouth (CRAW-DAD)[2] represents the effort of the National Science Foundation (NSF) to create an archive which stores wireless tracing data from many international contributors. The need for such a high capacity datastore has spawned from the data starvation that plagued research in wireless networks, as well as the limitations of synthetic models which were used as a replacement for real life user traces. CRAWDAD comes to the aid of researchers by hosting the contributed traces and developing better tools for collecting and post-processing data (e.g. anonymizing user traces).

Based on the fact that tracing experiments and studies related to them are extremely difficult to set up (as shown in Sect. 2.4), CRAWDAD is aimed at solving problems that are automatized: anonymizing the captured data in order to preserve privacy, or creating development tools for traces such as parsers.

The CRAWDAD initiative supports the human dynamics community as it understands the importance of data captured from live wireless networks in identifying and understanding the real problems, in evaluating possible solutions for said problems and also in evaluating new applications and services.

2.3 Social Tracer and HYCCUPS

In order to have mobility traces for an academic environment in certain conditions and containing a specific set of features, we performed two tracing experiments at

---

[2] http://crawdad.cs.dartmouth.edu/

our faculty. For each of these traces, we implemented an Android application that was deployed on the participants' smartphones for the duration of the experiment. This section presents the two applications and the tracing experiments performed.

2.3.1 Social Tracer

For our initial trace (which we called UPB 2011), we implemented an application entitled Social Tracer[3] in Android [28]. The participants in the tracing experiment were asked to run the application whenever they were in the faculty grounds, as we were interested in collecting data about the mobility and social traces in an academic environment. Social Tracer sent regular Bluetooth discovery messages at certain intervals, looking for any type of device that had its Bluetooth on. These included the other participants in the experiment, as well as phones, laptops or other types of mobile devices in range. The reason Bluetooth was preferred to WiFi was mainly the battery use [29]. For example, in four hours of running the application on a Samsung I9000 Galaxy S with discovery messages sent every five minutes, the application used approximately 10 % of the battery's energy. The period between two successive Bluetooth discovery invocations could be set from the application, ranging from 1 to 30 min (the participants were asked to keep it as low as possible, in order to have a more fine-grained view of the encounters).

When encountering another Bluetooth device, the Social Tracer application logged data containing its address, name and timestamp. The address and name were used to uniquely identify devices, and the timestamp was used for gathering contact data. Data logged was stored in the device's memory, therefore every once in a while participants were asked to upload the data collected thus far to a central server located within the faculty premises. All gathered traces were then parsed and merged to obtain a log file with a format similar to the ones from Haggle. Successive encounters between the same pair of devices within a certain time interval were considered as continuous contacts, also taking into account possible loss of packets due to network congestion or low range of Bluetooth.

The UPB 2011 tracing experiment was performed for a period of 35 days at the University POLITEHNICA of Bucharest in 2011, between November 18 and December 22. There were a total of 22 participants, chosen to be as varied as possible in terms of study year, in order to obtain a better approximation of mobility in a real academic environment. Thus, there were twelve Bachelor students (one in the first year, nine in the third and two in the fourth), seven Master students (four in the first year and three in the second) and three research assistants.

---

[3] http://code.google.com/p/social-tracer/

2.3.2 HYCCUPS

In order to get more relevant data regarding a mobile device user's behavior, we implemented a new tracer, called HYCCUPS, which is an Android application designed to collect contextual data from smartphones. The application runs in the background and can collect traces for multiple features. These features can be classified by the temporality of acquisition into static or dynamic, or by the semantic interpretation into availability or mobility features.

Moreover, static properties can be determined at application startup and are comprised of the device's traits, while dynamic features are momentary values acquired on demand. On the other hand, availability features represent values pertaining to the overall computing system state, while mobility features describe the interaction of the device with the outside world. We chose to collective an extensive dataset for future use.

As such, the features that the HYCCUPS Tracer can collect are as follows:

- **Minimum and maximum frequency**: static properties describing the bounds for Dynamic Voltage/Frequency Scaling (DVFS).
- **Current frequency**: momentary value of the frequency according to DVFS.
- **Load**: the current CPU load computed from */proc/stat*.
- **Total memory**: static property of the device describing the total amount of memory available.
- **Available memory**: momentary value which represents the amount of free memory on the device (bear in mind that, in Android, free memory is wasted memory).
- **Out of memory**: asynchronous event notifying that the available memory has reached the minimal threshold and, in consequence, the Out Of Memory (OOM) Killer will stop applications.
- **Memory threshold**: the minimal memory threshold that, when reached, triggers the OOM events.
- **Sensor availability**: static property which conveys the presence of certain sensors (e.g. accelerometer, proximity).
- **Accelerometer**: the accelerometer modulus is a mobility feature which characterizes fine grain movement (if available).
- **Proximity**: proximity sensor readings (if available).
- **Battery state**: the current charging level (expressed in %) and also the current charge state.
- **User activity**: availability events representing user actions that trigger opening/ closing application activities.
- **Bluetooth interactions**: momentary beacons received from nearby paired devices (similar to what Social Tracer does).
- **AllJoyn interactions**: interactions over WiFi modelled using the AllJoyn framework [30]. AllJoyn is an open-source peer-to-peer software development framework which offers the means to create ad hoc, proximity-based, opportunistic inter-device communication. The true impact of AllJoyn is expressed

through the ease of development of peer-to-peer networking applications provided by: common APIs for transparency over multiple operating systems; automatic management of connectivity, networking and security and, last but not least, optimization for embedded devices.

- **WiFi scan results**: temporized wireless access point scan results.

Tracing is executed both periodically, with a predefined timeout, as well as asynchronously on certain events such as AllJoyn interactions or user events. This chapter concentrates on dynamic mobility features represented by the last three tracing features from the above list.

Therefore, the second tracing experiment, entitled UPB 2012, lasted for 65 days, in the spring of 2012 and also took place at the University POLITEH-NICA of Bucharest. A total of 66 volunteers participated varying in terms of year and specialization: one first year Bachelor student, one third year Bachelor student, 53 fourth year Bachelor students, three Master students, two faculty members and six external participants (only from office environments). The experiment implied an initial startup phase, also called pairing session, when all attendants were asked to meet and pair all devices for Bluetooth interactions.

The participants were asked to start the HYCCUPS Tracer each weekday between 10 a.m. and 6 p.m. as we assumed this was the interval in which most participating members were attending classes or work. As expected, the volunteers in our experiment did not always respect the instructions with conscientiousness, so on occasion they didn't turn the application on. Nonetheless, the results proved that the captured tracing data was sufficient for our needs.

2.4 Benefits and Limitations

The main reason for developing and using tracing applications such as the ones previously presented instead of synthetic mobility models spawns from the need for better mapping onto real life situations. As previously stated, trace models follow a heavy-tailed distribution with spikes and bursts, making the random walks on graphs model and other such models obsolete.

The major benefit of tracing applications is the use of a custom data model in order to relate to real situations, real problems and optimized solutions for said issues. However, this can also lead to a pitfall: if the data model is not correctly designed at the start of the experiment, the entire outcome of the analysis can be biased.

Among the potential challenges of setting up our tracing experiments, we dealt with the following:

- Finding volunteers representative to our goals was not such an easy task as it may seem. For example, if we would have chosen all participants from the same class, then our results would have been biased because we would have been limiting our targeted scope to a partition of our community graph instead of reaching the entire collective. Moreover, all of the candidates for the experiment needed to have Android devices capable of tracing our data model: Bluetooth connectivity, WiFi connectivity, sensors etc.

- The design and development of the tracing application needed to take into account compatibility with multiple types of viable Android devices of variate versions. Furthermore, when developing the tracers, we were obliged to take into account the additional overhead of our applications, as most participants complained about the supplementary power consumption.

- The installation effort of the tracer was tremendous due to issues such as Bluetooth pairing: all of the participants' devices needed to pair to each other in order for us to be able to trace their interactions.

- Last, but not least, we were confronted with the human factor of such experiments: the lack of conscientiousness of our volunteers. Due to the participants not running the tracing application as instructed, the collected data was incomplete. Furthermore, this affected the analysis of said results, as we needed to deploy measures to deal with uncertainty.

3 Techniques for Data Analysis

This section presents several techniques for analyzing the data collected in a mobility trace. We begin by looking at the distribution of encounters and contact times, and then go on to present how we can verify if a trace exhibits contact predictibility in the shape of a Poisson distribution. Finally, we study the limits of predictability of the mobility and behavior of mobile users with regard to wireless access points.

3.1 Contact Times Distribution and Time of Day Dependence

The first step in analyzing a mobility trace should be looking at the distribution of contact and inter-contact times. Approximating these distributions using heavy-tailed or power law functions (as shown in Sect. 2.1) can help in the creation of synthetic mobility models, but more importantly it can aid in the development of routing and forwarding algorithms suitable for the specific network that is being analyzed. Apart from the contact and inter-contact times described in Sect. 2.1, two other metrics defined in [31] are the any-contact time (ACT) and inter-any-contact time (IACT). These are similar to the previous metrics, except that they are computed with regard to any node in the trace, not per pairs of nodes. Thus, the ACT is the time in which any internal or external node is in range with the current observer, while the IACT is the period when the current device does not see anyone in range. The former metric specifies the time window in which a node can forward messages to other participants in the network, whereas the latter is the opposite: the period when a node doesn't have any contacts at all.

The distribution of contacts with internal or external devices also highlights various characteristics of a mobility trace, such as the contact opportunities with given nodes, or the possibilities of using external devices for transporting data to a certain destination. In addition to analyzing the contact times and distribution, one should take into account the dependence of a trace on the time of day. Knowledge about this dependence can prove to be truly useful, especially in situations like the ones we analyzed and presented in Sect. 2.3, where there is generally little activity in the network during the times of day when the students aren't at the faculty. This means that a routing or forwarding algorithm should take advantage of those times of day when there are many contacts, in order to reduce the effect of time periods with few encounters. We analyze the UPB 2011 and UPB 2012 traces in terms of contact times, contact distribution and time of day dependence in Sect. 4.1.

3.2 Contact Predictability

An important challenge in mobile networks is knowing when and to which node should a message be passed, in order for it to reach its destination as fast as possible. Therefore, it would be important if we were able to predict the future behavior of a node in such a network, in regard to its encounters and contact durations. We propose a way to predict this behavior by analyzing a node's past encounters and approximating the time series as a Poisson distribution.

Since the nodes in an academic trace are students and professors from the faculty, we believe that their behavior is predictable. This should happen because the participants have a fixed daily schedule and interact with each other at fixed times in a day. For example, a professor and his students interact when the students attend the professor's class, which happens regularly each week. Likewise, two students from the same class would interact at almost all times when they are at the faculty.

Thus, we attempt to prove this supposition by analyzing the traces in terms of predictability. The first metric we use is the total number of encounters between a node and the other nodes. The number of encounters mainly specifies the popularity of a node, since the more encounters a node has, the more popular it is. The second metric is the contact duration of every encounter of a node in a given time interval. Similar to the number of encounters, it suggests the popularity of a node, but also its mobility. If a node has many encounters in a time interval, but all the encounters are short in terms of duration, it means that the node is very mobile and it doesn't stay in the same place for long periods of time.

In order to verify if a node's behavior in the opportunistic network is predictable, we use Shannon's entropy, which is basically a measure of predictability (the lower the entropy, the higher the chances are of a prediction being successful). When the entropy is 0, it means that a node's behavior is 100 % predictable. The formula for entropy is $H(X) = - \sum_{i=1}^{n} p(x_i) \log p(x_i)$, where X is a discrete random variable with possible values in the interval $x_1, ..., x_n$ and $p(X)$ is a probability mass function for X. A first possibility would be to compute $p(X)$ as the probability of encountering node N at the next time interval. However, the sum of probabilities in this case would not be 1, because a node might be in contact with more than one other node at a given time. Thus, we split the entropy computation into two parts: predicting that the next encounter will (or won't) be with node N, and predicting if a contact will take place at the next time interval. In theory, combining these two values will result in a prediction of the time of an encounter with a given node. Moreover, they also have a use on their own and not necessarily together.

We start by computing the entropies for contacts with a given node N. The probability function in this case is given by the ratio between the total number of times node N was encountered and the total number of contacts with any other nodes during the experiment. The second entropy function is given by the ratio between the number of time units a node was in contact with another node for, and the entire duration of the experiment. Based on these two entropies, we can decide whether it is possible to predict the node that the next encounter will be with, and whether there will be a contact at the next time interval. However, for now we will only focus on the second scenario.

Because there are only two values for this prediction (having a contact or not having a contact at the next time interval), a node's behavior regarding future contacts can be modelled as a Bernoulli distribution, which is a particular case of a binomial distribution. However, simply knowing if there will be a contact at the next time interval is not enough for a good opportunistic routing algorithm, since we need to know how many contacts there will be, in order to decide if the data packet should be forwarded, kept, or forwarded but also kept. The Bernoulli distribution does not offer such information, so we propose using the Poisson distribution, because it expresses the possibility of a number of events (in our case encounters with other nodes) to occur in a fixed time interval.

The probability mass function of a Poisson distribution is $P(N, \lambda) = \frac{e^{-\lambda} \lambda^N}{N!}$, where in our case $P(N, \lambda)$ represents the probability of a node having N contacts at a given time interval. In order to prove that a Poisson distribution applies to our traces, we use Pearson's chi-squared test [32], which tests a null hypothesis stating that the frequency distribution of mutually exclusive events observed in a sample is consistent with a particular theoretical distribution (in our case Poisson). We apply the chi-squared test for every node in the network individually and present the results in Sect. 4.2.

3.3 Predictability of Interacting with Wireless Access Points

This section presents a proposal for a methodology not just as a basis for studying already existing mobile traces which involve scanning of nearby wireless APs, but also as a set of guidelines for future tracing application developers to take into consideration when designing and developing tracers.

The basic principles for the methodology are inspired from the analysis conducted by Song et al. in [21]. In their paper, the authors study the limits of predictability in the mobility and behavior of mobile users over Cell towers. Here we try to formalize, adapt and enhance their analysis in order to map it onto wireless network traces. As such, we need to fill the gap between their analyzed context and ours, namely bridging the difference between the range of Cell towers and wireless access points.

Seeing that the main focus of this methodology is interaction with wireless APs we need to define a measure of sufficiency of the tracing data, namely observed interval sufficiency. This measure determines if the tracing data has converged to a point where it is sufficiently informed in order to perform additional operations on it. Moreover, we define the observed interval sufficiency as the minimum interval in which the discovery of access points converges. Recalling that we are dealing with WiFi networks in academic and office environments, we can assess that the surroundings of such a tracing experiment are limited and, as such, patterns are visible sooner than in mobile networks (e.g. GSM). This should limit the tracing interval to several months or weeks, rather than a year.

Also an important factor of our analysis is the number of subjects involved in the experiment, as well as their conscientiousness (or control over the tracing application; we will emphasize more on conscientiousness later in this section). As such, we have empirically discovered that a minimum of 10–20 users are necessary in order to provide statistical correctness of the analysis.

Naturally, the next step in such an analysis is to formalize the interactions between users and wireless APs. The reader should be aware that the tracing experiments that this methodology is aimed at must contain the temporized results of wireless AP scans. As such, we define a virtual location (VL) as the most relevant access point scanned by a user during an hour. Taking into consideration that multiple APs can be scanned by a user during an hour, we need to define a heuristic of choosing the most relevant one. We represent VLs as Basic Service Set Identifiers (BSSIDs), since Service Set Identifiers (SSIDs) are prone to name clashes, fact which may influence our study.

Although it may seem unintuitive at first, choosing hours to be the analysis' temporal step has many reasonable explanations. First of all, tracing applications which gather wireless scan results might use different timing intervals and we considered an hour should be viewed as a maximum value. Furthermore, the object of our study refers to academic and office environments in which an hour is usually the unit of work. We propose the use of two VL-choosing heuristics:

1. **First Come First Served (FCFS)**: choose the first sighting of an access point as the most relevant VL. The purpose of this heuristic is to mimic a pseudo-random algorithm of picking VLs.
2. **Alpha**: choose the most outstanding virtual location by weighing both the number of sightings during an hour, as well as the average wireless signal

strength. Basically, we choose a VL as the access point that maximizes the following expression:

$$\alpha \times count(VL_i) + (1 - \alpha) \times average(signalStrength(VL_i)) \tag{1}$$

If FCFS describes a pseudo-random heuristic, Alpha offers more control over choosing access points; by tweaking the α factor we can guide the algorithm towards more realistic situations: when α is lower, signal strength is more important than the number of sightings, thus better mapping on a situation with reduced mobility (closed surroundings) where the are few access points and the signal strength is the most valuable feature. On the other hand, when increasing α we turn our attention towards situations with a high range of mobility; signal strength is only momentary, whereas sighting an access point multiple times shows a certain level of predictability.

Based on these two heuristics, we define a VL sequence as the result of splitting up the entire tracing interval into hourly intervals and generating a chain of VL symbols for each hour of the monitored period. Whenever the VL of a user is unknown for a segment, it is marked with a special symbol (e.g. '?'). These shortcomings in tracing data, also known as lack of conscientiousness, are approximated by means of the knowledge coefficient similar to the q parameter used by Song et al. [21], which characterizes the fraction of segments in which the location is unknown. Also similar to [21], we chose a lower limit of 20 % for our knowledge coefficient as we found it sufficient for our needs.

In total, a set of 12 sequences are to be generated for each user: 1 FCFS and 11 Alpha sequences (by sweeping the α value from 0 to 1 with a 0.1 step value). Based on the VL sequences, three measures of entropy for each user should be computed:

- S_{rand} is the entropy of a user i travelling in random patterns and is defined as:

$$S_{rand}(i) = \log N_i \tag{2}$$

where N_i is the total number of VLs that user i has discovered.
- S_{unc} is the entropy of spatial travelling patterns without taking into account the temporal component of an interaction (also named temporally uncorrelated entropy). It is defined as:

$$S_{unc}(i) = \sum_{j=1}^{N_i} -p_i(j) \times \log p_i(j) \tag{3}$$

where $p_i(j)$ is the probability of user i to interact with a specific VL_j.
- S_{est} is the estimated entropy computed by means of a variant of the Lempel-Ziv algorithm [33], which takes into consideration the history of passed encounters. By so doing, we correlate the temporal dimension with the VL interaction patterns. We have constructed an estimator which computes entropy as:

$$S_{est} = \left(\frac{1}{n}\sum_i \lambda_i\right)^{-1} \log n, \tag{4}$$

where n is the length of the symbol sequence and λ_i is the shortest substring that appears starting from the index i, but which is not present for indexes lower than i. S_{est} converges to the real entropy when $n \to \infty$ [34].

We consider that $S_{est}(i) \leq S_{unc}(i) \leq S_{rand}(i) < \infty$ [21] to be a reasonable assumption for each user i, as a participant taking random actions will be less predictable than another one frequenting VLs iregardless of time, and both are less invariable than a real user taking logical decisions.

4 Experimental Results

This section present the results obtained when applying the data analysis techniques presented in Sect. 3 on our two traces.

4.1 Contact Times Distribution and Time of Day Dependence

We begin our analysis with the contact times distribution and time of day dependence for the two traces presented in Sect. 2.3.

4.1.1 UPB 2011

In the UPB 2011 trace, there were 22 internal devices numbered from 0 to 21. The total number of contacts between two internal devices (i.e. internal contacts) was 341, while the number of external contacts was 1,127. There were 655 different external devices sighted during the course of the experiment, which means that in average each different external device has been seen about 2 times. External devices may be mobile phones carried by other students or laptops and notebooks found in the laboratories at the faculty. Some of these external devices have high contact times because they may belong to the owner of the internal device that does the discovery, therefore being in its proximity for large periods of time. However, external contacts are in general relatively short.

Figure 1a shows the distribution of contact and inter-contact times for the entire duration of the experiment for all internal devices. As shown in [31], the distribution of contact times follows an approximate power law for both types of devices, as well as contact time and inter-contact time. The contact time data series

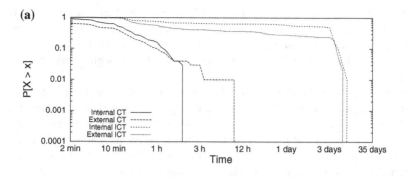

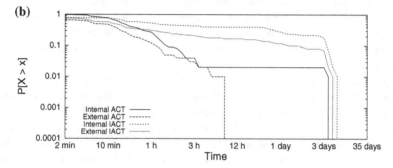

Fig. 1 Probability distributions of contact and inter-contact times (UPB 2011). **a** Contact (*CT*) and inter-contact (*ICT*) times. **b** Any-contact (*ACT*) and inter-any-contact (*IACT*) times

is relevant when discussing the bandwidth required to send data packets between the nodes in an opportunistic network, because it shows the time in which a device can communicate with other devices. As stated before, the number of internal contacts is 341, with the average contact duration being 30 min, which means that internal contacts have generally been recorded between devices belonging to students attending the same courses or lecturers and research assistants teaching those courses. External contacts also follow an approximate power law, with an average duration of 27 min. However, in this case there are certain external contacts that have a duration of several hours. This situation is similar to the one previously described, where these devices belong to the same person carrying the internal device. The inter-contact time distribution shows a heavy tail property, meaning that the tail distribution function decreases slowly. The impact of such a function in opportunistic networking has been studied in more detail in [1] for four different traces. The authors conclude that the probability of a packet being blocked in an inter-contact period grows with time and that there is no stateless opportunistic algorithm that can guarantee a transmission delay with a finite expectation.

Figure 1b shows any-contact and inter-any-contact times. As can be seen from the figure, they are greater than regular contact times, but the shape of the distribution is also a power law function. A conclusion that can be drawn from these charts is, as observed in [31], that contact times are bigger and intervals between contacts are smaller, so if a node wants to perform a multicast or to publish an object in a publish/subscribe environment it has a great chance of being able to do so.

Figure 2a shows the distribution of the number of times an internal or external node was sighted by other devices participating in the experiment. It can be seen that the maximum number of encounters of an internal device is 55 during the course of the 35 days of the experiment, whereas some internal nodes have never been seen. Most internal devices have been seen from 16 to 20 times. As for external devices, the majority of them have been encountered less than 5 times, with 534 of them having been sighted only once. There are few exceptions, as three external devices have been encountered more than 16 times. The conclusion is that there is a large number of nodes available in such an environment that can be used to relay a message, meaning that there is a lower chance of traffic congestion.

Figure 2b presents the number of times specific pairs of devices saw each other. It shows that the maximum number of contacts between two internal nodes or an internal and an external node is 17. Generally the number of contacts with external devices is larger than the number of contacts with internal devices. The maximum number of internal devices spotted by a participant is 17, whereas some participants have only encountered external nodes. Most of the internal devices have been in contact with between 10 and 15 other internal devices. As shown previously, the total number of external devices encountered during the 35 days of the experiment is far greater than that of the internal devices. There are six participants that have encountered between 15 and 50 external devices and five that have been in contact with more than 50 external nodes. The maximum number of different external devices spotted by a single participant is 197.

Figure 3a shows the distribution of inter-contact times for both types of devices for three time intervals. They are chosen between 8 AM and 8 PM because that is when students or teachers are at the faculty, and this experiment in not concerned

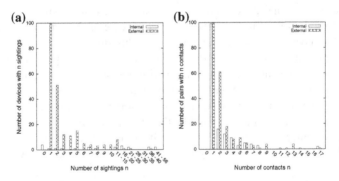

Fig. 2 Distribution of the number of sightings of a device and the number of contacts between pairs of devices (UPB 2011). **a** Sightings of a device. **b** Contacts between pairs of devices

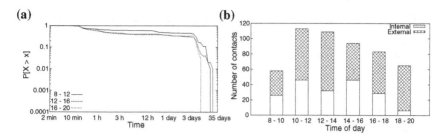

Fig. 3 Time of day dependence (UPB 2011) **a** Inter-contact times for three time periods. **b** Contacts for six time periods

with what happens in the rest of the day. The 8 a.m.–8 p.m. interval has been split into three parts, corresponding to three main time periods of the working day: morning (8 a.m.–12 p.m.), noon (12–4 p.m.) and afternoon (4–8 p.m.). As we can see from the figure, the three plots are very similar, following the same approximate power law function. This is different from [31], where daytime periods have a greater power law coefficient than night periods. This happens because we are only interested in periods when there are classes.

Figure 3b shows the percentage of contacts that take place in six two-hour intervals between 8 a.m. and 8 p.m. It can be seen that most contacts (113) happen between 10 a.m. and 12 p.m. and the smallest number of contacts in a two-hour interval (58) is recorded between 8 and 10 a.m. External contacts have a distribution similar to the one for all contacts, which shows that the faculty is populated the most between 10 a.m. and 2 p.m. This can also be explained by the fact that at noon students usually have lunch at the cafeteria, so they meet in a common place.

4.1.2 UPB 2012

The total number of internal contacts for the UPB 2012 trace was 12,003, which is far greater than for UPB 2011, showing not only that there were more participants in this experiment, but that they were more conscientious in terms of turning their tracing application on when they were at the faculty. Unfortunately, the tracing application did not export information regarding external contacts, so in this section we will only analyze encounters with internal devices. Out of all the internal contacts, 13 % of them were registered on Bluetooth and 87 % on AllJoyn.

We analyzed the distribution of contact and inter-contact times for this trace as well, and the results are shown in Fig. 4a. It can be seen that the curve of the distribution of contact times for internal devices is very similar to the corresponding one from the first trace. However, since there are more devices in this situation, the average contact duration decreases to about 14 minutes. The inter-contact time distribution, also shown in Fig. 4a, exhibits a heavy tail property just like the one from UPB 2011. The average duration between two contacts for this

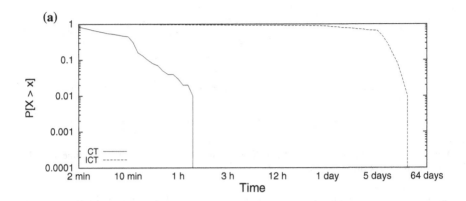

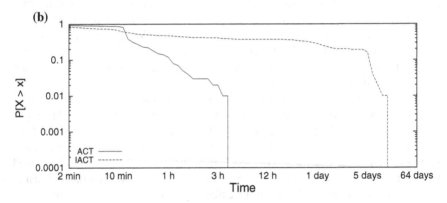

Fig. 4 Probability distributions of contact and inter-contact times (UPB 2012). **a** Contact (*CT*) and inter-contact (*ICT*) times. **b** Any-contact (*ACT*) and inter-any-contact (*IACT*) times

trace is 4.5 h, but it has to be taken into account that this value is computed simply by subtracting contact start and finish times. This means that if a contact takes place one day, and another happens the next day, the inter-contact time will be more than 8 h. Both the contact and the inter-contact time distributions follow an approximate power law. Figure 4b shows any-contact and inter-any-contact times as well, which also follow an approximate power law, with the times higher than for regular contacts. The inter-any-contact times are very high in this case because there are no external nodes in this trace, which means that the time between any contacts is only computed using internal nodes. In the UPB 2011 trace, the inter-any-contact time computation used the external nodes as well, which appeared more often, even if only for short periods of time.

Figure 5a shows the distribution of the number of times a device participating in the experiment was seen by the other participants. The maximum number of encounters of a certain node is 680, which is a lot higher than for the UPB 2011 trace. This means that some nodes in this trace are more popular and that there are more encounters and thus more possibilities of exchanging information. Out of the

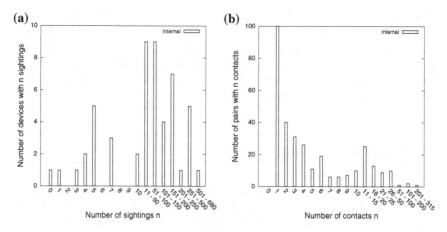

Fig. 5 Distribution of the number of sightings of a device and the number of contacts between pairs of devices (UPB 2012). **a** Sightings of a device. **b** Contacts between pairs of devices

chosen nodes, only one has not been encountered at all during the experiment, and the majority of devices have been sighted between 51 and 100 times, which is a clear improvement over the previous trace. The average number of times a device has been encountered by other nodes in the experiment is 106, which means almost twice a day.

Figure 5b outlines the number of times specific pairs of nodes have encountered each other. Most of them have met each other only once (100), but there are plenty of pairs of devices that have been in contact more than once (going to as much as 315 encounters between a pair of devices). The average number of encounters between the same two devices is 50 (much higher than for the UPB 2011 trace), which is a sign that devices met relatively often. This conclusion is useful in implementing a routing algorithm for opportunistic networks, because it means that this trace is closer to a real life situation than UPB 2011.

Figure 6a shows the distribution of inter-contact times per time interval. The three time intervals were the same as the ones chosen in Sect. 4.1.1: 8 a.m.–12 p.m., 12–4 p.m., 4–8 p.m. Again, the three plots are very similar to each other,

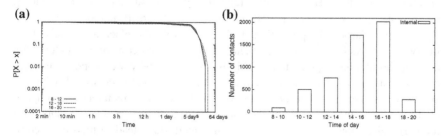

Fig. 6 Time of day dependence (UPB 2012). **a** Inter-contact times for three time periods. **b** Contacts for six time periods

following the same power law function. Figure 6b shows how many contacts happened in six two-hour intervals from 8 a.m. to 8 p.m. Unlike the UPB 2011 trace, the most contacts happened between 4 and 6 p.m. This is an indicative of the fact that most participants in the experiment had classes in that time interval. Many contacts were also recorded between 12 and 2 p.m., which was lunch time, when students may meet in the cafeteria.

4.2 Contact Predictability

As we have shown in Sect. 3.2, we propose approximating a node's future behavior as a Poisson distribution, and we show the results in this section. We performed our analysis on the UPB 2012 trace presented in Sect. 2.3 because it has a larger number of participants and encounters for a longer period of time than UPB 2011.

Figure 7a shows an example of a random node's behavior in terms of encounters with other nodes. It can clearly be seen that there is a weekly pattern, i.e. that on Tuesdays, Wednesdays and Thursdays the node has regular encounters with roughly the same nodes. The number of contacts in a day may differ, but this generally happens because there are short periods of time when the nodes weren't in contact or because of the unreliability of the Bluetooth protocol (thus yielding bursty contacts). The figure shows (for now just on a purely intuitive level) that there is a certain amount of regularity (and thus, predictability) in the behavior of the participants in the UPB 2012 experiment. Figure 7b presents contact durations per day for the same node as before. Just as in Fig. 7a, it can be seen that on Tuesdays, Wednesdays and Thursdays the contact durations are similar.

As we previously said in Sect. 2.3, we use two entropy functions: one for predicting that the next encounter will (or won't) be with node N, and the other for predicting if a contact will take place at the next time interval. Figure 8 shows the cumulative distribution functions for the two entropies. It can be observed that having a contact at the next period of time is mostly predictable, because the entropy is always lower than 0.35. However, predicting the node that will be seen at the next encounter is not so easily done based solely on the history of encounters, and this is shown by the high entropy values (as high as 4.25, meaning that a node may encounter on average any one of $2^{4.25} \approx 19$ nodes).

Since the entropy for predicting if a contact will take place at the next time interval is always lower than 1, the behavior of a node in terms of encounters with other nodes is highly predictable. This is the reason why we proposed using a Poisson distribution. The time interval chosen for applying the Poisson distribution and the chi-squared test was one hour. We tried to choose this interval in order to obtain a fine-grained analysis of the data. Choosing a smaller interval (such as a minute) and estimating the next contact incorrectly may lead to missing it completely. When we have an interval such as an hour, we can predict that in the next

(a)

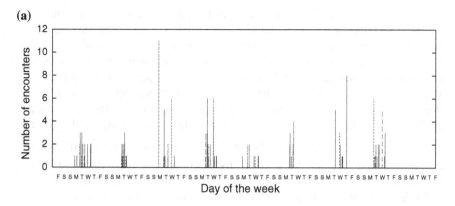

(b)

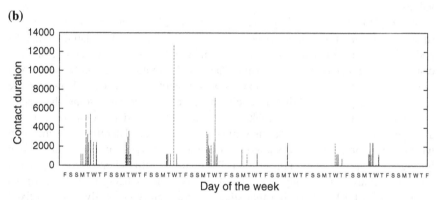

Fig. 7 Total encounters and contact durations per day for a random node. **a** Encounters. **b** Contact durations

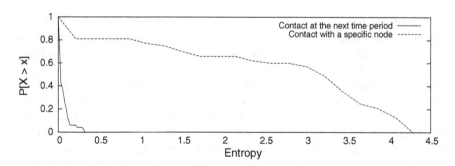

Fig. 8 Entropy values for predicting contacts and the time of contact

hour there will be a certain number of contacts with a higher rate of success, and the opportunistic routing algorithm can be ready for those contacts in the respective hour.

The first step of the chi-squared test was to count the frequency distribution of contacts per hour for the entire duration of our traces. The λ parameter can either be included in the hypothesis or it can be estimated from the sample data (as it was in our case). We computed it using the maximum likelihood method by averaging the number of encounters per hour over the entire experiment. Knowing λ, we were then able to find out the probability for having N encounters at the next time interval according to the Poisson distribution. Using this probability, we finally performed the chi-squared test for the time interval according to the formula $\chi^2_{k-p-1} = \sum_k \frac{(f_o - f_e)^2}{f_e}$, where f_o is the observed frequency, f_e is the expected frequency (computed using the Poisson distribution), k is the number of classes (which depends on the way the number of encounters is distributed for each node) and p is the number of parameters estimated from the data (in this case 1, the λ value).

We used the 0.05 level of significance for proving the hypothesis by using a chi-squared table, and the results can be seen in Fig. 9a (chart 1). We also included the nodes that have not had any encounters in the "Accepted" category, since a distribution with only zeros is a valid Poisson distribution. As can be seen from Fig. 9a, only 20.75 % of the hypotheses were accepted in this case. However, we have observed previously that a node's encounter history has a somewhat repetitive pattern for days of the week, so we then attempted to compute λ as the averaged number of contacts in the same day of the week. Therefore, we ended up with a larger number of chi-squared hypotheses to prove, but also with a much finer-grained approximation of the data. The results for this situation can be seen in Fig. 9a in chart 2, with only 29.65 % of the hypotheses being rejected. Still we went one step further, knowing that students at a faculty generally follow a fixed schedule in given days of the week and thus we computed the maximum likelihood value as an average per hour per day of the week. Thus, the results obtained were very good, with only 2.49 % of all the hypotheses rejected, as shown by chart 3 in Fig. 9a.

The results from charts 1–3 in Fig. 9a are computed for the total number of encounters in an hour. However, if the HYCCUPS application misbehaved at some point in the experiment and instead of logging a long contact between two nodes, logged a large number of very short contacts, the results of applying a Poisson probability may be wrong. Because of this situation, we also applied the chi-squared tests described above using only unique contacts. Therefore, the number of contacts in an hour is equal to the number of different nodes encountered in that hour. The results are shown in the final three charts in Fig. 9a. For the first test case (with λ computed over the entire experiment, as seen in chart 4 from Fig. 9a), 75.47 % of the hypotheses were rejected. In the test that uses the average per day of the week (chart 5), 24.26 % of all chi-squared hypotheses were rejected and finally just 1.31 % of distributions were not Poisson according to the chi-squared test for computing the maximum likelihood value per hour of a weekday (chart 6).

(a)

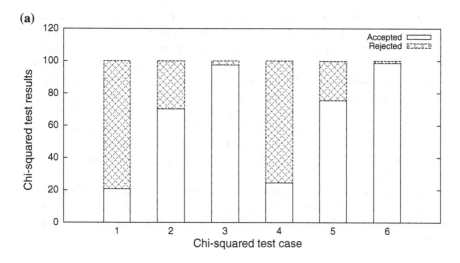

(b)

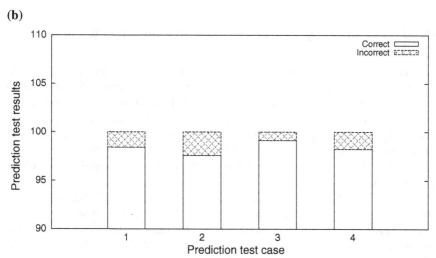

Fig. 9 Chi-squared test results and prediction success of the Poisson distribution; for the chi-squared tests, datasets 1, 2 and 3 are computed using the total number of encounters and varying the max likelihood (*1* for the entire experiment, *2* per weekday, *3* per hour of a day of the week), while datasets 4, 5 and 6 are computed using unique encounters; for the prediction success, datasets 1 and 2 are computed using the total number of encounters (*1* the next to last week, *2* the last week) and datasets 3 and 4 are computed using unique encounters. **a** Chi-squared. **b** Prediction success

In order to prove that these results aren't valid for the UPB 2012 trace only, we also ran them on UPB 2011. The results for the λ-per-hour test with unique contacts are even better than for the current trace, since only 0.11 % of the hypotheses were rejected.

To further prove our assumptions, we eliminated the last two weeks from the UPB 2012 trace and computed the Poisson distribution probabilities for each hour per day of the week on the remaining series. We compared the value that had the highest Poisson probability (i.e. the most likely value according to the distribution) with the real values. If the Poisson predictions were to be correct, then the two values should be equal. The results of this test for both total and unique contacts are shown in Fig. 9b. It can be seen that for total encounters 97.59 % of the Poisson-predicted values are correct for the next to last week (chart 1), and 98.42 % for the last (chart 2). When taking into account individual encounters, the predictions are even better: 98.24 % for next to last week (chart 3) and 99.14 % for the last week (chart 4).

We have shown in this section that, by knowing the history of encounters between the nodes in a mobile network in every hour of every day of the week, we can successfully predict the future behavior of a device in terms of number of contacts per time unit.

4.3 Predictability of Interacting with Wireless Access Points

Our initial premises for the following experiment are that synergic patterns in academic and office environments are subject to repeatability. As opposed to previous more generic studies [21, 35], we focus towards environments where human behavior can be predictable, and try to understand the physical laws governing the human processes. Our work, from this perspective, is somewhat similar to [36]. However, we cover a more generic space in determining the social and connectivity predictability patterns in case of academic and office environments.

As expected, the social aspects of mobile interactions have influenced tracing data, as participants tend to interact more with users in their community or social circle. Academic and office environments are naturally grouped into social communities, in our case groups of students, faculty members and office colleagues. The Faculty of Automatic Control and Computer Science at the University PO-LITEHNICA of Bucharest is structured as follows: there are four years for Bachelor students split up into four groups of about 30 persons each and ten Masters directions with about 20 students each. By running the MobEmu emulator [37] with k-CLIQUE [38] on our tracing data, we have computed the UPB 2012 communities which are illustrated in Fig. 10a for AllJoyn interactions, respectively in Fig. 10b for Bluetooth contacts.

In computing the communities we have varied the two k-CLIQUE parameters, namely the contact threshold and the community threshold, as follows:

- **Contact threshold = 3,600 s, community threshold = 8**: this configuration proved to be too restrictive as we ended up ignoring interactions and even omitting nodes from the communities.

(a)   **(b)**

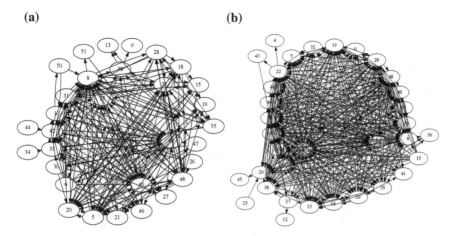

Fig. 10 Detected communities with a contact threshold of 1,200 s and a community threshold of 6. **a** PAllJoyn. **b** Bluetooth

- **Contact threshold = 600 s, community threshold = 4**: as opposed to the previous configuration, the current one is placed at the other extreme being too permissive as we obtained an almost full-mesh community.
- **Contact threshold = 1,200 s, community threshold = 6**: this is the appropriate balance between the previous two configurations as can be observed in Fig. 10a and b.

As expected, there is a high degree of connectivity considering we usually obtain one large community. This is easily explained by the spatial restraint, as almost all participants are students of the same school and therefore interact on the grounds of the university. However there is a slight difference, as interacting over Bluetooth tends to isolate stray mini-communities as can be seen in Fig. 10b. We believe that the key factor in this separation is range, as Bluetooth is designed for shorter ranges (about 5–10 m) while WiFi APs have ranges up to 30–40 m.

After ascertaining the social structures in our experiment, it is high time we explored the predictable behavior of participants while interacting with peers. We take into consideration both Bluetooth and AllJoyn interactions.

As such, we analyze and compare the tracing data for both types of synergy. We observe that AllJoyn interactions occur much more often than those on Bluetooth, respectively WiFi encounters cumulate up to 20,658, while Bluetooth sums up only 6,969 which amounts to only 33.73 % of the latter. We believe that such results are reflected by the low range of Bluetooth which was also observed in the community analysis.

We study the hourly interactions of individuals on a daily basis and as such we compute the probability that an individual interacts at least once each day at the

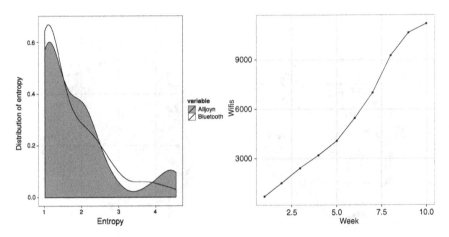

Fig. 11 Distribution of entropy of interacting for bluetooth and AllJoyn, and of number of discovered access points for each week in the UPB 2012 trace. **a** Entropy. **b** Discovered APs

same hour with any other peer. Figure 11a shows the entropy of hourly interactions. As can be seen, AllJoyn hourly interactions peak almost as low as Bluetooth. We point out that the comparison between the two peer-to-peer solutions actually comes down to a compromise between low range versus low power saving as more powerful radios lead to much faster battery depletion. In this experiment, we choose to further analyze WiFi interactions as their exceeding rate of interactions offers higher statistical confidence.

During the UPB 2012 experiment, a total of 6,650 access points were discovered; Fig. 11b shows the distribution of distinct APs discovered for various weekly intervals. As can be observed, 10 weeks are sufficient for the number of discovered APs to converge and, as such, we can state that the most frequented wireless network devices have already been detected. Also noteworthy, most participants have limited mobility as they meet few access points; these restricted travel patterns favour interacting, as individuals are clustered into communities situated in closed surroundings in the range of a few preferential wireless APs.

By applying the proposed methodology on the UPB 2012 trace, we obtain VL sequences with 368 symbols (8 h × 46 weekdays), each symbol corresponding to an outstanding VL for a specific hour. Unfortunately, the lack of conscientiousness of volunteer participants has left its imprint on the tracing data, as by applying such a knowledge coefficient we trimmed down more than half of the participants.

Figure 12 illustrates the distributions of entropy $P(S_{rand})$, $P(S_{unc})$, respectively $P(S_{est})$ for FCFS and Alpha(0.7) and, as expected, we found that the inequality $S_{est} \leq S_{unc} \leq S_{rand}$ holds for our experiment as well.

Figure 13 illustrates a comparison of the distributions for the three proposed entropies on all VL sequences. As expected, FCFS presents one of the most

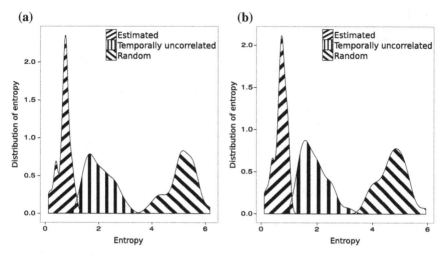

Fig. 12 The inequality of entropies for $S_{rand}, S_{unc}, S_{est}$ for the UPB 2012 trace. **a** FCFS. **b** Alpha 0.7

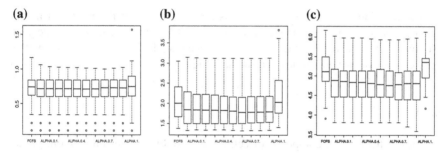

Fig. 13 Comparison of entropy distributions for all sequences for the UPB 2012 trace. **a** Estimated entropy. **b** Temporally uncorrelated entropy. **c** Random entropy

skewed distributions for each of the proposed entropies; this proves that pseudo-random simulations tend to suffer from unrealistic traits. Most surprising, the cases for Alpha(1) also show that in the UPB 2012 trace signal strength was a tie-breaker for choosing VLs; one interpretation could be that the users involved in the experiment have a low range of mobility and travel in surroundings limited to school and offices. What can also be observed is that Alpha around 0.7 generates results close to normal distributions.

Based on the results obtained by applying the proposed methodology, we can state that the wireless behavior of users in the UPB 2012 trace is subject to predictability, as a real user can be pinpointed to one of $2^{0.68} \approx 1.6$ locations, whereas a user taking random decisions will be found in one of $2^{4.94} \approx 30.7$ locations.

Fig. 14 Distribution of
number of discovered access
points for each week in the
rice trace

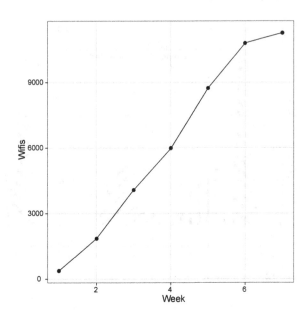

The remainder of this section presents the application of the proposed methodology on two external traces accessed from CRAWDAD, namely Rice and Nodobo. These external traces have also been studied in order to show the applicability of our guidelines and methodology.

4.3.1 Rice

The Rice[4] trace set is composed of cellular and WiFi scan results from the Rice community in Houston, Texas; 10 subjects have participated in the tracing experiment that lasted for 44 days, from 16 January 2007 to 28 February 2007. During the experiment, a total of 6,055 wireless access points have been discovered and Fig. 14 shows the distribution of discovering APs and, as can be seen, the 8 weeks are almost sufficient for convergence.

In consequence, the proposed methodology can be applied in order to study the predictability of interacting with wireless APs. The distributions of entropy $P(S_{rand})$, $P(S_{unc})$, respectively $P(S_{est})$ for FCFS and Alpha(0.7) are illustrated in Fig. 15.

Furthermore, Fig. 16 illustrates all of the distributions for the three measures of entropy on all generated sequences. As opposed to UPB 2012, the distributions of the estimated entropy are heavily skewed, but consistent; this may be a consequence of the knowledge factor. Most surprisingly, although the Rice trace set

---

[4] http://crawdad.cs.dartmouth.edu/rice/context/

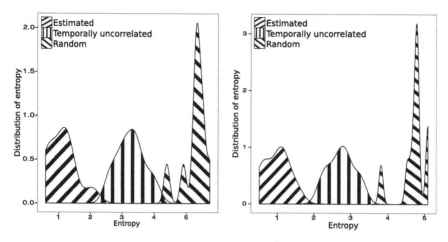

Fig. 15 The inequality of entropies for S_{rand}, S_{unc}, S_{est} for the Rice trace. **a** FCFS. **b** Alpha 0.7

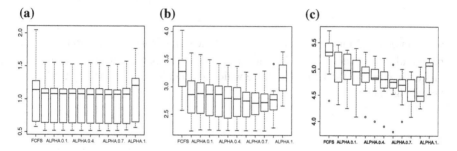

Fig. 16 Comparison of entropy distributions for all sequences for the rice trace. **a** Estimated entropy. **b** Temporally uncorrelated entropy. **c** Random entropy

contains data from only 10 users, all of them have a high degree of collected knowledge; all users have a knowledge factor of over 60 %. This increased informational gain may also affect the Random entropy as can be seen in Fig. 16c: each generated sequence is generally different from the others. This further proves that, in real life, random heuristics are not able simulate human behavior.

As a resemblance with the previously presented UPB 2012 analysis, signal strength also is a tie-breaker in choosing the outstanding VL; as such, it seems that while tracing wireless access points the quality of an AP is more important than the number of sightings. Also, in both traces, FCFS seems to have the same behavior: the distributions are skewed and the peaks are higher, but they still do not reflect the worst case scenario.

By applying the proposed methodology, we show that the users which collected the traces in the Rice experiment are subject to repeatability. Furthermore, a real

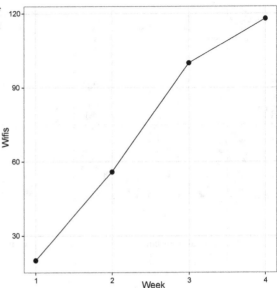

Fig. 17 Distribution of number of discovered access points for each week in the Nodobo trace

user can be pinpointed to one of $2^{1.6} \approx 3.03$ locations, whereas a random user can be found in one of $2^{4.9} \approx 29.85$ locations. The estimated entropy seems quite higher than that of the UPB 2012 trace which also could be explained by the higher knowledge factor.

4.3.2 Nodobo

The Nodobo[5] trace set was collected by means of a social sensor software suite for Android devices (also dubbed Nodobo); the experiment involved 21 subjects and lasted for 23 days (it actually lasted for a longer period, but we chose this subset for it was longest contiguous interval) from 9 September 2010 to 1 November 2010.

In applying the methodology, the Nodobo trace has triggered more than one warning as, not only is the tracing period insufficient, but there aren't sufficient users with a knowledge factor over 20 %. As the guidelines from the methodology point out, such a trace cannot be studied for predictability; as can be seen in Fig. 17, the distribution of the discovered nodes does not converge. As a comparison with the previous two traces which accumulated up to more then 6,000 discovered wireless APs, the Nodobo trace discovered only 153. The reader should bear in mind that the low knowledge factor is not necessarily influenced by the

---

[5] http://crawdad.cs.dartmouth.edu/strath/nodobo

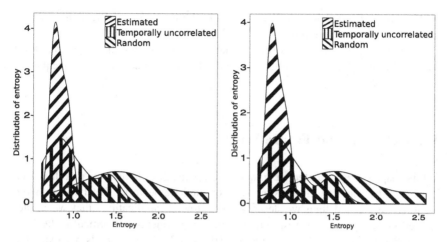

Fig. 18 The inequality of entropies for $S_{rand}, S_{unc}, S_{est}$ for the Nodobo trace. **a** FCFS. **b** Alpha 0.7

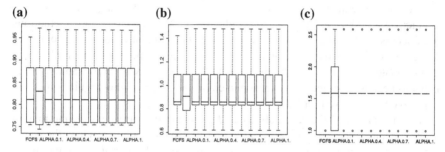

Fig. 19 Comparison of entropy distributions for all sequences for the Nodobo trace. **a** Estimated entropy. **b** Temporally uncorrelated entropy. **c** Random entropy

lack of wireless APs in the vicinity of mobile users, but more by the lack of conscientiousness of the volunteers involved in the experiment.

To further validate the methodology and guidelines, we attempted to analyze the predictability of interacting with wireless access points in Nodobo. Figure 18 illustrates the distributions of entropy $P(S_{rand})$, $P(S_{unc})$, respectively $P(S_{est})$ for FCFS and Alpha(0.7); we can state that the goal inequality of predictability does not hold. Furthermore, when attempting to compare the distributions for all sequences (as illustrated in Fig. 19), the insufficiency of both the tracing interval and of the sample size impact heavily on the statistical analysis, the results being inconclusive.

The proposed methodology applied on Nodobo has shown that this trace set is insufficient to determine any measure of predictability of wireless behavior. The guidelines prove to be efficient in filtering the trace set before the statistical

analysis is performed in positive cases (like UPB 2012 and Rice), but also in
negative cases (i.e. Nodobo).

We have also shown in this section that, after eliminating the traces that don't
have sufficient information, we are left with data that confirms Song's inequation
$S_{est} \leq S_{unc} \leq S_{rand}$.

5 Conclusions and Future Work

In this chapter, we presented techniques and applications to analyze mobility data.
We began by highlighting existing projects and frameworks that have gathered
mobility traces, and then we described two tracing applications that we implemented ourselves and the resulting mobility experiments performed at the University POLITEHNICA of Bucharest in 2011 and 2012. Afterwards, we extracted
the benefits of performing such experiments rather than using mathematical
mobility models, as well as several limitations and challenges brought by this
approach.

We then presented ways in which mobility data can be analyzed in terms of
contact distribution, as well as the predictability of encounters and interactions
with access points. We showed that the future behavior of a node in an opportunistic network in terms of the number of contacts in the next time interval can be
approximated as a Poisson distribution, with high levels of predictability. This
happens because contacts in an academic environment are highly regular, since the
participants have fixed daily schedules.

We also analyzed the repeatability and predictability of access interactions in
academic and office environments based on two separate points of view: the group
view and the individual view. Furthermore, we applied a distributed community
detection algorithm and found that confined surroundings lead to the creation of
large highly-adhesive communities. However, the wireless communication media
can have an important influence as low ranged solutions, such as Bluetooth, tend to
isolate loosely-coupled micro-communities. As for the individual's perspective
over interactions, we focused more on AllJoyn since Bluetooth interactions
occurred three times more rarely than the latter. Furthermore, we proposed a
methodology and a set of guidelines to be used in analyzing the predictability of
interaction between mobile users and wireless access points based on the study of
Song et al. [21]. By applying the methodology on three cases (UPB 2012, Rice and
Nodobo), we proved that mobile users have a predictable wireless behaviour if the
trace sets are complete, correct and sufficiently informed.

For future work, we plan to explore availability and usage patterns and further
correlate them with the current mobility and interaction patterns. By doing so, we
will be able to build an accurate detector for smart mobile collaborations based on
machine learning techniques trained with the tracing data. We believe that
studying the predictability of human behavior based on real mobile user traces can
prove to be the key to intelligent mobile collaboration in opportunistic networks

comprised of smartphones, that will eventually lead to less power consumption and which will be able to harness the full potential of contextual data by distributed context aggregation and detection.

References

1. A. Chaintreau, P. Hui, Pocket switched, networks: real-world mobility and its consequences for opportunistic forwarding, in *Technical Report, 2006* Computer Laboratory, University of Cambridge, Feb 2005
2. R-C. Marin, C. Dobre, F. Xhafa, A methodology for assessing the predictable behaviour of mobile users in wireless networks, in *Submmited at INCoS-2012 Special Issue Concurrency and Control* (Wiley, New York, 2013)
3. R-I. Ciobanu, C. Dobre, Predicting encounters in opportunistic networks, in *Proceedings of the 1st ACM Workshop on High Performance Mobile Opportunistic Systems, HP-MOSys '12*, New York, NY, USA, 2012. ACM, pp. 9–14
4. T. Camp, J. Boleng, V. Davies, A survey of mobility models for ad hoc network research. Wirel. Commun. Mobile Comput. (WCMC) **2**(5), 483–502 (2002). Special issue on Mobile Ad Hoc Networking: Research, Trends and Applications
5. M. Musolesi, C. Mascolo, Mobility models for systems evaluation, in *Middleware for Network Eccentric and Mobile Applications, Chapter 3*, ed. by B. Garbinato, H. Miranda, L. Rodrigues (Springer, Berlin Heidelberg, 2009), pp. 43–62
6. A. Barabási, The origin of bursts and heavy tails in human dynamics. Nature **435**(7039), 207–211 (May 2005)
7. P. Gerl, Random walks on graphs, in *Probability Measures on Groups VIII, volume 1210 of Lecture Notes in Mathematics*, ed. by H. Heyer (Springer, Berlin Heidelberg, 1986), pp. 285–303
8. D.B. Johnson, D.A. Maltz, *Dynamic source routing in ad hoc wireless networks, in Mobile Computing*, vol. 353 (Kluwer Academic Publishers, Dordrecht, 1996)
9. A. Doci, L. Barolli, F. Xhafa, Recent advances on the simulation models for ad hoc networks: real traffic and mobility models. Scalable Comput. Pract. Experience **10**(1) (2009)
10. A. Doci, W. Springer, F. Xhafa, Impact of the dynamic membership in the connectivity graph of the wireless ad hoc networks. Scalable Comput. Pract. Experience **10**(1) (2009)
11. K.A. Hummel, A. Hess, Movement activity estimation for opportunistic networking based on urban mobility traces, in Wireless Days (WD), 2010 IFIP (Oct 2010), pp. 1–5
12. A. Vázquez, Exact results for the Barabási model of human dynamics. Phys. Rev. Lett. **95**(24), 248701+ (Dec 2005)
13. J.G. Oliveira, A. Barabási, Human dynamics: Darwin and Einstein correspondence patterns. Nature **437**(7063), 1251 (Oct 2005)
14. A. Vázquez, J.G. Oliveira, Z. Dezsö, K.I. Goh, I. Kondor, A. Barabási, Modeling bursts and heavy tails in human dynamics. Phys. Rev. E 73(3), 036127+ (Mar 2006)
15. A. César, R. Hidalgo, Conditions for the emergence of scaling in the inter-event time of uncorrelated and seasonal systems. Phys. A Stat. Mech. Appl. **369**(2), 877–883 (Sept 2006)
16. Z. Dezsö, E. Almaas, A. Lukács, B. Rácz, I. Szakadát, A. Barabási, Dynamics of information access on the web. Phys. Rev. E (Stat. Nonlin. Soft Matter Phys.) **73**(6), 066132+ (2006)
17. B. Gonçalves, J.J. Ramasco, Human dynamics revealed through Web analytics. CoRR, abs/ 0803.4018 (2008)
18. B. Gonçalves, J.J. Ramasco, Towards the characterization of individual users through web analytics, in *Complex (2), volume 5 of Lecture Notes of the Institute for Computer Sciences, Social Informatics and Telecommunications Engineering*, ed. by J. Zhou (Springer, New York, 2009), pp. 2247–2254

19. J.P. Onnela, J. Saramäki, J. Hyvönen, G. Szabó, D. Lazer, K. Kaski, J. Kertész, A. Barabási, Structure and tie strengths in mobile communication networks. Proc. Natl. Acad. Sci. **104**(18), 7332–7336 (May 2007)

20. J.P. Onnela, J. Saramäki, J. Hyvönen, G. Szabó, M.A. de Menezes, K. Kaski, A. Barabási, J. Kertész, Analysis of a large-scale weighted network of one-to-one human communication. New J. Phys. **9**(6), 179 (2007)

21. C. Song, Q. Zehui, N. Blumm, A. Barabási, Limits of predictability in human mobility. Science **327**(5968), 1018–1021 (Feb 2010)

22. R.-C. Marin, C. Dobre, F. Xhafa, Exploring predictability in mobile interaction, in *2012 Third International Conference on Emerging Intelligent Data and Web Technologies (EIDWT)* (IEEE, 2012), pp. 133–139

23. J. Su, J. Scott, P. Hui, J. Crowcroft, E. De Lara, C. Diot, A. Goel, M.H. Lim, E. Upton, *Haggle: seamless networking for mobile applications, in Proceedings of the 9th International Conference on Ubiquitous Computing, UbiComp '07* (Springer, Berlin, Heidelberg, 2007), pp. 391–408

24. M. Pitkänen, T. Kärkkäinen, J. Ott, M. Conti, A. Passarella, S. Giordano, D. Puccinelli, F. Legendre, S. Trifunovic, K. Hummel, M. May, N. Hegde, Thrasyvoulos spyropoulos, SCAMPI: service platform for social aware mobile and pervasive computing. SIGCOMM Comput. Commun. Rev. **42**(4), 503–508 (Sept 2012)

25. S.M. Allen, M. Conti, J. Crowcroft, R. Dunbar, P.P. Lió, J.F. Mendes, R. Molva, A. Passarella, I. Stavrakakis, R.M. Whitaker, Social networking for pervasive adaptation, in *Second IEEE International Conference on Self-Adaptive and Self-Organizing Systems Workshops, 2008. SASOW 2008*, Oct 2008, pp. 49–54

26. A. Chaintreau, P. Hui, J. Crowcroft, C. Diot, R. Gass, J. Scott, Impact of human mobility on opportunistic forwarding algorithms. IEEE Trans. Mobile Comput. **6**(6), 606–620 (June 2007)

27. P. Hui, J. Crowcroft, E. Yoneki, BUBBLE rap: social-based forwarding in delay-tolerant networks. IEEE Trans. Mobile Comput. **10**(11), 1576–1589 (Nov 2007)

28. R.-I. Ciobanu, C. Dobre, V. Cristea, Social aspects to support opportunistic networks an academic environment, in *Ad-hoc, Mobile, and Wireless Networks, volume 7363 of Lecture Notes in Computer Science, Chapter 6*, ed. by X.-Y. Li, S. Papavassiliou, S. Ruehrup (Springer, Berlin, Heidelberg, 2012), pp. 69–82

29. E. Ferro, F. Potorti, Bluetooth and Wi-Fi wireless protocols: a survey and a comparison. Wirel. Commun. IEEE **12**(1), 12–26 (Feb 2005)

30. Qualcomm Innovation Center Inc: Introduction to AllJoyn. HT80-BA013-1 Rev. B (2011)

31. P. Hui, A. Chaintreau, J. Scott, R. Gass, J. Crowcroft, C. Diot, Pocket switched networks and human mobility in conference environments, in *Proceedings of the 2005 ACM SIGCOMM workshop on Delay-tolerant networking, WDTN '05*, New York, NY, USA, 2005. ACM, pp. 244–251

32. A. Stuart, K. Ord, S. Arnold, *Kendall's Advanced Theory of Statistics, Classical Inference and the Linear Model, volume Volume 2A (2007 reprint)*, 6th edn. (Wiley, New York, 1999)

33. J. Ziv, A. Lempel, Compression of individual sequences via variable-rate coding. IEEE Trans. Inf. Theory **24**(5), 530–536 (Sept 1978)

34. I. Kontoyiannis, P.H. Algoet, Y.M. Suhov, A.J. Wyner, Nonparametric entropy estimation for stationary processes and random fields, with applications to english text. IEEE Trans. Inf. Theory **44**(3), 1319–1327 (May 1998)

35. A. Noulas, S. Scellato, R. Lambiotte, M. Pontil, *A Tale of Many Cities Universal Patterns in Human Urban Mobility*, Mascolo, Oct 2011

36. M. Kim, D. Kotz, S. Kim, Extracting a mobility model from real user traces, in *Proceedings of the 25th IEEE International Conference on Computer Communications INFOCOM vol.* 2006, pp. 1–13 (2006)

37. R-I. Ciobanu, C. Dobre, V. Cristea, D. Al-Jumeily, Social Aspects for Opportunistic Communication, in *2012 11th International Symposium on Parallel and Distributed Computing (ISPDC)*, June 2012, pp. 251–258
38. P. Hui, E. Yoneki, S-Y. Chan, J. Crowcroft, Distributed community detection in delay tolerant networks, in *Proceedings of 2nd ACM/IEEE international workshop on Mobility in the evolving internet architecture, MobiArch '07*, New York, NY, USA, 2007. ACM, pp. 1–8

Memory Support Through Pervasive and Mobile Systems

Mauro Migliardi and Marco Gaudina

Abstract Memory is one of the most important components of human self-consciousness. As such, it has always been the subject of scientific studies and is one of the human functionalities with the most ancient examples of prosthetic devices. In this chapter we will describe the evolution of memory prosthetic devices from the paper agenda to the modern context-aware and mobile systems, and we will describe how our system, with its capability to autonomously recognize locations where a user may efficiently perform a previously defined task, represents a significant innovation with respect to the state of the art. Finally, we will describe our experimental results and we will provide some concluding remarks.

1 Introduction

Memory is one of the most important components of human self-consciousness and intelligence. As such, it has been the subject of study in biology, cognitive science, as well as in the psychology. A first level categorization, usually divides memory into short term and long term memory. Although both a clear separation and a precise discontinuity between these two are still somewhat controversial, there is a general agreement that different functions exists. Long term memory, as its name suggest, is intended to store information over a long time, while short term memory acts as a scratch-pad for a temporary recall of the information under process. Precise details about how the information is transferred from short term

M. Migliardi (✉)
Centro Ingegneria Piattaforme Informatiche (CIPI), University of Genoa and University of Padua, Via Opera Pia 13, 16145 Genoa, Italy
e-mail: mauro.migliardi@cipi.unige.it

M. Gaudina
Circle Garage, Piazza Colombo 2/13, 16100 Genova, Italy
e-mail: marco.gaudina@circlegarage.com

F. Xhafa and N. Bessis (eds.), *Inter-cooperative Collective Intelligence:*
Techniques and Applications, Studies in Computational Intelligence 495,
DOI: 10.1007/978-3-642-35016-0_9, © Springer-Verlag Berlin Heidelberg 2014

memory to long term memory, as well as the complete mapping of the different
memory related functions onto the different sections of our brain, are still under
debate. However a general consensus exists on aspects such as the importance of
repetition and semantic associations, the fact that long term memorization efficiency may differ greatly from subject to subject and the fact that memorization
capabilities are influenced by the emotional state and age of the subject [1–3].

When a subject experiences difficulty in storing information in long term
memory, a common effect is a disruption of his capability to efficiently follow an
activity plan. Things to do tend to come to mind unordered, in the wrong moment
and in the wrong place. This has two negative effects:

- it reduces one's personal efficiency;
- it generates a deep sense of frustration that produces further stress and may even
  worsen the situation.

When this phenomenon gets to the extreme, it turns into a sort of "Activity
Thrashing", i.e. a situation in which the subject cannot manage to complete any
task due to the fact that while he tries to pursue the completion of the task at hand,
his attention is continuously captured by the need to complete other tasks that
require to move into a different context. Thus the subject ends up moving around
without completing anything. Anyway, even in less extreme versions, this phenomenon has a negative impact on the capability of subjects to perform and
conclude tasks as planned and tends to be extremely stressful.

In modern, western society, life expectation has dramatically increased while
the birth rate has similarly shrunk; therefore the average age is steadily rising [4,
5]. If we combine this effect with the fact that the average stress level is growing
too, then it is easily predictable that the impact of memorization problems is bound
to grow in the near future [6] and, as a consequence, that the importance of
memory support application will grow too [7, 8].

1.1 Evolution of Memory Prosthetics Devices

The paper agenda has always been considered as the first line of defense against
memory difficulties; however, the pervasive introduction of information technologies opens new and interesting possibilities. More in details, the development
frontier for memory support systems today is characterized by three extremely
significant capabilities. The first one is the capability to be *context-aware*, i.e., the
capability of systems to change their behavior in relationship with the situation and
position (in space and time) of the user which the service is related to. The second
one is the capability to retrieve information from the World-wide-Web at any time
and thus to efficiently turn that information stream into a knowledge base for the
system. The third one is the capability to adopt novel input/output modes that
greatly improve usability by overcoming the constraints of the traditional touch-
typing terminal. Taking advantage of these capabilities, the evolution trend in

memory support systems is hence from the pocket agenda that has a completely passive interaction with the user toward a completely different object. Such a new object provides help in the form of much more sophisticated reminders that it generates not only at the right time but also in the right place and in the right situation trough interaction channels that may be adapted to suit the user current needs. The evolutional trend, thus, is to turn from simple data collectors as the paper agenda toward effective *Active Personal Information Manager*. It is our opinion that Personal Information Manager needs to become active to effectively stave off the memory problems deriving from stress and growing age and to prove our point we have developed a system that is capable of capturing user needs and tasks to be carried out expressed in pseudo-natural language, and, through the combination of some reasoning with internet based Geographical Information Systems, to autonomously provide to the user contextualized hints about when and where he can fulfill those needs or complete those tasks.

2 Context Aware Systems

Communication between humans is extremely efficient and it is extremely more efficient than the communication between humans and computers. Several factors contribute to this efficiency: the richness of the language that humans share, the mutual knowledge and the implicit knowledge of everyday situations. When humans talk with other humans they are able to use implicit information related to the current situation, namely context, to increase the bandwidth of the communication. Unfortunately, the ability to leverage this type of information is completely absent in the communication between humans and computers. During a traditional human computer interaction, users have a very narrow channel and a limited possibility to generate inputs for a computer. Consequently, computers are not able to take advantage of the human-computer dialog context. Improving the capability of a computer so that it may take advantage of context would significantly increase the efficiency of the communication in human-computer interaction, and it would make possible to implement computational services with a higher degree of efficiency and usability.

The concept of sensitivity to a context (*Context Awareness*) has been introduced in the ICT field for the first time by Schilit [9]. He was referring to the capability of computer to relate to the ambient it was inserted into. Through the use of sensors, a computer could have the "perception" of what is happening around itself.

Sensors for measurable quantities, such as temperature or movement, are able to provide ambient related information and could be used inside a computer program via *sensor fusion* techniques. These techniques, in fact, may be used to produce a model for the identification and characterization of the context where the system is positioned. Hence, it is possible to use information about the surrounding ambient as the starting point to generate models with the aim to adapt the behavior of an application to a particular event or situation.

In almost 2 decades, the "context awareness" concept has slowly moved from being limited to the discovery of the surrounding environment to become a very important and influential topic in many fields of information technology.

As an example, Kaltz et al. [10] extended the context concept to Web Engineering techniques proposing new referring models for developing a web application related to the proposed context. This, in practice, translates into adaptation of the contents with respect to the position of the user. In fact, from the position of the user it is possible to formulate several different considerations regarding cultural aspects of the web navigator. This allows proposing content that is continuously adapted to the needs of the web navigator and increases significantly the dynamicity and interactivity of the system as it is perceived by the user.

Recently, Zainol and Nakata [11] showed that it is necessary to define the ontology of a context to extend its validity and to allow an automatic reasoning engine navigating inside it. In particular, the definition of an ontology allows to model a generic context that could be applied not only in a specific way but can be extended, adopting the triple *Extrinsic Context, Interface Context, Intrinsic Context*, to every situation where there is the necessity to implement a relation between the position of the system/user and the generation of an event.

Furthermore, in the last years, the steady improvement of mobile devices has provided users with terminals that may leverage integrated GPS systems and radio cells triangulation to achieve cheap but reliable positioning. This enabled the mobile device to evolve from an entity in relation with the surrounding environment to a proxy for relations between the user and his environment.

Taking into account these new trends, a good definition of what is and what is not "context" is the one given by Abowd et al. [12]:

> Context is any information that can be used to characterize the situation of an entity. An entity is a person, place, or object that is considered relevant to the interaction between a user and an application, including the user and applications themselves.

From the definition above, if a piece of information can be used to characterize the situation of a participant in an interaction, then that information is a context. Adversely, if the information does not affect the user or the application for the purpose of the task, it is not a context.

Abowd et al. [12], suggests the existence of four primary contexts that indicates the types of information necessary to characterize a situation:

- Location context provides information about where the entity is localized.
- Identity context provides information about who the entity is.
- Activity context provides information about what is happening.
- Time context provides information about when the entity is.

In the next section we leverage the above taxonomy to provide some examples of real application and show how memory support applications evolved with the adoption of context awareness.

3 Application Contextualization in Memory Support Systems

Writing down appointments and notes into a paper agenda has always represented the first line of defense against the mounting tide of the information overload. However, the agenda's main characteristics are the impossibility to contextualize the information contained and the availability of the agenda itself: to guarantee that no appointment is lost, it is necessary to perform repetitive (almost continuous) checks to the agenda. If this is not the case, the presence of an important appointment will go undetected and the chance to buy a needed object (e.g. some milk) will be lost despite the fact that we are standing in front of a dedicated shop (e.g. the milk shop).

3.1 The Introduction of Time Context

A first level of support to the contextualization of the information present in an agenda has been provided by the fact that every digital system contains, for its intrinsic nature, a clock that allows it to keep track of the current time. This capacity to contextualize in time any kind of information allows the development of applications that actively suggest that a deadline or an important appointment is approaching. This feature made the "electronic calendar/agenda" one of the most diffused desktop applications. Consider, as an example, that the calendar application has been present since the very first version of *X Windows*.

However, the intrinsic dependency of this kind of application from the immediate availability of a switched on computer limited enormously its capability to surrogate the paper agenda in the role of memory support system. In fact, until the massive diffusion of notebook computer took place, there was no real convenience in substituting the paper agenda with the electronic one.

The massive adoption of laptop computer in the nighties reduced the obstacle to the adoption of electronic versions of traditional agenda by weakening the previously very tight relationship between a computer and a desktop to put it on. This, in turn, fostered the evolution of *Personal Information Management* application toward new levels of sophistication. In particular, the integration of the calendar with the address book and the e-mail-box, together with the possibility to share data among the participants into a working group in a reliable, selective way, have represented the main enabling conditions for the pervasive diffusion of *Group Information Management* applications that we have today.

Nonetheless, even with the advent of laptop computer, the necessity to have the pc turned on still represented a major drawback of the complete substitution of the paper agenda with an electronic one. In fact, while the form factor of laptop/notebook computers is capable of guaranteeing a satisfactory level of independence from the need for a fixed working position such as an office desk, the inability of those same devices to stay always switched dictates that, in order to get a reminder, a user needs to actively perform an action on the device in a way that is even more cumbersome than the one needed to check for appointments in a paper agenda.

It is only with the massive spread of smartphones devices that took place during the last decade that the electronic agenda has finally overcome all the shortcomings previously described. In fact, the smartphone has both a form factor that makes it completely portable and the capability to stay always switched on. This allows such a device to both provide support to the user without the need for him to be proactive and to be available to store new information immediately, without long bootstrapping processes.

3.2 The Introduction of Location Context

An important technological enabler is the presence of positioning hardware on all the smartphones of the current generation. The most common way to provide a reliable geographical positioning is to combine the presence of a GPS system on the smartphone with the capability to measure the power of the signal coming from different antennas to triangulate the device position. Furthermore, to ease the task of identifying the user position even when the GPS signal is not available (e.g. indoor) Google made a map of the WiFi networks present on the territory and smartphone devices are able therefore to use this information to increase the precision of the localization when the satellites signal is poor. With the addition of this hardware and software components, it is possible to integrate location context awareness into memory support systems and applications.

An example of a memory support application available on the smartphone platform and capable of complete awareness to time and identity context is Astrid [13].

Astrid (see Fig. 1) is a quite diffused Android application that is capable of inserting in a personal database user defined activities and to generate reminders at user specified times. A significant feature of Astrid is its capability to synchronize data with Google Calendar [14]. Portability and ease of interaction are often contradictory goals and the most conveniently portable device, the smartphone, has strong limitations in the human-computer interaction channels, e.g. small screen, small keyboard, etc. For these reasons, the integration of complete web based interfaces with smartphone specific applications provides a significant added value. This kind of integration, in fact, provides both agile support to users in mobility and rich interfaces that the same users may leverage in more stable environments (e.g. at home, in the office, etc.).

As noted previously, the native form of Astrid is capable of context awareness only in relation with time context and identity context; however, there is a plugin that connects Astrid to another Android application providing a limited level of location context awareness. The level of awareness is limited because a user has to

Fig. 1 Interface of the
context aware smartphone
application Astrid

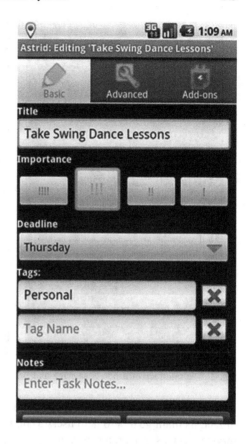

manually pre-define all the positions that constitute a location context, nonetheless, with the help of this add-on we obtain an application capable or reminding a user of a task not only when the deadline approaches but also when the right location is at hand.

Some applications that introduce awareness of the location context disregard time context information that had already successfully exploited in past systems. Two examples of this typology of application that can be somewhat assimilated to memory support systems are *Hotels near me* [15] and *Loopt* [16] (see Fig. 2).

Hotels near me is an application that leverages the position context of the user to help him in the task of finding an hotel nearby. Forcing the memory support concept, this type of application can be seen as a support tool if, as an example, the user forgets to book a hotel in advance. However, it is obvious that its primary functionality is instead to cope with unexpected situations. The type of functionality provided together with a very successful campaign aimed at getting large discounts from hotel's companies have guaranteed a major success to this application.

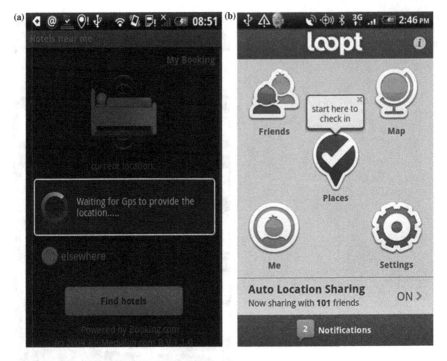

Fig. 2 The interface of **a** hotels near me and **b** loopt

Loopt is quite similar to the previous application but has a more generic target. In fact, its main goal is to help users to find not just hotels but many categories of services such as restaurants, bar, money dispenser and others. If we compare *Loopt* to the previous application, we can clearly see that in this case there is a wider range of use cases; nonetheless this application too is closer to a tool designed to cope with the unexpected than a real memory support system.

3.3 Web Integration

An example of a complete integration between web application and mobile device is constituted by the Google suite with web applications [17] and the corresponding components installed on every Android device [18].

In this suite we can find **Google Calendar**, without any doubts one of the most diffused personal calendar and perfectly integrated with the remaining part of the suite. The interface allows adding tasks with complete descriptions. These tasks may come both in the form of temporally located appointments (e.g. meet the CEO at 5 p.m.) and in the form of to-dos (e.g. write the memo before Wednesday). Furthermore, the interface is extremely flexible: it allows defining tasks with a minimum level of details to cope with the needs of users in mobility, and to refine this definition adding details such as location, a complete description and a list of participants at any time; it is important to notice that this kind of refinement is not constrained to be performed through the web-interface, on the contrary, even if it is by far less convenient, if the need arises a user can fill up all the information directly through the mobile interface. From the notification point of view, the suite offers a good level of integration between web-based and mobile-based interfaces. In fact, several different notification modalities are offered: a user can choose to receive an email, to show a popup dialog on the web interface, to get a sound and visual notification on a smartphone, or to receive a text message (sms) on a traditional mobile phone.

Even if it offers a very good level of integration between web-based interfaces and mobile-based interfaces, the suite still suffers from the inability to leverage information about the user location context. In fact, while it is possible to define a location for an appointment or a task, the hints that the system is capable of providing to the user are completely based on time context alone.

3.4 Integration of Time and Location Context

Two examples of integration not only of the mobile interface and the web based interface, but also of the integration of information acquired from different context typologies, namely the spatial context and temporal context, are *Remember the Milk* [19], *ReqAll* [20], the *Reminders* feature in *IOS5* [21] and *GeoStrings* [22].

Remember the Milk is a clear example of the integration of several different contexts. The system is capable of interacting with the user via several different channels (e-mail, sms, popup, etc.); personal tasks are contextualized in time through associated deadlines but they are also geo-localized. This geo-localization, though, requires the user to manually identify a location where the task can be efficiently carried out and to manually associate this location to the task. The system is then capable of generating a hint to the user both when the deadline expires and when the user gets to the place associated to the task. As an example, consider a task that defines the need to buy milk before 4 p.m. at a specific milk-shop. The system will notify the user either if it is 4 p.m. or if the user goes to that specific milk-shop.

ReqAll is an application that aims at registering all the tasks that a user has to complete in a certain place, so that it may remind the user what he has to do when he gets to a specific location at a specific time. It can be integrated with some of the most diffused calendar manager such as Outlook or Google Calendar and it allows sharing reminders with other users. The mobile application is capable of generating geo-localized alarms associated with time contextualized to-do lists.

Reminders of IOS5, similarly to the previous examples, allows users to define reminders that will be fired up either at a given time or when the user gets to the user defined location. The system leverages the GPS system embedded in the iPhone.

GeoStrings is an application for WebOS mobile devices. It allows users to define one or more locations where a specific reminder has to be fired up. The availability of multiple locations is a significant refinement over the previous systems, however, all the locations have to be manually picked up by the user.

The integration of both human-computer interaction channels and different types of context provided by the two above described systems represents a major advancement for memory support systems; however, both systems still lack the capability to explore the environment surrounding the user to search for places representing alternatives to the one selected by the user. Thus, while the system may remind a user about what he wants to do when he gets into a specific place, the system can neither remind the user that he has to go to that place nor it can discover that the user does not need to go there to complete the task.

3.5 Toward a New Generation of Memory Prosthetic Devices

Although the examples provided in this section represent very interesting use cases for the integration of human-computer interaction channels and different typologies of context, they lack any form of autonomy and depend completely on the capability of the user to proactively and precisely define the time and location context in which notification events have to be generated.

As we stated in the introductory section of this chapter, it is our opinion that, in order to (1) provide an efficient support to human memory, (2) prevent the negative effects that both stress and aging have on memory, and (3) enhance the capability to efficiently follow a plan, there is the need for a system that synergistically combines the capability to provide to users hints about tasks that need to be completed with the capability to autonomously search for locations that lend themselves to the completion of the memorized tasks. Furthermore, in order to catch all the hints the user expresses about his needs, human-system interaction should take place in a way as close as possible to human-human interaction.

The introduction of these capabilities, namely (1) the capability to autonomously seek locations apt to the task at hand and (2) the capability to accept input in pseudo natural language are two of the most significant innovation of our system with respect to the previously described state of the art.

Our system is capable of:

- memorizing activity plans described by the user in pseudo-natural language [23, 24];
- reasoning on and classifying the user needs to match them with Points of Interest (POIs) available in the user environment;
- tracking the user movements to perceive his/her proximity to identified POIs;
- reminding users of those planned activities at the right time and in the right place.

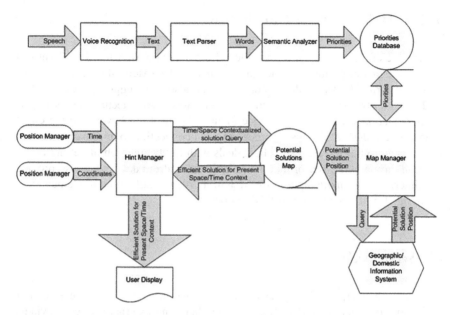

Fig. 3 The functional architecture and data flow of the complete support system for memory impaired subjects

In Fig. 3 we represent a logical, data flow based scheme of such a system.

A user provides a speech input depicting his need. Speech is converted into text by a speech recognition engine (e.g. the one provided by android through Google servers) and produces text. The text in parsed to identify words. Words are submitted to a semantic analyzer that grasps their meaning and produces Priorities objects encapsulating Tasks to be performed. These priorities/tasks are submitted to a reasoning system that generates a list of POIs categories in which the priority/ task may be efficiently performed. The list of POI categories is submitted to a map manager that queries a set of Geographical Information System (GIS) providers to get actual locations of POIs. These locations are stored in a database of potential solutions for the user's priorities/tasks. When the user's client provides a change of context (in principle it could be both space and time, but our current implementation focuses on space), the hint manager queries the potential solution database to see if there is any priority/task that can be efficiently fulfilled/performed in the user's vicinity. If it finds one, then it generates a notification to the user.

To test the effectiveness of this approach, we have implemented a prototype system. In the next section we describe its architecture.

4 The Architecture of Our System

To achieve our goal, we had to map the logical data flow shown in Fig. 3 onto an actual system architecture. The actual system was also extended to include events besides priorities/tasks, so that it could be used also as a simple agenda.

The first, obvious choice, was to use Client-Server Architecture. The server is implemented as a REST Web Service in Java Application Server, and the client is an Android device. Expensive processing and connection to third party provider is delegated to the server, while client deals with lightweight job of getting user's location through GPS sensor that is available on Android device and also provide user interface for input/output to the system. Further detail on the implementation is provided in the remaining of this section.

4.1 Server Side

In a very abstract, high level decomposition, the structural components of our server that take care of interaction with the client and external service providers are depicted in Fig. 3.

The activity starts from a simple HTTP requests that will be handled by a servlet engine, that, in our current implementation, is a Tomcat Server. The Servlet engine, through its internal mechanisms, will dispatch a servlet for every client request. Each servlet has its own address space, and has access only to its own resources. The REST Parser component is responsible of detecting HTTP method (GET, POST, DELETE, PUT), header, cookie, and also parsing XML body when needed. This will be our only component whose methods can be accessed directly (although remotely) from the client.

Inside the server, three additional major Manager Components are responsible for different tasks (see Fig. 4). The Event Manager is responsible for managing all

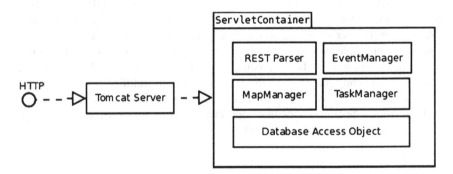

Fig. 4 High level structural abstraction of the components that provide the interface to the client inside the server

access to events object, which involves creation, deletion, selection of specific events based on date, selection based on title, etc.

The Task Manager does the same for task objects; the Map Manager encapsulates the task of querying Geographical Information Providers (local or external GIS systems) and is responsible for grabbing POI locations that will be used by the hint manager component of the system (see also Fig. 3).

In order to provide flexibility for future developments, every Manager is implemented using a Subscriber-Provider Design Pattern. A third party provider will be integrated as a subscriber to the apt Manager instance. The adoption of this design pattern makes it easy for us to add or delete another service provider in the future. As an example (see Fig. 5), let's consider the Event Manager and Google Calendar as a service provider for registering Events. In the case of absence of the Google Calendar subscriber, be it for lack of connectivity or for an active decision of the user to disable that service, the system will not generate updates to the Google Calendar account, however it will still continue to run with its own local database. In the same analogy, creation of any future subscriber will just need to extend a specific subscriber class and implement the required methods defined by the interface.

Each of the three managers provides the logic of the execution, and provides access to its own Database Access Object (DAO). Event Manager only has access to Event DAO, Task Manager to Task DAO, etc. DAO provides access to the database in a primitive low level manner (CREATE, DELETE operation on tuples). Our first implementation used the db4o Java library [25], which can save and retrieve simple Java object into a file. This choice speeded up our development process as compared to using contemporary relational database such as SQL, in fact we could avoid the definition of a stable data model as every object could be simply and automatically turned into a line in the database.

For databases that do not need to have a Manager implementation, e.g. the Session Database that keeps track of user session data, we do not create a

Fig. 5 Publisher-subscriber
design pattern

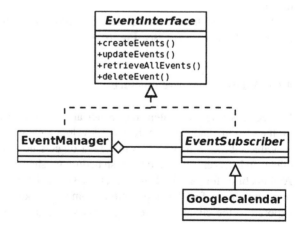

Fig. 6 Three-tier
architecture

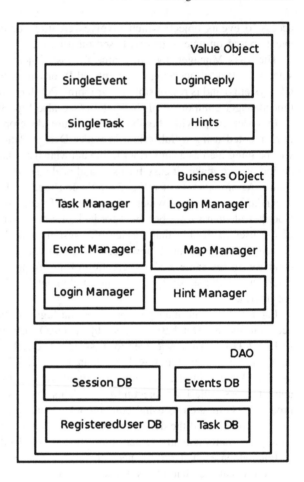

correspondent Manager class but we only provide a Session DAO to access the
database.

At system level, we have a 3-tier architecture of Value Object, Business Logic,
and Data Access Object (see Fig. 6).

4.1.1 A Three Tier Architecture

By decomposing our system into 3-tier architecture, we decouple the complexity
of the system and support further development and expansion of the project.
A Value Object consists of simple Java objects needed for transactions between
client and server. It will be marshaled/unmarshaled to XML by standard Java
Architecture for XML Binding (JAXB). A Business Object contains logic
implementation for the application. Some methods in this layer are indirectly
exposed to the client via interface classes that are implemented using RESTeasy

API. This interface enables communication to client using RESTful web service, which is supported by Android phone. This way, the client can just send a simple HTTP message, and object transfer is done by using XML.

Data Access Object layer provides abstraction to database using db4o library [25]. We save registered users, events and tasks in database. Suppose, for example, that in the future we found that the communication between client and server is too heavy on sending event object, we could just change the fields on Event Value Object, without changing the underlying business logic. Furthermore, if we want to change our database implementation from local file to cloud based database (a possibility supported also by our simple key-value design), we could do it easily by changing the Database Access Objects for the database. This choice to decouple into three-tier architecture follows the standard J2EE design pattern; hence, it is also possible for the application to be deployed in Java Application Server other than Tomcat, such as JBoss and Glassfish.

Unfortunately, the first simple performance and stability tests of the system revealed that the db4o component was a weak point of the system both for its low performance and for its low stability. Thus we immediately had to develop an adaptation layer to connect to a true relational database. The design and implementation of this adaptation layer required us to study how all the different objects that were present inside the server could be mapped into database tables so that the original structure and functionality of the server could be saved.

This process led us to the definition of an Entity-Relationship model for all the persistent objects in the server and then to the definition of a Java interface (IFDatabaseManagement) containing all the operations needed to interact with the relational database. Before implementing the interface, we selected the actual database: our choice was to adopt MySQL. The selection of the actual product completed, we proceeded to implement the actual class (MySQLDBManager). The design of a generic interface allows for a complete independence of the system from the actual DBMS, thus, if our current choice (namely MySQL) would turn insufficient in the future, we could simply substitute it by producing a new specific implementation of the interface IFDatabaseManagement.

Thread safety during execution of methods in DAO has been achieved in two steps. First, in order to provide a unified method of thread creation, we use a Factory Method Design Pattern to access the database. Furthermore, since we only have a single instance of DAO for each database, we need to restrict new instance creation from Java Virtual Machine. We do this by implementing each DAO class as a singleton. This way, every servlet that wants to get data from the database needs to call a single instance of Factory that produces a new client thread to access its method. Every Database Factory is initialized once during the starting of Tomcat server, and is closed only when the Tomcat Server is shut down. The locking mechanism adopted is the standard synchronized keyword of Java which adopts an object instance as the mutex lock.

4.1.2 Reasoning Engine

In the current implementation of our system, the inference engine that derives from the user's priority/task the POIs categories to search in the GIS is composed of an ontology and an ontology reasoner. As a first step we have modeled the relationships between types of priority/task and places where they can be fulfilled/performed in an ontology written in OWL using Protégé 4.1. At present, our ontology consists of three classes (item, action and location) with reciprocal object relation: canBeFoundIn, canBePerformedIn, locatedIn, performedIn. After the definition step, the ontology is populated with data of item-location relation. Using OWL API 3 and HermiT reasoner, we can then search in our ontology if the item or action requested by the user to satisfy a priority or perform a task exists, and infer the locations that can fulfill this priority or task.

Security design in a RESTful web service is quite a challenge for mobile application, since REST protocol is basically a stateless protocol. As a first step to mitigate security problems, we adopted the HTTPS protocol to secure communication between client and server. Having secured the communication, we have created our own session database which saves a user token. Token creation happens only once during the user's first login. The Android device saves this token inside the Account Manager, so that the user does not need to login again every time, and only uses session-id. This approach, however, makes the system prone to attacks capable of stealing the token either from the client or from the server. In future implementations we may adopt a mechanism of regeneration and exchange of fresh session-ids to mitigate this factor. The clients will have to renew their session-id after a defined period over, so old session-id (that might have been acquired by malicious user) will no longer be valid.

Restriction of access in Tomcat server is done by configuring local access policy. This way, every servlet will only have access to files that are accessed by DAO.

In order to provide a means for the user to personalize the result, we added Hint Manager that processes the data provided by the GIS servers. In the current implementation, our GIS server is Google Local Business Search. When results from the GIS server are collected, they are passed to a Hint Manager which filters them according to the preferences expressed by the user. The Hint Manager is a pipe and filter design pattern, which parameters are retrieved from the client. The final result from Hint Manager is then sent to the client. Upon receiving the result, a notification will appear on Android screen.

4.2 Client Side

On the client side, i.e. the Android smartphone, one of the most stifling limitations was the one connected to battery duration. In fact, heavy network traffic together with continuous usage of the GPS drained the batteries in a very short time. To reduce the level of energy consumption, we decided that both queries to the server and user position assessments had to be performed in discrete instants, while network and GPS could be switched off at all other times. To achieve this result, we studied an algorithm capable of forecasting the next instant in which the client needs to assess the user position and query the server for changes. The algorithm is not designed to cope with emergency situations involving dramatic changes in user's speed such as the case of an accident.

The system defines two parameters to provide hints to the user and download hints on the surrounding environment from the server. The first one is the maximum distance at which a POI will trigger a notification to the user. The second one is the radius of the zone inside which POIs will be downloaded into the client. Obviously, the value of the second parameter must be greater than the value of the first one, otherwise we require the system to notify the user about POIs whose existence it does not know locally and we force it to query the server continuously. The difference between the radius of inner (notification) ring and the radius of the outer (local caching) ring define a circular space around the user (see Fig. 7) that we may call the slack ring. The slack ring defines a space where the user may move without triggering the need for the client to query the server for additional map and POIs information as they are all already in the client local memory. It is a classic case of performance trade-off where one trade-off dimension is the amount of local memory dedicated to caching and the other dimension is the amount of

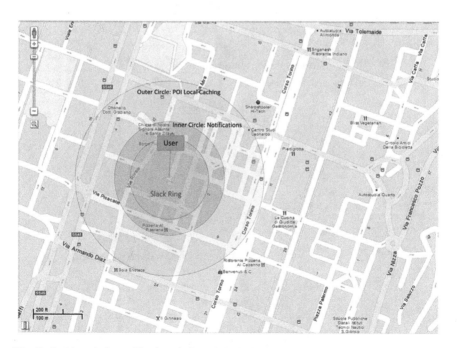

Fig. 7 Caching circle, notification circle and slack ring

bandwidth and thus battery consumed; in fact, it is obvious that a rather large slack ring requires the user to move further away before new POIs data are needed. Minimizing queries to the server reduces the battery consumption caused by radio transmission; however, keeping the GPS and the application active all the times still drains the battery too quickly. Thus, we decided to look for a way to switch on position checking intermittently and we ended up with two different solutions.

The idea behind the first solution is to allow a user to navigate inside the slack ring before checking if new notifications are due. To do so, we identify the current user velocity vector and we calculate the time that it will take for the user to get to the border of the slack ring. Then, we switch off the GPS and we suspend the application for the calculated amount of time. When the time expires, we switch back on the GPS and we check the user position and we identify his current velocity vector. If he is still inside the slack ring, the application may need to add new notifications but there is no need for querying the server. Thus the system may use the new velocity vector to calculate the new switch off time and repeat the process. When the system wakes up and finds that the user has moved outside the slack ring, then he needs to query the server for data in a new outer ring centered on the current user position, identify the current velocity vector of the user, notify the user for the POIs that have entered the inner ring, calculate the next switch off period and go to sleep again.

In Fig. 8 we may see an example of this. In t0 the client downloads the POIs to populate the outer ring and notifies the user for the POIs in the inner ring (the one

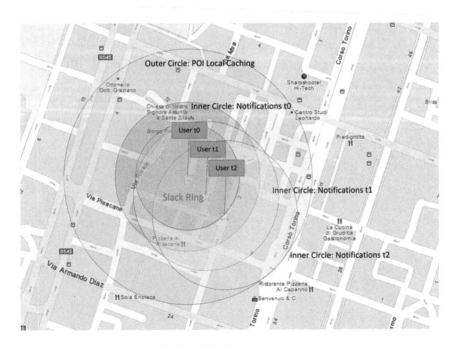

Fig. 8 Example user movement inside slack ring

identified in Fig. 8 by the label *Inner Ring: Notitifications t0*). Then he identifies the velocity vector of the user, calculates the instant t1 at which the user is expected to cross the boundaries of the slack ring and goes to sleep for a time equals to t1 − t0. In t1 the client wakes up and checks the user position; as he is still inside the slack ring (the user may have stopped for a while or he slowed down his pace), there is no need for new data from the server, it is sufficient to notify the user for the POIs that have entered his inner ring that is now identified in Fig. 8 by the label *Inner Ring: Notifications t1*. Then the client identifies the new user velocity vector, calculates a new time t2 in which the user is expected to cross the boundaries of the slack ring and goes to sleep for a time equals to t2 − t1.

If the current speed of the user is lower than 2 km/h (a leisurely stroll) the max between his previous measured speed and the threshold speed is used, so that the system never sleeps indefinitely. In t2 the clients wakes up and checks for the user position; this time the user is outside the slack ring, thus the client needs to download data from the server to build a new outer circle, then he notifies the user for the POIs that have entered his inner circle (identified in Fig. 8 by the label *Inner Circle: Notification t2*), calculates a new slack ring and starts again the process as in t0.

The adoption of this mechanism allows switching off the GPS and putting the application to sleep, thus greatly reducing the impact of our system on the smartphone's battery. However, this mechanism is not perfect as it has some drawbacks.

The first obvious drawback is due to the fact that switching of the application for a time interval introduces a granularity in the capability of the system to notify the user. In fact, when the application on the client is off, a POI that enters the *Inner Circle: Notification* will not be immediately signaled to the user; on the contrary, the notification will take place only when the application on the client will come back on. In Fig. 9 we see an example of this behavior. The POI Enoteca Sola is of interest for the user. In t0 the POI is outside the *Inner Circle: Notification t0*, so the system generates no notification. At time t0+ the POI enters the *Inner Circle: Notification t0+*, however, the user won't get any notification until time t1 when the application on the client wakes up again.

The granularity of notification may be the cause for even worse misbehavior though, as Fig. 10 shows. Once again the POI Enoteca Sola is of interest for the user but it's outside the *Inner Circle: Notification t0*, so no notification is generated. The POI enters *Inner Circle: Notification t0+*, however, when the application on the client wakes up in t1, the POI has moved outside *Inner Circle: Notification t1* thus the user will never be notified that he had a chance to complete a task at the POI Enoteca Sola.

The granularity of notification is controlled by the size of the circles and by the speed of the user: at any given user's speed larger circles imply larger granularity while at any given circles' size larger user's speed means smaller granularity. Thus, it is possible to reduce the granularity by simply reducing the size of the circles, but this also shortens the intervals of GPS and application switch off and so reduces the amount of battery saved.

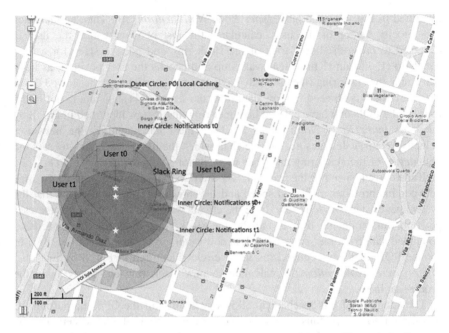

Fig. 9 Example of delayed POI notification due to app and position checking switch off

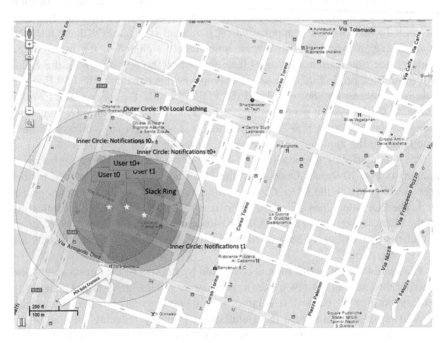

Fig. 10 Example of missed POI notification due to app and position checking switch-off

It is important to notice that, while the user's speed of movement is mainly due to his means of transportation, only the slowest mean of transportation (i.e. walking) is guaranteed to deprive the user of a power source. On the contrary, a user moving on a car or a taxi-cab will be able to recharge his device. Furthermore, according to recent studies conducted in a major Italian city the average speed for surface buses is definitely low, between 10 and 12 km/h [26]. Thus, if a user is deprived of an energy source, his low speed allows the system to save power with a significant increase of the check granularity in fact a lower speed guarantees that the distance covered by the user will be small even if the system adopts a significant granularity for the checks. At the same time, if a user moves at higher speed then the system may reduce the granularity check because the faster battery depletion could be countered by the capability of recharging the device on the means of transportation.

Another drawback of introducing a position check granularity of duration inversely proportional to the measured user speed is the fact that the system measures instantaneous speed and doesn't take into any account the acceleration (it's a purely proportional control, there is no differential component). This could take to wrong estimations and may delay positioning checks and notifications beyond the intended. Nonetheless, it is once again very important to notice that the situation in which saving battery is crucial is just when the user is moving on foot. In such a situation, the speed variations are limited and even going from a leisurely pace to a full sprint will produce only a limited misbehavior in the system.

Nonetheless, in future developments we plan to take into account the distance the user moved through in previous intervals to correct the instantaneous speed with an averaged value, in this way an integral control factor will be added.

5 Experiments and Users' Feedbacks

To test and evaluate our prototype, the ideal experiment would have been to provide to a statistically significant group of users our system and to compare the results obtained in daily use with the results of a control group with no memory support system. Unfortunately, this was not possible because of both budgetary and time restrictions, thus we had to devise a different experiment that, although not capable of providing full statistical evidence of the validity of our system, could anyway provide some useful insights both on the perception the users had of our system, of its strong points and of its weaknesses.

The experiment we decided to carry out was as such. First, we gathered a group of volunteers chosen in such a way that none had been previously involved in the project and only half of them had a technical background. We divided them in two cohorts of almost equal size (namely 16 and 15) and as similar as possible for age range and availability of technical background. The first cohort was equipped with our memory support system, while the second, acting as a control group, had the smartphone with no helper application but for mere movement tracking purposes.

The experiment consisted in providing to the user a course in a urban setting to which the user was familiar (either his hometown or the city where he had been living for no less than 4 years) and a set of tasks (ten tasks) to carry out while walking along the course in a constrained amount of time (no more than 2 h).

We selected the tasks so that they were all possible to complete with limited diversions from the devised course. Each user had 10 min to memorize the list of task before starting to walk. During the walk, we tracked down each user movement and we plotted his path using data from the GPS system of the smartphone. After the walk, each user had to fill a questionnaire about his "user experience" with the system. In Tables 1 and 2 you can see the questions we proposed to the users. The experiment was carried out in four different locations, namely Genova, Treviso (see Fig. 11), Bassano del Grappa and Castelfranco Veneto (see Fig. 12).

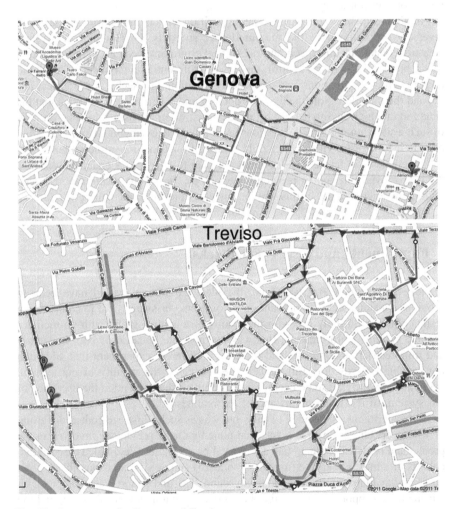

Fig. 11 Course maps for Genova and Treviso

In Table 1 we see how the users answered to the questions that required a quantitative (1–10) evaluation, the first half of the table deals with more general aspects of the application while the second half focuses on notifications and notification delivery mechanisms. As you can see in line one, the evaluation of the user experience provided by the application on the smartphone and the system in general was quite good and we had a strong concentration of scores very close to the average value. In line two, we see that the evaluation of the usefulness of the application was even higher (almost 9 out of 10) even if the scores were a little bit more spread around the average value.

The question we show in line three deals with the capability of the application to find solutions in such a way so that the planned course of the user is not heavily disturbed.

The score is still quite good showing a very low level of perceived disturbance, however, this parameter does not represent an effective evaluation of the

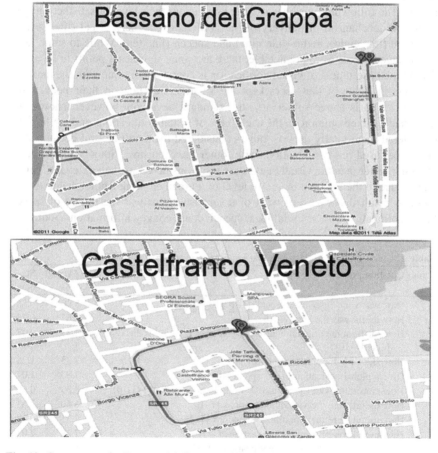

Fig. 12 Course maps for Bassano del Grappa and Castelfranco Veneto

intelligence of the system; as a matter of fact simply swamping the user with hints for all the available POIs allows the user to choose POIs close to his planned course (as the evaluation shows) but does not implies any intelligently selective behavior on the system side. Line four deals with the user interface quality. In this case, the scores provided by the users were still satisfactory but implied that use of the application required a certain level of technical proficiency.

This fact needs to be taken into strong account for future development of the system, even more so because the age range of the users involved in our experiments was from 24 to 45, so a population that is usually familiar with high-tech devices. In line five, we can see that the global evaluation of the application and system was very positive and that the scores were quite concentrated on the average value.

The second half of Table 1 deals specifically with the notifications provided to users by the smartphone client. In line 6 the first question is about how relevant to the expressed need the notification were. All the users were very satisfied as the score is almost 9 and the standard deviation shows a very concentrated set of scores. The following four lines ask the users about their level of satisfaction with a specific mechanism of delivery of the notification, namely visual (line seven), vibration (line eight), audio (line nine) and speech (line ten). While to some level the obtained scores were predictable, there are some significant points to be made. The limited usefulness of visual notification is completely in line with the fact (see line one in Table 2) that users kept the smartphone client in their pocket, the average value is above sufficiency, the standard deviation, however, shows that users had a mixed opinion on the usefulness of visual notifications and while some considered them totally useless, others considered them still helpful because they could be used in combination with other mechanisms. Sound notifications, although better than the visual ones, were also considered scarcely relevant. This is mainly due to the fact that the noise level of a urban environment tended to drown them. A result that was somewhat of a surprise for us was the failure of speech based notification. In our initial idea, speech could convey information about what

Table 1 Users' answers to questions that required a quantitative evaluation

Table	Question posed	Average	St. dev
1	How was your "user experience" [1–10]	8.00	0.58
2	How useful was the app [1–10]	8.67	0.75
3	How far from the original course did you steer because of the system's hints [1–10]?	2.33	0.47
4	How simple is the interface [1–10]	7.33	1.25
5	Give a global evaluation of the app [1–10]	7.83	1.07
6	How consistent with "your needs" were the notifications? [1–10]	8.83	0.69
7	How useful were the visual notifications [1–10]	6.50	2.63
8	How useful were the vibratile notifications [1–10]	9.33	0.75
9	How useful were the audio notifications [1–10]	7.17	2.03
10	How useful were the speech notifications [1–10]	5.17	2.11

Table 2 Users' answers to question that required a simple choice

Line #	Question posed	Unanimous answer
1	The smartphone was mainly in your pocket or in your hand (pocket-hand)	Pocket
2	Did you use the map in the app? (yes/no)	yes
3	Was it useful? (yes/no)	yes
4	Would you buy the app? (yes/no)	yes

kind of task was possible to complete and on the details of the possible destination without forcing the user to focus on the client smartphone display. However, all the users had quite negative a reaction toward speech notifications and when asked to elaborate on this they described the speech notification as too slow and they underlined the fact that the number of POIs around made the speech a very annoying, almost continuous droning. This result could be corrected if the system were capable of pre-selecting only a limited number of POIs among those available or in a country based environment where POIs were few, however, at the current level of development of the system, the experiments made quite clear that speech notifications were not a useful option. Another very interesting result of the questionnaires was the high level of success of vibration based notifications (see line eight of Table 1). The average score was close to the maximum and the standard deviation shows a quite tight distribution of scores. This result lead us to the decision of developing a more sophisticated vibration based notification system as described in [27].

Although it provided to us some very interesting insights, the proposed experiment is not capable of testing the actual capability of the system to contrast the stress or age induced forgetfulness of tasks during a normal day. In fact, the short duration of the experiment and the absence of stress inducing stimuli on the users produced only a limited washout effect in the memorized list of tasks. Furthermore, we realized too late that the chosen number of tasks, namely ten, allowed users to adopt memory tricks such as associating a task to each finger to avoid forgetting. Nonetheless, even in a limited space-time frame, our system enhanced users efficiency in terms of number of remembered tasks, time needed to perform the tasks, distances walked to perform the tasks. We collected travelling data for each user using GPS coordinates coming from the mobile phones and in Fig. 13 we show the resulting graphs for the average travelling distances of the live testing application. From this graph we can argue that the application really helps users in efficiently completing their to-do activities list, however we cannot confirm that the comparison between no-hint and hint data is statistically relevant due to the low number of users (less than 20 for each group, as previously stated) involved in this experiment session. For sure we can observe that the value of the distance that the users travelled to complete the tasks is almost always lower when our memory support system is available and, in total, our system allowed a saving of 25 %. Furthermore, in all the experiments sessions we observed a higher

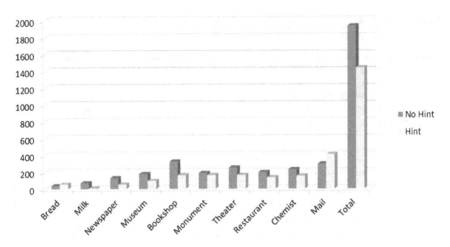

Fig. 13 Graphs for the two different phases of a live testing session. In *dark color* average travelling distances with no-hint and in *light color* average travelling distances with the use of the proposed application

percentage of completed tasks as well. In fact while the users with our memory support system always completed all the tasks, the average number of completed tasks for the control group was only eight, thus our system scored an increase in average task completion efficiency of 20 %.

6 Ontology Management and Enrichment

In this section we will focus on the different levels of efficiency that the system demonstrated in relation with the richness and completeness of the part of the ontology related to the tasks the user had to complete.

It was clear, as a matter of fact, that not all the tasks received the same level of support from our system. In some cases the suggestions allowed an almost immediate (both in time and space) completion of the task while in some other cases longer intervals (both in time and spaces) occurred.

Through the analysis of the system behavior and the profiling of the data flows, we managed to identify the source of this phenomenon as the different levels of completeness of the ontology that the system uses to classify a user need and to relate it to one or more site category.

This fact showed us that, in order to be efficient in finding solutions for all the needs of the users, our system needs a very complete and continuously evolving ontology. In general, the problem of ontology enrichment is a hard one [28] and is usually tackled either by leveraging the support of experts task-forces [29–31], or through statistical modeling of large text corpora [32–34]. However, in order to be effective, our system cannot afford the long training needed to build a statistical model; furthermore, the large text corpus onto which training would be performed is not available yet (although experiments are advancing). On the other side, it is impossible to build a task force of experts because it is not possible to identify a single, clearly defined domain inside which the user needs will be expressed.

Our first consideration was to find a solution to the situation in which the words contained in a user expressed need could not be matched in any of our ontology relations. We had to possible approaches:

- Use ConceptNet5. The project from MIT [35], aims at populating its ontology trough continuous scan of existing structured databases.
- Use a collaborative, social-network like approach. Users can collaborate and increase the "answering" capabilities of the system suggesting locations related to matching queries.

The first solution is a bold move toward web indexing. However, we decided to adopt the second solution and to give more importance to the users' direct opinion than to internet links or forum derived statements.

Thus, in order to achieve our goal we decided to adopt a social approach. Actually, the aspect of social integration is very common and diffused in several systems and web based applications. Most of the modern internet applications have a strong bias toward social interactions. It is well known that the data mined from the social interaction patterns of users can be leveraged for business purposes, we claim that social interactions can be a very useful tool for the enrichment of ontologies. Services such as Foursquare [36], are completely based on the social interaction among users. More in details, Foursquare provides to the users a possibility to interact with other users through their relationship with real world locations. A user suggests a particular POI increasing or decreasing the importance of such POI. As a consequence, the system can assemble a map of what users like or dislike and it can grow a map of preferences developing the possibility to increase its capacity of "suggestion". This kind of functionality is extremely interesting for the process of enriching the ontology on top of which our system's reasoning capabilities are based. Thus, we decided to implement a mechanism capable of leveraging the user experience and his/her social relations to generate new entries for our ontology.

When a user inserts a task, the system analyzes the phrase and recognizes the verb. If the verb has no associated meaning in the current ontology then the system asks to the user (See Fig. 14) to select the kind of action that better defines his task, i.e., search for something (a need to acquire an object), do something (a need to perform an action) or go somewhere (a need to reach a location).

Once the type of need has been identified, then the system searches the ontology to see if it knows where that kind of need can be satisfied. If no match is found, then the social ontology interface (see Fig. 15) is proposed to the user.

Through this interface the user may define associations between items and locations categories or actions and location categories. It is also possible to strengthen or weaken associations that other users have proposed (see Fig. 16) or

Fig. 14 Screenshots from the
categorization of a verb in a
task description

to add personal POIs to a specific category (see Fig. 17). In this way users can enrich the set of relationships the system uses and thus enhance its capability to understand users' needs.

Obviously, the system has to cope with unreal suggestions such as buying milk at the tobacconist. Thus we have introduced a voting mechanism that leverages the users' capability to judge the real value of the proposed addition to the system's ontology. The voting system requires that a majority of the users have voted positively a relation before the relation moves from the state in which it is suggested as "under vote" to the state in which it is incorporated into the ontology and used transparently to the users. For this reason, the possibility for a wrong relation to enter the ontology is minimized, nonetheless, to allow recovery from unexpected anomalies, administrative tools allow direct pruning of wrong or suspect relations from the ontology itself. It is important to notice that the user that originated a relation is not disclosed to the user who is voting the relation, thus our system does not need to cope either with inter-users trust problems or with user's privacy problems.

Furthermore, associations that indicate that an action can be performed or an object can be found in several different categories of places do not represent a conflict. On the contrary, they simply provide the information that a task can be

Fig. 15 Social ontology interface

Fig. 16 Interface to vote an
association between an item
and a POI category

Fig. 17 Interface for the addition of new POIs

completed in several different categories of places and the system will be able to provide a richer set of hints.

In our system, we do not introduce a system for the collaborative provision of POIs in a social database as it is done in Foursquare; on the contrary, what we do in our system is mainly to allow user cooperatively add new relations between actions and categories of locations or objects and categories of locations in order to enrich the system ontology. This different approach has a significant positive effect, in fact, we do not introduce in the ontology single locations, but associations between actions and categories or object and categories; thus the size of the ontology grows at a much slower pace. Besides, when we allow a user to introduce a new POI in the system database we ask him if he wants to keep it in a personal database that will be searched only on his behalf, or if he wants to publicly propose that location. In this second case, the same voting mechanism adopted for filtering new relations is adopted. A POI that receives a majority of positive votes is moved into the public database and used transparently to users; a POI that receives a majority of negative votes is left only in the user private database and never suggested to other users. The POI database may grow at a much faster pace than the ontology, in future versions of the system we may adopt a nonSQL database to cope with excessive growth.

The system internally keeps track of which user introduced each relation and each POI, a knowledge that may represent a sort of profiling of the users. However these data are not disclosed to users and do not represent a public source of information.

7 Conclusions

Memory is a defining function for human beings. Its efficiency though is negatively influenced both by stress and by age and when a subject experiences difficulty in storing information in long term memory, a common effect is a disruption of his capability to efficiently follow an activity plan. In fact, things to do tend to come to mind unordered, in the wrong moment and in the wrong place with the negative effects that it reduces one's personal efficiency and it generates a deep sense of frustration that produces further stress and may even worsen the situation.

Recent advancements in mobile devices and context aware systems may be leveraged to stave off these effects through the development of memory support systems. In the first part of this chapter we have described the evolution of memory support systems and provided some examples of applications. However, even if in the last decade there have been some significant improvements in the functionalities provided by memory support systems, it is our opinion that, in order to (1) provide an efficient support to human memory, (2) prevent the negative effects that both stress and aging have on memory, and (3) enhance the capability to efficiently follow a plan, there is the need for a system that synergistically combines the capability to provide to users hints about tasks that need to be completed with the capability to autonomously search for locations that lend themselves to the completion of the memorized tasks. Furthermore, in order to catch all the hints the user expresses about his needs, human-system interaction should take place in a way as close as possible to human-human interaction. Thus, we have designed a system capable of:

- memorizing activity plans described by the user in pseudo-natural language;
- reasoning on and classifying the user needs to match them with POIs available in the user environment;
- tracking the user movements to perceive his/her proximity to identified POIs;
- reminding users of those planned activities at the right time and in the right place.

While there are several systems capable of reminding users when they reach a specific, previously user defined location, the capability of our system to autonomously reason on user needs and POIs categories to provide original suggestions about locations that may be unknown to the user is one of the main innovations with respect to the state of the art.

In the second part of this chapter we have described the architecture of the system, our experiments with real users and how the results of said experiments lead us to further development. More in details, while the users had a generally positive reaction to our system, we were able to see that the ontology that represented the core of the reasoning autonomy of the system itself needed to be as complete and as up-to-date as possible. Thus, to pursue this goal, we developed a social-network like mechanism that allows users of the system to collaboratively enrich the ontology and update the relations it contains. The final result shows very good properties and in capable of collecting user needs, autonomously scan the user's neighborhood for locations where those needs can be fulfilled and finally deliver to the user suggestions about which tasks he can currently complete in a time and location contextualized manner.

References

1. M.S. Gazzaniga, *The Cognitive Neurosciences III*, 3rd edn. (Bradford Books, 2004), Nov 1, 2004
2. G.M. Wittenberg, J.Z. Tsien, An emerging molecular and cellular framework for memory processing by the hippocampus. Trends Neurosci. **25**(10), 501–505 (2002). (1 Oct 2002)
3. A.M. Magarinos, B.S. McEwen, Stress-induced atrophy of apical dendrites of hippocampal CA3c neurons: comparison of stressors. Neuroscience **69**(1), 83–88 (1995). (November 1995)
4. J.C. Chesnais, Fertility, family, and social policy in contemporary Western Europe. Population Dev Rev **22**(4), 729–739 (1996)
5. World Health Organization, Global Health Observatory Data Repository (2012), http://apps.who.int. Accessed on 31 July 2012
6. G.H. Bower, The Psychology of Learning and Motivation: Advances in Research and Theory, vol. 22. (Academic Press, 1988)
7. AAL, (2010), http://www.aal-europe.eu/. Accessed on 31 July 2012
8. universAAL, (2010), http://www.universaal.org. Accessed on 31 July 2012
9. B. Schilit, N. Adams, R. Want, Context-aware computing applications, in *IEEE Workshop on Mobile Computing Systems and Applications (WMCSA'94)*, Santa Cruz, CA, US, 1994, pp. 89–101
10. J.W. Kaltz, J. Ziegler, S. Lohmann, Context-aware web engineering: modeling and applications (PDF). Revue d'Intelligence Artificielle **19**(3), 439–458 (2005)
11. Z. Zainol, K. Nakata, Generic context ontology modelling: a review and framework, in *Second International Conference on Computer Technology and Development (ICCTD)*, 2–4 Nov 2010, pp. 126–130
12. G.D. Abowd, A.K. Dey, P.J. Brown, N. Davies, M. Smith, P. Steggles, Towards a better understanding of context and context-awareness, in *Proceedings of the 1st International Symposium on Handheld and Ubiquitous, Computing*, 1999, pp. 304–307
13. Astrid., Astrid todo list, http://weloveastrid.com/. Accessed 20 Apr 2012
14. Google, Google Calendar, http://www.google.com/support/calendar/. Accessed 20 Apr 2012
15. Blue Media, Blue Media Labs site: Hotels Near Me, http://blumedialab.com/. Accessed 20 Apr 2012
16. Loopt, Loopt uncovers the need-to-know info about your city, https://www.loopt.com/. Accessed 20 Apr 2012
17. Google, Google Applications suite, http://www.google.com/apps/intl/en/business/index.html. Accessed 20 Apr 2012
18. Google, Android Operating System official website, http://www.android.com/. Accessed 20 Apr 2012
19. Remember the Milk, http://www.rememberthemilk.com/. Accessed 20 Apr 2012

20. reQall, http://www.reqall.com/. Accessed 20 Apr 2012
21. Apple Corp., IoS 5, http://www.apple.com/ios/ios5/features.html#reminders. Accessed 31 July 2012
22. Hedami, GeoStrings, http://www.hedami.com/geostrings.html. Accessed 31 July 2012
23. M. Bates, Models of natural language understanding. Proc. Nat. Acad. Sci. U.S.A **92**(22), 9977–9982 (1995). (24 Oct 1995)
24. J. DiCarlo, Retrospective: What we learned from ubiquity - blog of mozilla labs developer. http://jonoscript.wordpress.com/2010/01/20/retrospective-what-we-learned-from-ubiquity/. Accessed 31 July 2012
25. Versant, db4objects API reference : native java, NET and mono open source object database, http://developer.db4o.com/Documentation/Reference/db4o-7.12/java/tutorial/index.html. Accessed 20 Apr 2012
26. Comune di Milano, Velocita' Mezzi Pubblici di Superficie, http://www.comune.milano.it/portale/wps/wcm/jsp/fibm-cdm/
FDWL.jsp?cdm_cid=com.ibm.workplace.wcm.api.WCM_Content/AreaC_RisultatiAttesi/4f919b00491b71cfb70dffae291725ad/
PUBLISHED&cdm_acid=com.ibm.workplace.wcm.api.WCM_Content/
Velocit%20mezzi%20pubblici%20di%20superficie%20al%2030%20giugno%202012%20-%20705eb6804c028fd488348f28d42ade19/705eb6804c028fd488348f28d42ade19/
PUBLISHED. Accessed 31 July 2012
27. M. Migliardi, M. Gaudina, A. Brogni, Enhancing personal efficiency with pervasive services, in *Proceedings of the Sixth International Conference on Broadband and Wireless Computing, Communication and Applications* (Technical University of Catalonia, Barcelona, Spain, 2011), 26–28 Oct 2011. doi: http://dx.doi.org/10.1109/BWCCA.2011.27
28. G. Pilato, A. Augello, G. Vassallo, S. Gaglio, Sub-symbolic semantic layer in cyc for intuitive chat-bots, in *Proceedings of International Conference on Semantic Computing, ICSC 2007*, Irvine, California, USA, 2007, pp. 121–128
29. G. Berio, M. Harzallah, Towards an integrating architecture for competence management. Special Issue: Competence Management in Industrial Processes, guest editors X Boucher, E Bonjour, N Matta, Computers in Industry **V58**(2), 2007 (2007)
30. E. Ermilova, H. Afsarmanesh, Modeling and management of profiles and competences in VBEs. J. Intell. Manuf. **V18**, 561–586 (2007)
31. J. Hodik, J. Vokrinek, J. Biba, P. Becvar, Competences and profiles management for virtual organizations creation. *Lecture Notes in Computer Science*, vol. 4696 (2007), pp. 93–102
32. K. Neshatian, M.R. Hejazi, Text categorization and classification in terms of multi-attribute concepts for enriching existing ontologies, in *Proceedings of the Second Workshop on Information Technology and Its Disciplines*, 2004, pp. 43–48
33. A. Faatz, R. Steinmetz, Ontology enrichment with texts from the www, in *Proceedings of the Second Semantic Web Mining Workshop at European Conference on Machine Learning and Principles and Practice of Knowledge Discovery in Databases*, 2006
34. G. Stumme, A.H.B. Berendt, Semantic web mining: state of the art and future directions. Web Semant.: Sci. Serv. Agents World Wide Web **4**(2), 124–143 (2006)
35. MIT Research Project: ConceptNet 5. http://conceptnet5.media.mit.edu/. Accessed 20 Apr 2012
36. Foursquare, Foursquare home page, http://www.foursquare.com. Accessed 20 Apr 2012

Viable Service Business Models Towards Inter-cooperative Strategies: Conceptual Evolutionary Considerations

P. Dohmen, N. Kryvinska and C. Strauss

Abstract Around half a decade ago, first scholars and practitioners started to think about the basic characteristics and marketing opportunities of services. These efforts underwent multiple conceptual transitions throughout the years, ultimately resulting in a new conceptualization called 'Service-Dominant Logic' (S-D Logic). Thus, the main goals of this work are twofold: on the one hand, a retrospective view on the development of viable service models is given, on the other hand, a detailed investigation of the S-D Logic evolution, into the context of inter-cooperative strategies and resources planning, is provided. Consequently, this chapter provides a retrospective view on the development of the field of service science and its viable service business models. It reviews how—commencing in the late 1970s—the field started to be developed into the contemporary era where a service-dominant (S-D) Logic has gained much scientific attention. The most important developments and debates regarding the (non-) usefulness of a differentiation strategy between goods and services in this period are highlighted. In addition, some illustrative examples regarding scientific application of the contemporary S-D Logic perspective are given.

P. Dohmen · N. Kryvinska · C. Strauss
Department of eBusiness School of Business, Economics and Statistics,
University of Vienna, Vienna, Austria
e-mail: paul.dohmen@univie.ac.at

C. Strauss
e-mail: christine.strauss@univie.ac.at

N. Kryvinska (✉)
Secure Business Austria (SBA), Vienna, Austria
e-mail: natalia.kryvinska@univie.ac.at
URL: http://www.sba-research.org/

F. Xhafa and N. Bessis (eds.), *Inter-cooperative Collective Intelligence:*   273
Techniques and Applications, Studies in Computational Intelligence 495,
DOI: 10.1007/978-3-642-35016-0_10, © Springer-Verlag Berlin Heidelberg 2014

1 Chronological Development of S-D Logic Model

The term 'service' has been subject to various interpretations during the last decades. Scientific debates, originally strongly dominated by the marketing discipline [1, 2], have been going on regarding the conceptual borders of services in comparison to goods.

The distinction between goods and services has its origin in the discipline of marketing. The [3] concisely describe of how the discipline of services marketing developed in the period between 1970 and 1990.

Consequently, this chapter provides a chronological overview regarding the distinction between goods and service business models, showing how services became an increasingly relevant scientific phenomenon. Figure 1 shows our findings of the main developments in this respect.

Tables 1 and 2 show the acceleration factors that contributed to the development of services marketing as they outlined them, divided into important events and important publications. Although others already 'planted the seeds', they describe that 1977 can be considered as a breakthrough year for services marketing. In this year, [4] published his article, in which it was first recognized and specified that the intangibility of services posed special challenges for the marketing. Fisk and colleagues outlined how some years before in 1969, Johnson already explicitly asked the question 'Are goods and services different?', thereby launching the debate that soon became the roots of an entirely new discipline [5]. In 1983, Christopher Lovelock continued this view by arguing that above the differences between goods and services, services could be divided into five heterogeneous groups, followed by his book on service marketing the year thereafter [6, 7].

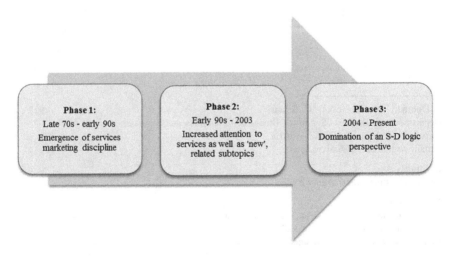

Phase 1:
Late 70s - early 90s
Emergence of services marketing discipline

Phase 2:
Early 90s - 2003
Increased attention to services as well as 'new', related subtopics

Phase 3:
2004 - Present
Domination of an S-D logic perspective

Fig. 1 The trace towards and development of service-dominant logic

Table 1 Factors contributing to the development of the services marketing discipline: major events (adapted from [3])

Year	Event
1977	Marketing Science Institute established the consumer services marketing research program and published is first report from the program
1981	American Marketing Association held the first national conference on services marketing in Orlando, Florida
1984	American Marketing Association established a separate services marketing division
1985	Arizona State University established the First Interstate Center for services marketing
1985	American Marketing Association held is first faculty consortium on services marketing at texas A&M University
1988	University of Karlstad, Sweden, hosted the first Quality in Services (QUIS) symposium, a multi-disciplinary, multinational research conference held every other year
1990	Universite d'Aix-Marseille, France, sponsored the first international research seminar in service management. Researchers from multiple disciplines and countries presented papers on various service-organization issues

One year later, Zeithaml and colleagues published an article, in which they aimed to "... offer a conceptual framework summarizing the unique characteristics of services, the problems stemming from these characteristics, and the strategies suggested as appropriate to overcome the problems" (cf. [8]). In this work, they also introduced and treated the IHIP framework extensively, which will be discussed later on. As described by Fisk et al. [5], in the years thereafter one witnessed an 'explosive growth of books, journal publications, conference proceedings and dissertations' regarding the nature and treatment of services, in which attention was paid to 'new' topics like service quality, service design, customer retention and relationship marketing. Table 3 shows an overview by Fisk et al. [5] of the number of articles about services that were published in the period between 1953 and 1992.

In 2004, Vargo and Lusch [9] proposed to consider all goods as services, and urged scholars and researchers to take a more service-dominant view. Because, in fact, every exchange or economic action fundamentally results in a form of service provision, they recommended abandoning strategies where it is distinguished between services and goods. Despite the fact that this view proved to possess internal consistency as well as to bring about interesting perspectives, Stauss [10] describes this shift as a so-called *Pyrrhic victory*, where an apparent victory (acceptance of the proposed view) might in practice equate a severe loss (the abandoning of a complete, relevant field of research).

Thus, why broadening the services perspective and abandoning this field's relevant lines of enquiry and research contributions could be necessary is exhibited in the form of six (potentially) problematic implications [10], as follows:

1. The all-embracing, broader definition associated with the view of Vargo and Lusch is undesirable, since such generalizations cause the opposite of additional insights: "A general definition of service that includes virtually everything defines virtually nothing" (cf. [10]).

Table 2 Factors contributing to the development of the services marketing discipline: major publications (adapted from [3])

Year	Publication
1974	George and Barksdale: marketing activities in service industries, *Journal of Marketing*
1976	Levitt: the industrialization of service, *Harvard Business Review*
1977	Kotler and Conner: marketing professional services, *Journal of Marketing*
1977	Shostack: breaking free from product marketing, *Journal of Marketing*
1978	Chase: where does the customer fit in a service operation?, *Harvard Business Review*
1978	Thomas: strategy is different in service industries, *Harvard Business Review*
1980	Berry: services marketing is different, *Business*
1981	Levitt: marketing intangible products and product intangibles, *Harvard Business Review*
1983	Lovelock: classifying services to gain strategic marketing insights, *Journal of Marketing*
1985	Parasuraman, Zeithaml and Berry: a conceptual model of service quality and its implications for future research, *Journal of Marketing*
1985	Solomon et al.: a role theory perspective on dyadic interactions: the service encounter, *Journal of Marketing*
1985	Zeithaml, Parasuraman and Berry: problems and strategies in services marketing, *Journal of Marketing*
1987	Shostack: service positioning through structural change, *Journal of Marketing*
1988	Parasuraman, Zeithaml and Berry: SERVQUAL: a multiple-item scale for measuring customer perceptions of service quality, *Journal of Retailing*
1988	Zeithaml, Berry and Parasuraman: communication and control processes in the delivery of service quality, *Journal of Marketing*

Table 3 Overview of "service" articles published in the period between 1953–1992 (adapted from [5])

Source	Number of "service"
Journal of Services Marketing	84
International Journal of Service Industry Management	41
Harvard Business Review	34
Service Industries Journal	30
Journal of Marketing	25
Business Horizons	22
Industrial Marketing Management	22
Journal of Retailing	19
Journal of the Academy of Marketing Science	16
European Journal of Marketing	15
Journal of Business Research	15
Journal of Professional Services Marketing	15
Sloan Management Review	15
Journal of Business Strategy	13
Academy of Management Review	10
Quarterly Review of Marketing	6
Journal of Consumer Marketing	5
Journal of Consumer Research	5
Journal of Marketing Research	5

2. Since production of physical goods differs manifestly from production of services, characterizing all goods as services simply because customer value is created with goods is highly undesirable, from a theoretical as well as from a practical perspective.
3. There are inseparable services (though not all), where production and consumption necessarily have to take place simultaneously [11]. Equating goods and services would imply that this important differentiation opportunity for such services goes unrecognized.
4. The same argumentation counts for the use of the term 'relationship'. Blithely using this term to refer to goods' transactions would deny the specific importance it has in a service context, where relationships with customers are far more than just transactional [10].
5. Modern service economy is not solely undergoing a process of 'servitization', but shows trends in opposite direction as well, with production oriented logics being applied to service industries.
6. So-called service specific knowledge developed because services' characteristics were so distinct to those of goods, posing specific challenges. Losing this distinction will therefore imply that dealing with challenges on the basis of relevant recommendations and contributions from the past will no longer be possible [10].

Since both streams that are described above might have put forward valid arguments, the truth might actually be 'somewhere in between'. Facts are that service industries have been rapidly expanding and in recent decades, the economy has been shifting towards a service-based economy [12–14]. The expressions 'servicizing products' and 'productizing services' illustrate that boundaries between goods and services are blurring indeed. Products are increasingly offered together with value adding services, which indicates that there is a nested relationship between goods and services in which so-called bundled solutions are offered [13, 44]. Hence, differentiating between goods and services becomes increasingly complex.

The next section shows how the Intangibility-Heterogeneity-Inseparability-Perishability (IHIP) framework was used in order to delineate services from goods, and why so many authors considered this framework from a skeptical point of view.

2 The Role of Intangibility-Heterogeneity-Inseparability-Perishability Framework in the Separation of G-D and S-D Logics

As highlighted in the previous section, it is pointed towards crucial distinctions regarding the characteristics of services when defending the field of service research. In the discussions, it was often referred either explicitly or implicitly to the IHIP framework, which is an abbreviation of four typical service characteristics: intangibility, heterogeneity, inseparability and perishability [2, 11]. The next paragraphs will explain this framework in more detail, together with arguments as to why a critical view towards these factors might be important.

2.1 IHIP Characteristics and Its Critics

During the second phase as represented in Fig. 1, it started to become clear that the boundaries between goods and services were blurring. Nevertheless, services kept often being differentiated from goods in the literature. The most important framework in this respect was the IHIP framework. Although it is acknowledged in the literature, that there has never been any scientific justification for its characteristics, the framework has been widely used [12].

The characteristics of the IHIP framework were already mentioned half a century ago by Regan [15]. In a thorough evaluation regarding the origins of this framework, Lovelock and Gummesson [16] elaborate on how, after various literature reviews and numerous citations, this framework has been accepted as a key source of wisdom in the service (marketing) literature. The framework basically consists of following elements: *intangibility* refers to the fact that services are not like physical goods that can be perceived with basic human senses, implying a high level of experience: "The idea is that services are activities that cannot be touched, for example, the service of an opera performance" (cf. [12]). Two other characteristics result from this: *perishability*, which implies that services (activities and processes) cannot be stored, as well as *inseparability*(earlier in this chapter called 'simultaneity of production and consumption'), which implies that services are produced and consumed at the same time [17]. Inseparability exists between production, delivery and consumption, and typically takes place with the presence of the customer [12]. Finally, a distinct characteristic of services is their *heterogeneity*, which means that services "differ regarding their quality across time, organizations and people" (cf. [17]). Because services are usually performed by humans and not by machines, heterogeneity results from the non-existent need to standardize them, in contrast to (automized) production of standardized, homogeneous goods [12].

The impact of the IHIP framework during the 1980s and 1990s, is emphasized in detail by Edgett and Parkinson. In the introduction of their article in 1993, where they state: "It is now generally accepted that the marketing of services is sufficiently distinctive from the marketing of physical products to deserve separate treatment. ...the majority of scholars now accept that the debate is over." (cf. [18]). These words show that, at least in the marketing discipline, consensus existed regarding a separate treatment of goods and services during this time.

The Refs. [8] and [18] review the literature in order to discover what studies addressed the different characteristics from the IHIP framework. Table 4 shows the number of IHIP related studies identified by only [8] (1st column), only [18]

(2nd column) and the overlapping findings, mentioned by both of them (3rd column). (*For detailed information regarding their findings please see the original work by both authors*).

A critical attitude towards the validity and applicability of these characteristics again brings us back into the debate that was described above: do goods really differ from services? Lovelock and Gummesson [16] argue that this claim only holds for certain types of services, thereby rejecting the IHIP characteristics as a contemporary valid framework. They call for a critical stance, even scepticism towards the framework, arguing that not all characteristics of the IHIP framework are applicable to all types of services, because there are "sufficient exceptions to discredit the claim of universal generalizability" (cf. [16]). Vargo and Lusch [2] argue that due to the inaccuracy of the IHIP definitions and contradictory implications, the IHIP framework fails to delineate goods and services adequately. They conclude that "a strategy of differentiating services from goods should be abandoned" (cf. [2]). Hence, in the remainder of this chapter it is must be taken into consideration that, whether goods and services can be distinguished, is an underlying discussion that has been an important dispute in the past decades. In this respect, an interesting—but of course also debatable—recommendation seems to be the one by Edvardsson et al. [11], who choose to strike a balance between the different opinions: "We should not generalize the characteristics to all services, but use them for some services when they are relevant and in situations where they are useful and fruitful" (cf. [11, 44]).

2.2 Implications for the Definition of Services

As a consequence of the debate above, it follows that defining the term 'service' is a complicated task. Not that there is a lack of definitions, on the contrary. Sixteen experts were asked to come up with a definition of service in the research of [11]. Although similarities could be identified in the independent responses, for example the keywords 'performance' and 'processes' were mentioned in about half of the responses, it turned out that all definitions were on an abstract level, open to interpretation at an operational level. In fact, almost all experts responded that it does not make much sense to define services in one or two lines. Some authors referred to services as 'deeds, processes and performances', which, in turn, is maybe the best way to describe services, albeit in a very general way [11]. Maglio

	Period	Number of findings [8]	Number of findings [18]	Number of joint findings
Table 4 Synthesis of studies dealing with IHIP between 1963–1993 (adapted from [8, 18])	1963–1973	1	5	3
	1974–1978	6	5	12
	1979–1984	11	31	12
	1985–1993	–	38	–

and Spohrer follow the definition proposed by Vargo and Lusch, in which services are considered to be the application of resources for the benefit of another" (cf. [19]). Although these definitions might relatively well capture the meaning of services, it remains important to keep in mind that the concept of services is hugely debated, and therefore every definition is subject to interpretation within each respective specific context [2].

An important additional remark regarding terminological issues related to service perspectives is whether one is talking about 'service' or 'services'. According to Vargo [20] and Vargo and Lusch [22], the plural 'services' is typically a Goods-Dominant representation (see Sect. 3 for more detailed information), whereas the singular 'service' reflects the Service-Dominant perspective. Hence, 'service economies' depict economies in which the resources of one party are used for the benefit of another. In such situations, "the locus of value creation moves from the 'producer' to a collaborative process of co-creation between parties" (cf. [20, 21]), as will be explained more detailed in the next paragraph.

The next sections will look at how services have become the basis of a whole new perspective in research and business, as represented with phase 3 in Fig. 1.

3 Service Science Perspectives: Service-Dominant versus Goods-Dominant logic

3.1 Resource-Based Definition of Service Logic

In line with the previously sketched debate regarding the role of services, Lusch et al. [23] distinguish two conceptually differing logics that might underlie the field of service science. On one hand, there is a goods-dominant logic ('G-D Logic'), on the other hand there is a service-dominant logic ('S-D Logic'). G-D logic is based on the essence "that economic exchange is fundamentally concerned with units of output that are embedded with value during the manufacturing process" (cf. [20]). In contrast, in S-D logic the focus of value creation "moves from the producer to a collaborative process of co-creation between parties" (cf. [20]). As obvious proponents of the second logic, it must of course be realized that the mentioned authors belong to the group of scholars that propose to adhere to the approach where no differentiation is made between goods and services, as explained earlier. This should be kept in mind while reading the next section, where the main differences between the two logics according to Lusch et al. [23] are summarized.

G-D logic is mainly based upon an orientation toward so-called *operand resources*: "Operand resources are those that are acted upon; they are static and usually inert" (cf. [23]). Such resources are typically physical [25]. As a result, the production of (tangible) goods with these resources is the main focus of companies, since producing and selling those goods is how value is created. Customers have to be targeted more or less in a one-way direction, through promotion of these goods. Firms have a relatively static view of the world, with linear supply chains, utility-maximizing consumers and profit-maximizing firms [23, 24].

In contrast, S-D logic presumes an orientation toward *operant resources*: "Operant resources are often intangible (e.g., knowledge and skills) and are capable of acting on operand resources and even other operant resources to create value." (cf. [23]). These 'human' resources are not inherently valuable in this perspective, but become valuable after application of operant resources: they have to be turned into benefits [25]. This happens under interaction of firms and customers, which are actively involved in transforming the inputs into value.

In the literature, a threefold hierarchical classification among operant resources has been developed by [25]. First, basic operant resources, like for example the skills and knowledge of individual workers, form the building blocks of higher-order resources. Second, composite operant resources, combine two or more basic operant resources, thereby enabling firms to create market offerings more effectively and/or efficiently. Third, interconnected operant resources are similar to composite operant resources, with the only difference that they are based on lower order resources significantly interacting and reinforcing each other. Such interaction enables then additional positive effects for the organization. The rationale behind the hierarchy in general, is that if one moves up in the hierarchy, resources become increasingly connected as well as more difficult for competitors to acquire or develop, which implies that a firm possesses greater potential for sustainable competitive advantages [25]. Figure 2 graphically represents the overview of the described resources classification under G-D logic and S-D logic.

Hence, according to S-D logic, the customer is seen as a co-creating collaborative partner of the firm, responsible for the creation of value [23, 24]. Since 2004, S-D logic has become fundamental to service science, serving as its philosophical foundation [19]. Considering the increasing importance of this

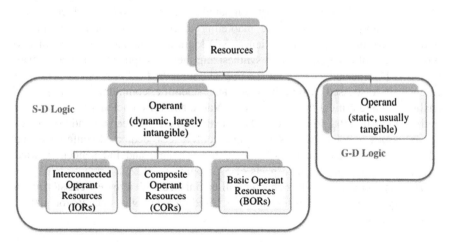

Fig. 2 Schematic representation of the distinction between operant and operand resources (adapted from [9, 25])

"emerging revolution in business and economic thinking in the twenty-first century", it is relevant to have a closer look at how this philosophy is used as a foundation in practical research (cf. [26]).

3.2 S-D Logic and General Theory

The Brodie and Saren [27] relate the development of S-D Logic to middle range as well as general theory. Middle range theory can have the important function of being a theoretical bridge between empirical findings and general theory, which is broader in scope and more abstract [27]. If S-D Logic based middle range theory is able to fulfill the function of being able to focus on a subset of phenomena relevant to the particular context at hand, empirical results can be translated to general theory. Because middle range theory development helps to show explicit links between the process of theory formulation and verification, the resulting general theory enjoys a stronger foundation and justification: "... middle range theories play an important role in the 'scientific circle of enquiry' in providing a bridge between general theory and empirical findings both in the context of discovery and the context of justification" (cf. [27]). In this way, S-D Logic has the chance to guide the development of an empirical research agenda and thereby develop from an emerging paradigm into a higher order general theory.

4 Case Studies of Empirical Applications of S-D Logic

In this section, several examples of empirical applications of S-D logic will be provided. It will become clear that compared to earlier periods, in which service research belonged mainly to the marketing discipline (as emphasized in the introduction), in contemporary science, S-D logic is at the basis of research in a broad variety of disciplines, often synthesizing these disciplines in a creative way [1, 28]. S-D logic and its position and impact within the field of service science are extensively discussed within business and scientific communities, often in a challenging and critical fashion (e.g. [1]). Vargo and Lusch noted 4 years after the initial publication that introduced S-D logic that "there has been a steady stream of special issues, special sessions at leading conferences, dedicated conferences, an edited book with contributions from 50 top scholars, and independent journal articles dealing with various aspects of S-D logic" (cf. [22]).

Thus, the remainder of this chapter will be finished with a review classifying an exemplary variety of such research that is based on S-D logic, since its foundation was developed in 2004.

4.1 S-D Logic, Branding and Networks

Under S-D logic, the function and role of brands might change. The Brodie [29] argues that brands are at the basis of value adding processes that are so important in the S-D perspective, thereby contributing to the value perception of the customer. New logic in this field "acknowledges that brand value is co-created between the firm and its stakeholders" (cf. [30]). The brands facilitate and mediate processes that are used for realizing experiences that drive co-creation of value [29]. Hence, branding in this context has to be seen as a dynamic and social process that is stretched over a broad context. As a result, in today's service-based economies, investing into strong branding relationships becomes increasingly important [30].

In general, relationships between social and economic actors are at the basis of networks. Such networks allow for quality interaction (e.g., in the form of high levels of trust), which serves in turn as the key ingredient for successful co-creation of value under S-D logic [31].

4.2 S-D Logic and Discontinuous Innovations

S-D logic might also help to explain discontinuous innovations. The [32] term innovations discontinuous if they (1) change how value is created and (2) significantly affect market size, prices, revenues or market shares. Employing S-D logic to the concept of innovations requires firms to look beyond the traditional focus on value-in-exchange. There are two reasons for this. First, in discontinuous innovations, the role of the customers changes into a co-creator of value. Second, the firm's value creation is changed in the case of discontinuous innovations. Value creation is now based on operant resources (as explained earlier) like skills, knowledge and competencies, whereas the customer acts as an integrator of these resources [32]. Hence, the S-D perspective also influences and changes (conventional) approaches to innovation.

4.3 S-D Logic and Co-Produced Knowledge

The Vargo and Lusch [9, 22] argue that in an S-D perspective, the customer is always co-creating value in an interactive way. At the same time, this jointly created knowledge between companies and customers serves as a fundamental source of competitive advantage [33]. The Blazevic and Lievens [34] show how S-D logic can contribute to the management of this knowledge. They investigated the role customers play in knowledge creation, and were able to identify three roles differing in the extent to which knowledge was actively coproduced: customers could act as passive users, active informers or bidirectional creators. Customers are acting as exchange partners for joint knowledge creation in all these three cases, however to a different extent and with different subsequent influences on the innovation tasks and processes [34]. Hence, their research underscores that the coproduction of knowledge, which is typical within an S-D context, occurs in diverse fashions and with dissimilar outcomes. The Payne et al. [33], therefore, stress the importance of knowledge as a key operant resource. A deep understanding of customer experiences and processes is crucial in order to understand the dynamics of successful knowledge co-creation under S-D logic.

4.4 S-D Logic and Social Construction Theories

Instead of focusing on the value-creation process as such, Edvardsson et al. [35] focused on the social setting in which the co-creation of value occurs. In their exploratory study, they applied key concepts from social construction theories to S-D logic. Social construction theories are used to interpret the social world and the behavior of actors within this environment. In the context of S-D logic, this means that by using such theories, understanding about service systems and value creation within these systems can be enhanced [35].

Figure 3 shows how [35] depict this situation. Two actors, the firm and the customer, act as resource integrators that mutually try to create value. Both parties are embedded in wider networks that play an important role in the service exchange, and can be considered as service systems [20, 22]. Because this service

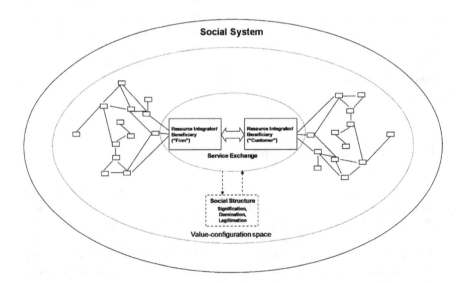

Fig. 3 S-D Logic and social construction theories [35]

exchange takes place within a wider social system, the actors draw upon the rules and resources ('social structures') within this context. Social structures enable and constrain, and hence, influence, the service exchange in an unobservable and implicit way: "Social structures are expressed through the norms, values and ethical standards that guide what is acceptable during interactions between individuals, which has implications for service exchange and value co-creation" (cf. [35]). This shows that consolidating social construction theories with S-D logic might also help to advance knowledge within and understanding of the dynamics associated with service-dominant thinking.

4.5 S-D Logic and Logistics Service Value

Successful co-creation of knowledge and mutual development of value propositions, are S-D logic based activities that might help to create logistic service value, according to Yazdanparast et al. [36]. The authors specify how, on the basis of S-D logic, logistics practitioners can increase the co-creation of value in such way that it may lead to competitive advantages. It is emphasized that particularly the field of logistics, typically characterized by its dynamically changing service offerings, might benefit from the application of S-D logic theory [36].

4.6 S-D Logic and Procurement

In their discussion on the implications of 'servitization' of procurement, Lindberg and Nordin [37] reach a two-sided conclusion when looking at the perspectives of industrial buyers. On the one hand, a clear movement towards S-D logic can be identified, since buyers acknowledge that they are increasingly looking for the purchase of entire solutions rather than only products. On the other hand, a counteractive movement towards objectification of services seems equally apparent: buyers are trying to objectify and (partly) standardize the services they are buying, which is more in line with G-D logic. Hence, regarding the interplay between procurement and S-D logic, they conclude that many buyers are "employing buying behaviour that is in line with both kind of logics" (cf. [37]).

4.7 S-D Logic and Service Experience

A key concept within the S-D Logic paradigm is that of a 'service experience'. This concept refers to how we, consciously or unconsciously, experience services. Following up on the statement by Vargo & Lusch that service experience is the foundation of all business, Helkkula shows that the way in which customers perceive and engage in the co-creation of value is a key question under S-D Logic [38].

Service experiences, often incorrectly reduced to only hedonic experiences, are relational, social and inter-subjective at the same time. Because under S-D Logic stakeholders are simultaneously involved in the co-creation of value, they continuously experience the social phenomenon called service experience [38]. The concept underlies an interesting paradox. On the one hand, in line with S-D Logic, services and the corresponding experiences are to be considered holistically rather than as process-and-outcome components [39]. On the other hand, the service experience is unique to each individual, due to the specificity of time, location, context and content of the experience [38]. As a central concept within S-D Logic thought, future research on the phenomenon is needed in order to develop the concept further and relate it more closely to other concepts like for example (the co-creation of) value.

4.8 S-D Logic and Competitive Advantage

The Lusch et al. [24] also propose to consider economic issues through an S-D lens. They come up with nine propositions about how to achieve competitive advantage on the basis of services. In an environment, in which stakeholders in the market place are viewed as operant resources [24], collaboration and knowledge application are at the ways in which value (and possibly competitive advantage) is created. Value creation is subjective here, and other stakeholders in the market act as co-creators of it. Hence, on the basis of this, they formulate how managers can achieve competitive advantage from an S-D point of view: "...the most fundamental implication is that firms gain competitive advantage by adopting a business philosophy based on the recognition that all entities collaboratively create value by serving each other" (cf. [24]) (for a detailed investigation on operant resources that are important in value creation, see [25]).

4.9 Cloud Computing as an Illustrative Example of S-D Logic

The earlier described trend towards the 'servitization' of goods might well be illustrated by the example of cloud services. Defined as "a computing mode in which applications, data and IT resources are provided as service for users over the Internet", cloud computing has become an interesting information technology provisioning approach for many firms (cf. [40], p. 145). Cloud computing has even proven to possess the potential to leverage increased value for all involved stakeholders in certain industries [41]. The locus of value creation no longer resides within the boundaries of the firm. Instead, co-creation of value typically takes place between all the actors in the network [42]. This typical S-D Logic characteristic indicates that cloud computing might be an excellent contemporary illustration of the emerging service dominated mind-set.

Cloud services must not be restricted to the simple moment of operative use. Such a narrowed view, treating cloud computing as a simple, uni-directional activity would correspond with G-D Logic [43]. This would be incorrect, since the full life cycle of cloud services have to be taken into account if one discusses the value creation that such technologies are able to leverage. Taking such an analytical perspective, clouds can be considered as platforms for value co-creation, in which the common effort of two or more service systems are combined. The Schmidt [43] describes how value creation can take place in each of the various interactive stages of the cloud-service lifecycle:

- First, cloud services have to be *integrated* into the client organization by granting the users access to the service and its resources;
- Second, the user has to provide data to the cloud so that the cloud service can be *configured*;
- Third, before the consumer can request the service, resources to be operated upon by the cloud have to be *assigned*;
- Fourth, the actual service *operation* takes place;
- Fifth, the customer can make sure resources passed over are returned;
- Sixth, the cloud services are removed from the consumer's computing environment, thereby being fully *disintegrated*.

In each of the above described interactions, co-creation of value occurs. Hence, it shows how people, technology, internal and external service systems jointly co-produce value throughout the lifecycle of a cloud service, thereby being an excellent illustration of the S-D Logic paradigm [43]. Future research on the specific dynamics and potential for maximized value creation is necessary in order to increase understanding of the way in which S-D Logic helps to leverage the benefits of cloud computing.

5 Conclusions and Future Work

S-D logic is posited as an open source evolution [20, 22] that is "always open for further elaboration, refinement and development" (cf. [35]) throughout its discourse.

Accordingly, the purpose of this article was to trace back the roots of S-D logic, including the important debate regarding a possible distinction between services and goods, as carried out during the last decades. Subsequently, it was exposed that since 2004, S-D logic has become an emerging foundation in the field of service science.

Besides, the selection of exemplary applications of S-D logic in recent literature given at the end of this work serves as an illustration of how S-D logic is still applied within the field of marketing but also in an increasing amount of other areas.

Future suggestions and recommendations regarding S-D logic must be grounded in attempts to either strengthening its foundations, thereby developing S-D logic as a solid, general theory, or in attempts to extend the service-centered mindset, ideally by collecting (further) empirical evidence. The further advice comprises a sheer endless amount of application areas, as revealed in this chapter. To recall, it was already shown that the foundational premises and core assumptions of S-D logic theory were studied in an increasing variety of areas (far beyond marketing), like supply chain management, logistics, (marketing) ethics, procurement, branding, organizational studies, social sciences, policy regulators, etcetera. A complementary shift towards more technical areas like (e.g., e&m-Business, e-Governance, e-Health, IT Service Systems, Business Services and Processes Management), where co-creation of value plays a central role, might also yield valuable insights. S-D logic possesses the potential to provide a reoriented perspective on the market, for all stakeholders involved [12, 44]. Ultimately, S-D logic might even be able to live up to its extremely ambitious role of, to say it in the words of its creators Vargo and Lusch, "providing a basis for reorienting theories of society and economic science" (cf. [20, 22]).

References

1. B. Stauss, International service research—status quo, developments, and consequences for the emerging services science. Serv. Sci. 57–70 (2007)
2. S.L. Vargo, R.F. Lusch, The four service marketing myths: remnants of a goods-based, manufacturing model. J. Serv. Res. **6**(4), 324–335 (2004a)
3. L.L. Berry, A. Parasuraman, Building a new academic field—the case of services marketing. J. Retail. **69**(1), 13–60 (1993)
4. G.L. Shostack, Breaking free from product marketing. J. Mark. **41**(2), 73–80 (1977)
5. R.P. Fisk, S.W. Brown, M.J. Bitner, Tracking the evolution of the services marketing literature. J. Retail. **69**(1), 61–103 (1993)
6. C.H. Lovelock, Classifying services to gain strategic marketing insights. J. Mark. **47**(3), 9–20 (1983)
7. C.H. Lovelock, *Services Marketing: Text, Cases and Readings* (Prentice-Hall, Englewood Cliffs, N.J., 1984)
8. V.A. Zeithaml, A. Parasuraman, L.L. Berry, Problems and strategies in services marketing. J. Mark. **49**(2), 33–46 (1985)
9. S.L. Vargo, R.F. Lusch, Evolving to a new dominant logic for marketing. J Mark. **68**(1), 1–17 (2004b)
10. B. Stauss, A pyrrhic victory: the implications of an unlimited broadening of the concept of services. Managing Serv. Qual. **15**(3), 219–229 (2005)
11. B. Edvardsson, A. Gustafsson, I. Roos, Service portraits in service research: a critical review. Int. J. Serv. Ind. Manage. **16**(1), 107–121 (2005)
12. E. Gummesson, R.F. Lusch, S.L. Vargo, Transitioning from service management to service-dominant logic: observations and recommendations. Int. J. Qual. Serv. Sci. **2**(1), 8–22 (2010)

13. K. Nam, N.H. Lee, Typology of service innovation from service-dominant logic perspective. J. Univ. Comput. Sci. **16**(13), 1761–1775 (2010)
14. R.C. Larson, Service science: at the intersection of management, social, and engineering sciences. IBM Syst J. **47**(1), 41–51 (2008)
15. W.T. Regan, The service revolution. J. Mark. **27**(3), 57–62 (1963)
16. C. Lovelock, E. Gummesson, Whither services marketing? J. Serv. Res. **7**(1), 20–41 (2004)
17. C. Sichtmann, I. Griese M. Klein, *Determinants of the International Performance of Services: A Conceptual Model*. (Diskussionsbeiträge des Fachbereichs WiWi der freien Universität, Berlin, 2007), pp. 1–20
18. S. Edgett, S. Parkinson, Marketing for service industries: a review. Serv. Ind. J. **13**(3), 19–39 (1993)
19. P.P. Maglio, J. Spohrer, Fundamentals of service science. J. Acad. Mark. Sci. **36**(1), 18–20 (2007)
20. S.L. Vargo, From goods to service(s): divergences and convergences of logics. Ind. Mark. Manage. **37**(3), 254–259 **(The Transition from Product to Service in Business Markets)**
21. S. Alter, Does service-dominant logic provide insight about operational IT service systems? *AMCIS 2010 Proceedings, Paper 138* 2010
22. S.L. Vargo, R.F. Lusch, Service-dominant logic: continuing the evolution. J. Acad. Mark. Sci. **36**, 1–10 (2008)
23. R.F. Lusch, S.L. Vargo, G. Wessels, Toward a conceptual foundation for service science: contributions from service-dominant logic. IBM Syst J. **47**(1), 5–14 (2008)
24. R.F. Lusch, S.L. Vargo, M. O'Brien, Competing through service: insights from service-dominant logic. J. Retail. **83**(1), 5–18 (2007)
25. S. Madhavaram, S.D. Hunt, The service-dominant logic and a hierarchy of operant resources: developing masterful operant resources and implications for marketing strategy. J. Acad. Mark. Sci. **36**(1), 67–82 (2007)
26. P.P. Maglio, S.L. Vargo, N. Caswell, J. Spohrer, The service system is the basic abstraction of service science. Inf. Syst. e-Bus. Manage. **7**(4), 395–406 (2009)
27. R.J. Brodie, M. Saren, J. Pels, Theorizing about the service dominant logic: the bridging role of middle range theory. Mark Theory **11**(1), 75–91 (2011)
28. R.J. Glushko, Designing a service science discipline with discipline. IBM Syst. J. **47**(1), 15–27 (2008)
29. R.J. Brodie, From goods to service branding: an integrative perspective. Mark. Theory **9**(1), 107–111 (2009)
30. M.A. Merz, Y. He, S.L. Vargo, The evolving brand logic: a service-dominant logic perspective. J. Acad. Mark. Sci. **37**(3), 328–344 (2009)
31. A. Fyrberg, R. Jüriado, What about interaction?: networks and brands as integrators within service-dominant logic. J. Serv. Manage. **20**(4), 420–432 (2009)
32. S. Michel, S.W. Brown, A.S. Gallan, An expanded and strategic view of discontinuous innovations: deploying a service-dominant logic. J. Acad. Mark. Sci. **36**(1), 54–66 (2007)
33. A.F. Payne, K. Storbacka, P. Frow, Managing the co-creation of value. J. Acad. Mark. Sci. **36**(1), 83–96 (2008)
34. V. Blazevic, A. Lievens, Managing innovation through customer coproduced knowledge in electronic services: an exploratory study. J. Acad. Mark. Sci. **36**(1), 138–151 (2007)
35. B. Edvardsson, B. Tronvoll, T. Gruber, Expanding understanding of service exchange and value co-creation: a social construction approach. J. Acad. Mark. Sci. **39**(2), 327–339 (2011)
36. A. Yazdanparast, I. Manuj, S.M. Swartz, Creating logistics value: a service-dominant logic perspective. Int. J. Logistics Manage. **21**(3), 375–403 (2010)
37. N. Lindberg, F. Nordin, From products to services and back again: towards a new service procurement logic. Ind. Mark. Manage. **37**(3), 292–300 (2008)
38. A. Helkkula, Characterising the concept of service experience. J. Serv. Manage. **22**(3), 367–389 (2011)
39. S. Schembri, Rationalizing service logic, or understanding services as experience? Mark. Theory **6**(3), 381–392 (2006)

40. Qing, L., & Chun, Y. (2010). Development trends of MIS based on cloud computing environment. *The Third IEEE International Symposium on Information Science and, Engineering*, pp. 145–148

41. A. Mladenow, E. Fuchs, P. Dohmen, C. Strauss, *Value Creation Using Clouds: Analysis of Value Drivers for Start-Ups and Small and Medium Sized Enterprises in the Textile Industry.* The 26th IEEE International Conference on Advanced Information Networking and Applications (AINA 2012), Fukoka, Japan, March 26–29 2012

42. G. Kontos, K. Kutsikos, A service classification framework for value co-creation. The case of the Internet of services, in *Proceedings of the 2011 Naples Forum on Service*, ed. by E. Gummesson, C. Mele, F. Polese. Service Dominant Logic, Network and Systems Theory and Service Science. Napoli, 2011

43. R. Schmidt, Value-co-creation in cloud-computing, in *Proceedings of the 2011 Naples Forum on Service*, ed. by E. Gummesson, C. Mele, F. Polese, Service Dominant Logic, Network and Systems Theory and Service Science. Napoli, 2011

44. S. Moussa, M. Touzani, A literature review of service research since 1993. J. Serv. Sci. **2**, 173–212 (2010)

Discovering the Hidden Semantics in Enterprise Resource Planning Data Through Formal Concept Analysis

Martin Watmough

Abstract Enterprise Resource Planning (ERP) systems contain massive amounts of data that is frequently under utilised. This chapter presents a method for discovering useful and semantic data in a practical manner that demonstrates the potential applications of this approach. Data extracted directly from a productive ERP system forms the use case and represents the challenges of handling large volumes along with the variations and noise associated with real data. ERP systems support the core and many other business functions. Enterprises invest significant resources into systems during implementation, ongoing maintenance and actual use. These systems control operations through to integrating with business partners and should be leveraged for any competitive advantage available. A frequently overlooked source of data are the user transaction logs as they offer an insight into the actual patterns of use; this is the use case for this chapter. Formal Concept Analysis (FCA) has been applied in order to discover semantics and relationships from transactional data. This chapter describes the method applied alongside examples of the analysis in order to support the theory and demonstrate a practical application. The findings from the analysis are discussed, and it is explicated how semantic discovery from ERP is a useful exemplar for the Internet of Things and ERP development.

1 Introduction

Modern Enterprises are complex and rely heavily on people and electronic systems to control and manage their operations. Enterprise Resource Planning (ERP) systems are central to many Enterprises as they provide an integrated and best-practice set of

M. Watmough (✉)
Sheffield Hallam University, Sheffield, United Kingdom
e-mail: Martin.Watmough@Ciber.com

F. Xhafa and N. Bessis (eds.), *Inter-cooperative Collective Intelligence: Techniques and Applications*, Studies in Computational Intelligence 495, DOI: 10.1007/978-3-642-35016-0_11, © Springer-Verlag Berlin Heidelberg 2014

processes coupled with control and governance. A significant problem is that market forces make constant change inevitable; systems must be adaptable to meet the information needs of the Enterprise. Mobile devices are now capable of communicating audibly and visually, almost on par with any fixed location device. They are also capable of functioning, sensing and communicating without human input; the potential range of information available to an Enterprise is creating significant challenges about how to handle and more importantly benefit from this new environment. The main constant is ERP as it traditionally represents the core processes of an Enterprise and forms a central repository to support critical functions such as the financial statement. Any new method of analysis requires a framework that is equally flexible to these needs.

The predominant trend within ERP solutions has been for the design and architecture to be created by process experts. This tenancy to manually design processes and reporting solutions makes it difficult to conquer the challenges of increasing data volumes, process diversity and the range of interactions.

Approaches such as Business Process Management (BPM) are being applied to provide Enterprises with more responsiveness through flexible and graphical modelling tools. These tools support process design in a flow chart format and the technical capability to create the programs necessary to operate and integrate with other systems. This ability to model processes quickly is enhanced by Service Orientated Architecture (SOA), an approach that is focussed on the integration and interoperability of systems. These approaches have the capability to address the needs of the Enterprise however there is a gap—the ability to understand and analysis highly variable processes. This highlights the need to employ techniques that discover information and semantics in order to gain knowledge, insight and pro actively apply it.

This view of ERP systems is representative of many organisations at the present time but it does not encapsulate the challenges and solutions currently being applied or developed. Debevoise defines the key drivers as visibility, organisation alignment and adaptation [1], a theme that is continued in this chapter and echoes the sentiments of exceptions becoming the normal mode of operation coupled with localised processes and varying data. This is not intended to suggest that core processes are not controlled, it is intended to highlight how processes can be refined locally to reflect the interaction with local partners. The term process within the context of this subject is used to describe a planned sequence of events in order to achieve an outcome and represents the combination of humans and systems. The manner of interaction is expected to differ due to the availability of information and the ability to communicate through a wide range of devices; this includes input from sensors, connected system or even automated decision making.

There is no shortage of data however identifying good information, managing and understanding it is an enormous challenge. Improving understand through semantics could provide better insight and improved decision making. Semantics and context are critical considerations, this is supported according to Wille by Peirce's pragmatism which claims that we can only analyse and argue within restricted contexts where we always rely on pre knowledge and common sense [2]. Semantics is, as McComb states : Semantics = Data + Behaviour [3], or put another way semantics is about meaning. Context considers time, location and historical dimensions, it may change for any given event or user perspective. The definition used here for semantics focusses predominately on the canonical meaning and is less time dependant than context. Making use of semantics and context is certainly a multilevel problem that is not restricted to a unique event or object, it is equally applicable to a sequence of events or a situation.

A method for discovering information and semantics is described in this chapter based on data collected directly from an ERP system employed by a real Enterprise; all data has been anonymised. User interaction with the system has been captured and processed using Formal Concept Analysis (FCA), as introduced by Rudolph Wille and Bernhard Ganter [4] as the mechanism for semantic discovery. The method applied is repeatable and not constrained to ERP as a data source, the examples are intended to be a useful exemplar for semantic discovery within the Internet of Things. Similarly, multiple user interactions with the system could be viewed as part of the Collective Intelligence within the organisation.

The aim of the following section is to analyse data sourced directly from an ERP system using techniques that are both mathematically sound and complementary to human cognition. For this reason FCA has been applied as it embodies many of the desired properties. Wille describes FCA's roots as being in philosophical logic with a pragmatic orientation and formalisation of concepts [2]. As discussed above the recognition of a situation, the capability of maintaining an understanding of the position and relationships through formal concepts in order to provide a context for the analysis is vital for understanding. Winograd and Flores as discussed by Devlin [5] developed systems to complement human communicative skills, this is undoubtedly the direction of this chapter, it is fundamentally an alliance between human cognitive processes and a sound mathematical technique, it is not targeting a solution to Artificial Intelligence.

An understanding of ERP is not imperative. Technical terminology is partnered with a meaningful description of the process and transaction however a brief overview of ERP is provided as background.

1.1 Enterprise Resource Planning Systems

ERP systems provide a transactional capability that forms a fundamental platform upon which the majority of today's organisations operate. ERP provides a detailed and structured mechanism for controlling and capturing operational data and a platform for analysis. ERP is not to be considered as an isolated system, in practice they form part of a complicated architecture communicating and interacting with many other systems. One important system aspect of this is Business Intelligence (BI); by definition this provides decision makers with valuable information and knowledge by leverage a variety of sources of data as well as structured and unstructured information [6]. The analysis described in this chapter is based directly on transactional data from ERP; the approach has the potential to form part of the Enterprise's Business Intelligence Suite.

ERP systems are typically based on a relational database, certainly this is where the majority of transactional activity is captured. Data is normalised to minimise redundancy and remove ambiguity which makes it useful for analysis, however, a great deal of process logic is embedded within the ERP system and not the database. Relational databases when used for on-line transactional processing (OLTP) are good repositories for detailed information. BI solutions typically transform this data so that it is suitable for on-line analytic processes (OLAP) by converting data into a format that is more applicable for fast analytic applications, frequently at a level where granularity is reduced. This focussed and efficient analysis tool comes at a cost in terms of transforming the data and maintaining its meaning, particularly when data is consolidated across systems or geographical areas.

ERP offers a relatively rigid set of data in a well-structured format, its operation relies on programmed logic that is not necessarily represented in the data. In addition to documents, objects, statuses and relationships ERP systems also capture a variety of log files including user tasks, time stamps and changes. All these in combination will potentially reveal otherwise latent semantics that can be of benefit to the Enterprise and form part of a successful BI application. ERP systems do not typically incorporate semantics and the stored data represents only a proportion of that available within an organisation. Non-integrated but complementary systems and humans form the repositories where the majority of data is stored. Correspondingly the majority of control mechanisms and procedures are not contained directly in the data but encapsulated in programs or dictated by human interaction with the system.

Business Process Management (BPM) is an approach to modelling processes that can address variations and logic problems. It can accomplish this because it incorporates a mix of metaphors including decisions and events into the model. A metaphor is a way of reducing the dimensions of the descriptions of a process to a more understandable and visible basis [1]. There is no argument against the need for metaphors, however, determining them and communicating them without loss of or mis-understanding is a difficult tasks.

Regardless of the form ERP takes, the core function and method of interaction is a transaction. A transaction is a fixed sequence of actions with a well-defined beginning and a well defined ending [7]. Transactions are a method of starting a function such as a report, data entry, browsing or virtually any other purpose the system is used for. The use of many transactions in a defined sequence is a process with an aim of meeting the goals of the Enterprise.

Discovering latent semantics and the ability to analyse large and complex data sets from different sources and formats in a timely manner, without requiring expert knowledge, is the main focus and motivation of this chapter. A number of analysis methods have been applied in order to discover semantics; these are described and evaluated including a review of the source data and resulting semantics.

1.2 Rationale

ERP is a compelling data source for this analysis due to the wide number of applications, industries, the sheer volume of data available and finally because it is highly structured and consistent. Data is one of the key components within ERP systems; it ranges from highly structured to virtually unstructured. In ERP the bias is towards structured data, if a view of data across the whole Enterprise is considered the bias would be towards unstructured data.

The principle of this research is to discover if ERP user data can reveal any useful semantics or information. Potential applications included user management, authorisations, process design, interface design and understanding general patterns of use. Cross referencing this with respect to a time period or geographical data may also reveal useful information. The data used in the examples include both content and structure that supports both of these considerations. Traditional BI solutions do not focus on user transactional data; therefore, this approach and data set presents an interesting dimension.

Devlin states that humans have tried to represent knowledge and understand the laws of thought for thousands of years and that we are still unable to explain exactly how our minds perform such feats [5]. He argues that our minds are intimately intertwined with the world around us, and that our feelings and perceptions, even our social norms, play crucial roles in the marvellous complex dance of human cognition. In the domain of ERP this is equally applicable, systems currently have limited contextual awareness but this is changing. The availability of inputs from sensors and mobile devices is increasing, if it can be harnessed a step change maybe possible. Decision making in systems is rule based and constrained therefore, at least at present, systems need to complement human capabilities.

Dreyfus and Dreyfus introduced a five stage model about human performance based on human skill acquisition [8]. Level one represents novice where rules are followed without context. The middle stages introduce a holistic understanding with some reliance on rules. Level five is where experts function, essentially through skill and being fully aware of context, with virtually no referral to rules. This represents an interesting choice of paradigm for future ERP systems; there is a strong argument for rule based systems that are capable of automating or supporting the decision making processes and representing Enterprises in an understandable form. Clearly a business would not want to operate at novice level, somewhere around the middle of this scale where rules are followed with some contextual knowledge would be desirable for systems however this is far from the optimum for humans. The converse view of this is that the sheer volume of data and complexity would prevent a human from operating at an expert level across the whole organisation as humans do not posses the communication capability to operate as a collective body, certainly not at any great speed or to the detriment of other activities. This is not intended as a sweeping statement across all areas and there will be many situations where simple rules are all that is needed, but the drive towards expert level will need a union of systems and humans.

2 Overview of Analysis

The following section contains an overview of the analysis techniques performed. Data has been extracted from an ERP system and processed into a format suitable for FCA. Details about the structure are contained in the following sections. All data used has been extracted directly from the database, there are no steps that could not be mechanised within the data preparation.

The basic steps performed are similar in nature and utilise a set of research tools for creating Formal Concepts and displaying lattices, these are FcaBedrock [9], In-Close [10] and Concept Explorer [11]. A brief overview is contained below however a comprehensive description has been produced by Andrews et al. [12].

The end result of applying FCA is a Concept Lattice, Fig. 1 provides an example. The object 'Yacht' has attributes 'Engine' and 'Sail' in contrast to 'Kayak' that does not. Phrased another way, a yacht has an engine and a sail. The lattice is read vertically with lower items sharing the attribute values of higher objects to which they are linked. A central notion of FCA is called a 'Galois connection', an intuitive way of discovering hitherto undiscovered information in data and portraying the natural hierarchy of concepts that exist in a formal concept [12]. FCA is useful as it can be used to analysis raw data, find relationships and produce a converted format suitable for displaying graphically as a lattice. In the example many types of boats could have been analysed in order to produce a simple view of relationships as highlighted in the lattice, this is how it can be a powerful technique when applied to large data sets where relationships are unknown at the outset. Formally this is isomorphism, corresponding or similar form and relations [13], a feature of the Galois connection.

It is possible to read the concept table used to create the lattice, see Fig. 2. The same relationships between the object 'Yacht' and attributes 'Engine' and 'Sail' are visible in table; the creation of this table is fundamentally what FCA achieves. Two terms are important to understand at this point—Intent and Extent. The extents are the objects such as 'Yacht' and 'Dingy' that share the attribute 'Sail'. Intents are the attributes such as 'Sail' that share the objects 'Yacht' and 'Dingy'.

Fig. 1 Simple lattice

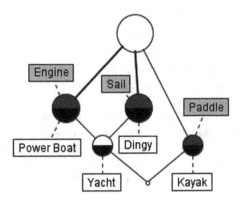

Fig. 2 Concept table for simple lattice

A	B	C	D
	Engine	Sail	Paddle
Yacht	X	X	
Dingy		X	
Power Boat	X		
Kayak			X

A method of restricting the output to only larger sets of intents or extents is supported by InClose, this applies a technique known as minimum support and restricts the content of the concept table. If a minimum support of two extents is applied and because 'Kayak' only has one attribute it would be excluded from the concept table, this is useful for identifying prominent relationships in large data sets.

FCA has been applied to a wide range of applications; this chapter does not represent a comprehensive review but an illustration of pertinent applications. The approaches described below benefit from the capability of FCA to analyse large data sets and the discovery of relationships through tabular or graphical analysis.

A common application of FCA is the classification of large data sets as in Andrews and McLeod study using FCA [14]. Groups of genes with similar expressions profiles were extracted and discovered, essentially identifying classes with similar features or properties.

Poelmans et al. applied FCA to business processes and data in order to discover variations and best practice in a medical care situation [15]. The ability to identify process anomalies and exceptions was identified but more importantly the advantageous use of FCA as a discovery process. This represented an attempt to analyse physical events and outcomes and is pertinent to the analysis within this chapter, where this chapter differs significantly is in the recognition of sequences of events.

A different application was applied by Poelman et al. for filtering out interesting persons for further investigation by creating a visual profile of these persons, their evolution over time and their social environment [16]. The interesting outcome from this analysis was how it is possible to discover relationships from unstructured data sets, a task that is virtually impossible to achieve manually in a practical period of time.

The closest realistic comparison to the analysis contained in this chapter is broadly defined as web usage mining; approaches such as Clickstream or Statviz [17] are two examples of both existing software and approaches. Clicksteam is typically a web browser based approach to log user clicks when navigating Internet sites. Results are logged and analysed for many reasons, common areas targeted include navigation, marketing and profiling users. Statviz is very similar but this specific software graphically represents the movement between pages and can incorporate measures such as popularity.

The approach taken in this chapter is based on a transactional system, similar to Statviz in that it is tracking movement between transactions that could be viewed as the pages navigated by the user. The underlying difference here is that web-based navigation is typically related to fulfilling a single task in its entirety where as from a transactional viewpoint there is not a clear definition. Transactional systems such as ERP form an integral part of the organisations daily function and users differ enormously in breadth of function, responsibility and understanding. Individuals typically use a limited set of transactions in order to support a sub-section of processes while handling the normal day-to-day exceptions and interruptions.

Transaction based systems such as ERP systems have been relatively static when compared to web based systems for two primary reasons: standardisation and maintenance. They are typically internal systems that are not heavily branded, do not feature tailored screens or have navigation aids that enhances the user's experience. Secondly, due to the complexity of modern systems, few organisations have the skills or knowledge to support or maintain such systems. Therefore the development of road maps and upgrade cycles by specialist vendors enable even small organisations to benefit from the latest technologies at the expense of standard interfaces.

Ontology is the study of categories of things that exist or may exist in some domain. When combined with logic an ontology provides a language that can express relationships about the entities in the domain of interest [18]. The basis of the analysis contained in this chapter is a bottom up discovery of formal ontologies by applying FCA. Lattices constructed by FCA methods are structured in a manner that supports ontology development, this was also indicated by Sowa as hierarchies of categories [18].

First order logic (FOL) deals with predicates, these being functions that maps its arguments to the truth values, the synonym of predicate being relationship [18]. The principle of FOL is beyond the scope of this chapter, suffice to say, it supports the description of relationships in an algebraic form; these in turn can support the formation of rules and inference. The combination of ontology and FOL based on the analysis in this chapter could be a step towards integrating this approach into a practical and real time applications.

The basic principle of ontology can be demonstrated using the example in Fig. 1. The categories of Engine, Sail and Paddle have been identified and could therefore be queried using logic. If a secondly lattice or ontology also contained these categories a query could also utilise this knowledge, a useful feature given the complex environments under consideration. For this purpose of this chapter the discussion will focus on FCA and lattices however the subject of ontology is an important consideration.

2.1 Data Preparation

The data for this analysis has been extracted directly from an organisation's productive SAP ECC 6.0 system. Therefore, it is a representative sample of real-world data; all data has been anonymised. SAP A.G. is one of the leading vendors for ERP systems and accounts for a significant proportion of the worlds transactions, this is estimated to be between 60–70 % [19, 20]. This is clearly a very large quantity and given the growth in data and communications a volume that is likely to increase. ERP systems are typically rigid systems that represent the core functions of an Enterprise, although with the development of advanced technologies and architectures ERP systems are slowly increasing in flexibility at significantly lower costs than in previous times. This does not imply that ERP only represent simple systems, they are employed across global organisations with complex and unpredictable behaviours caused by internal and external factors.

A useful set of data captured in SAP ECC 6.0 are the transaction logs of its users, this forms the primary data source and use case. It has been combined in part with other data contained in the system, this will be described in detail but an example is a lookup for the description associated with a transaction. In general this is all data available within SAP ECC 6.0 that can be extracted by query. The steps taken in this analysis are consistent, the primary differentiator being the manner in which data is prepared and the graphical manipulation of the lattice in Concept Explorer. The first example includes all steps in detail, subsequent examples only show the differentiating factors.

In order for a transaction to access or change data it performs a dialogue step, this essentially requests information from the Database Management System. These requests are logged and available for analysis by the Business Transaction Analysis tool that displays kernel statistical data for user transactions or background processing [21]. The basic data structure is a time stamped record of the transaction or program executed by a user. The basic constraint is that data is only written at the end of the dialogue step, essential after successful completion of the activity, however performance related data is also include which enables the calculation of the start time for the dialogue step. The raw data is based on dialogue steps to the database therefore an individual transaction may result in a number of dialogue steps as the user retrieves data and makes updates, however, it is simple to identify the first dialogue step and filter out any secondary steps.

All data in this section has been prepared to a 3-column CSV format, this represents triples in the form subject-predicate-object. The data was also restricted to one week of transactional activity within a single department. This structure enables the inclusion of additional data potentially from other sources without requiring it to be contained in the ERP system itself.

2.2 Analysing Transactional Activity

The aim of this analysis is to discover semantics from 'transactions by user in a time period'. This is intended to be an elementary question to test the method, it could be answered with a simple query but it is necessary to understand FCA in this context. The raw data was queried in order to produce a text file (.csv) as

Table 1 Example input file

Subject	Predicate	Object
171	USER	User 171
942	USER	User 171
(Repeat for all users)		
171	TRANSACTION	Report_1
942	TRANSACTION	Report_1
(Repeat for all user / transaction)		

shown in Table 1. Two queries have been combined to retrieve file contents from the raw data, the file contained approximately 35,000 rows with a run time of only a few seconds. The first section of the data describes the user reference with the user name, for anonymity the names have been replaced by 'User-number'. The second section of the data describes the user and transaction. The tool set applied does not offer the capability to scale this approach indefinitely; tools such as Triple Stores, relational databases design with the sole intention of holding data in this format, would be required to support scaling up to the volume and performance levels required for an entire organisation.

In this example the context is 'transactions by user in a time period', it does not include any chronological data. This text file (.csv) is read into FcaBedrock, Fig. 3, and processed to produce a context file (Fig. 4). Without any further processing the output files contains the data as represented in Fig. 5, the image is taken from Concept Explorer rather than the data file purely for presentation purposes. In its simplest form the task of creating a formal context file is complete.

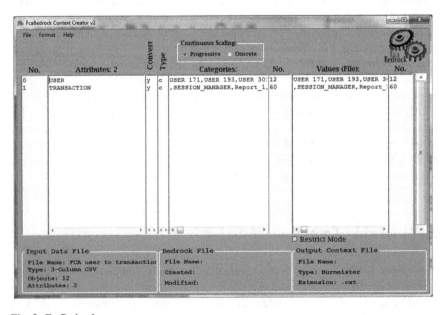

Fig. 3 FcaBedrock

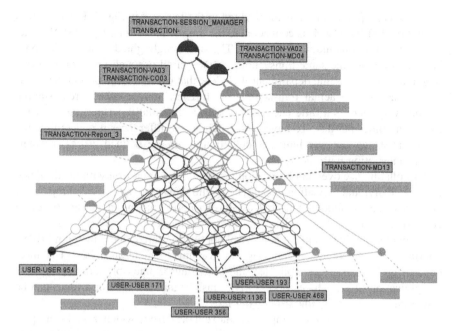

Fig. 4 Transaction based lattice

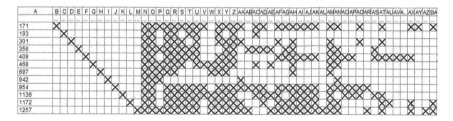

Fig. 5 Concept table for transaction based lattice

Figure 4 displays a lattice created with an attribute count reduced by 50 % within the context editor, in this example the attributes are the transactions. The initial lattice produced was complex and unreadable therefore simplification through reducing the number of attributes was attempted. Although this could be deemed as a visual improvement it has a significant drawback in that a proportion of attributes are simply removed, there is no ranking or measure of importance. With expert knowledge it is known that 'VA01' is a transaction to create a sales orders and therefore fundamental to the operation of the department in question, however, it has been lost from the analysis. This is where minimum support can be applied through the use of InClose to control the intents and extents therefore simplifying the lattice in a more controlled manner.

The second point to note is highlighted in the lattice, see Fig. 4. Reading from the bottom left 'User 954' is connected to the transactions 'Report_3', 'MD13' etc. as are other users including 'User 171'. The strand highlighted indicates that only half the users use the same transactions. Although all users all perform the same function within the same department it is clear that over the time period there was a difference in the actual transactions used. There maybe other circumstantial reasons why there is a difference as the team is essentially performing the same task many times per day. To have such a number of differences is concerning and further investigation may indicates deviations from the defined process or ineffective interaction with the system.

By collating use over an observation period or by comparison with the designed process, individual deviations could be highlighted and addressed. Objects or attributes can be hidden within Concept Explorer making it easier to reveal common patterns or differences.

From this simple example a number of applications are evident including BPM, interface design, process monitoring and training to mention a few. The most obvious problem is the complicated and bordering on unusable lattice produced. Simplification of the lattice is not a trivial task, significant data or a focus can be lost from the analysis without realising it.

Figure 5 represents the complete context table from which Fig. 4 was produced. The columns represent the transactions and rows the user, in simple terms, the 'X' marks an occurrence of user and transaction (the column headers have been reduced in width for clarity). The interesting aspect of this is the heavily populated groups particularly towards the middle of the table where many users are using the same transactions. One potential opportunity would be to understand why users are not using certain transactions, this may indicate a process feature or training requirement.

Figure 6 contains the same data as Fig. 4 manipulated into a form suitable for Excel. In this example, 'X' is replaced by '1' to support calculations, it is also

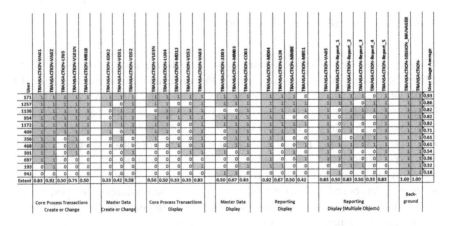

Fig. 6 Excel concept table with calculated values in excel

shaded for clarity. The column on the far right shows 'transactions usage by user'. By example, 'User 171' is performing 93 % of the transactional range within the department. The extent row along the bottom is reflected in the lattice and represents the range of users performing a transaction; this figure is the same as the extent value available in the lattice diagram. The core function of the department is known to be transactions 'VA01' and 'VA02' (create and change sales orders). These activities have high extents indicating virtually all users perform these transaction. A large data set could be analysed utilising minimum support to select only concepts with high extents. This is potentially useful and is demonstrated in subsequent examples where further attributes are introduced. Potentially this could include geographical, performance measures or job title/role attributes to mention a few.

An additional application of this tabular view is in the definition of the standard processes. Over a period all activities associated with a department or process step could be identified, capturing the actual process, supporting activities and providing feedback on the designed process. This point is explored later when transactional flow is considered.

2.3 Analysing Transactional Activity with Descriptions

The aim of this analysis is to discover semantics from 'transaction descriptions by user in a period'. Figure 7 is based on the same data set as Fig. 4. In this example the transactions have been replaced by their high level description, for example the transaction 'VA01' has been replaced by 'Create', see Table 2. In order to produce this file the query was modified to replace the transaction with its description. This data is available within SAP ECC as each transaction is held as a record including its description, associated program and other information. For the purposes of this example the first key word has been selected but with the addition of methods such as stemming and key word search this approach could be extended.

In this configuration, the lattice has taken a considerable step towards being more usable and a distinct hierarchy is starting to become visible, this is illustrated by a highlighted strand in Fig. 7. Although this analysis is based on the same data the use of general descriptions has caused a grouping effect and simplified the lattice. This is effectively applying metaphors and cognitive models in order to

Table 2 Example input file with descriptions	Subject	Predicate	Object
	171	USER	User 171
	942	USER	User 171
	(Repeat for all users)		
	171	TRANSACTION	Create
	942	TRANSACTION	Report
	(Repeat for all user / transaction)		

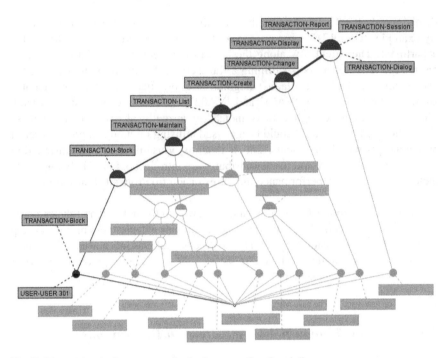

Fig. 7 Transaction lattice represent by basic transaction descriptions

understand and generalise complex processes or events. The highlighted strand indicates transactions that support the core process with sub processes or supporting activities appearing beneath them and not highlighted. An understanding of lattice construction and the ability to graphical interact and highlight strands is starting to make lattices an intuitive method of analysis.

2.4 Analysing Transactional Activity with Multiple Attributes

The aim of this analysis is to discover semantics from transactional activity using 'multiple attributes descriptions by user in a period'. Figure 8 is based on the same data set as Fig. 4. In this example, attributes have been added rather than replaced, see the date structure in Table 3. The USER attribute remains unchanged from the original example. The ACTION attribute has now been added to represent the description. The AREA attribute has been introduced to represent the business function. The final attribute is transaction date; TRANSACTION has been used to represent the date a transaction was used. As before, all data used has been extracted directly from the system.

This analysis has been restricted to a single user by using the 'restrict mode' in FcaBedrock, this essentially applies a filter. Because the AREA attribute has been derived from the transaction it is expected that they will align with each other.

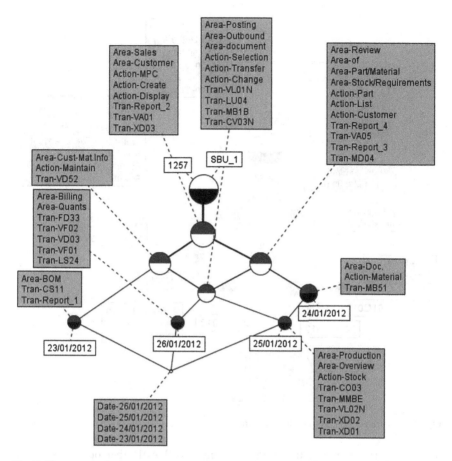

Fig. 8 User transactions

Table 3 Example input file containing multiple attributes

Subject	Predicate	Object
1257	USER	User 171
(Repeat for all users)		
VA01	ACTION	Create
(Repeat for all transactions / actions)		
VA01	AREA	Sales
(Repeat for all transactions / areas)		
Date	TRANSACTION	VA01
(Repeat for all dates / transactions)		

Using the same example as previously 'Create' and 'VA01' will share the same node. By visual inspection the lattice indicates that there is not a standard daily process, transaction 'VA01' is performed daily but the sub tasks vary. If it is

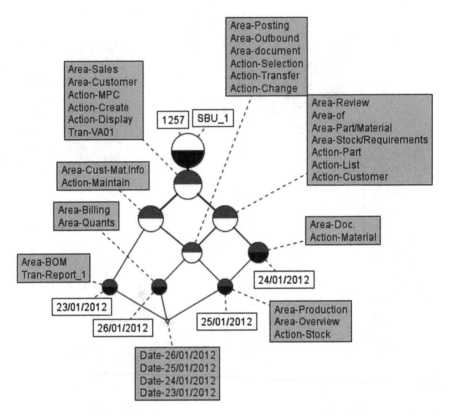

Fig. 9 User transactions limited to target transaction VA01

assumed that the core function of the department is 'VA01', then other tasks could indicate inefficiencies or distractions from the goal.

This capability to input multiple objects supports two features, firstly the ability to choose the level of detail displayed and focus the analysis. Figure 9 has all transactions hidden apart from the 'VA01'. The AREA and ACTION attributes remain for interrogation but the complexity caused by displaying all transactions is removed. The second feature is the continued maintenance of a context, a detailed focus is surrounded by a context in order to support a cognitive model.

2.5 Analysing Transactional Activity with Direct Comparison

The aim of this analysis is to discover transaction activity between 'users in a period' and support a detailed level of investigation. Two users have been compared side by side in order to demonstrate this point, see Fig. 10. The preparation steps are identical to Fig. 8 apart from focussing on a different user.

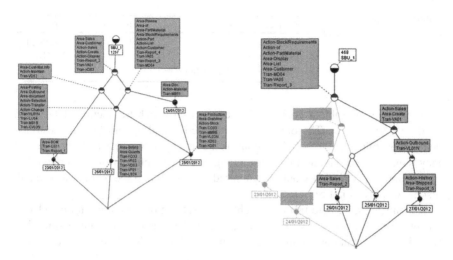

Fig. 10 User transactions for user 1257 *(left)* 468 *(right)*

A distinct difference can be observed when the 'create' node is highlighted, the user on the left shows that 'create' is a frequent used transaction and occurs every day. Conversely the user on the right used only display and reporting transactions for the first two days of the observation period.

Side by side comparisons of lattices has the potential to produce useful knowledge. The level of detail and granularity is very significant in the analysis. This example is highly detailed but the same method of comparison could equally be compared to groups of users or Enterprises. The ability to support large data sets, multiple attribute descriptors and graphical analysis is fundamental.

The ability to group and generalise features is fundamentally important to support a cognitive understanding for visualisation and comparison. Grouping could be used to combine nodes that share similar features, effectively promoting dissimilar features that may be at a different level of detail. Replacing attributes with alternative descriptions or values from the system or via ontologies is possible.

2.6 Analysing Transactional Sequence

The aim of this analysis is to discover semantics and patterns in transactional sequences. The analysis methods applied so far have focussed on the use of transactions over a time period; this section will focus on the sequence of transactions. The sequence of transactions closely represents how users actually interact with the system. It has the potential to be used for comparing planned and actual usage and highlight patterns of usage. This analysis could be useful in both BPM and Interface design.

Subject	Predicate	Object
VA01	FLOW	Report_108
Report_108	FLOW	MMBE
(Repeat for all transactions)		

Table 4 Example input file containing the sequence of transaction

The format in Table 4 contains the sequence in which transactions have been used by a single user. The object is the transaction and the attribute is the transaction that followed. The obvious point is that a transaction, particularly a report, may remain on screen and in use for a long period, therefore it may overlap other transactions. Any refresh would be captured but, just like if the report was printed, a static display is not represented.

It is intriguing to note from the highlighted section of the lattice in Fig. 11 that one of the core functions 'VA01—create sales order' has such a strong relationship with five reports and one maintenance transaction. This clearly indicates a need to

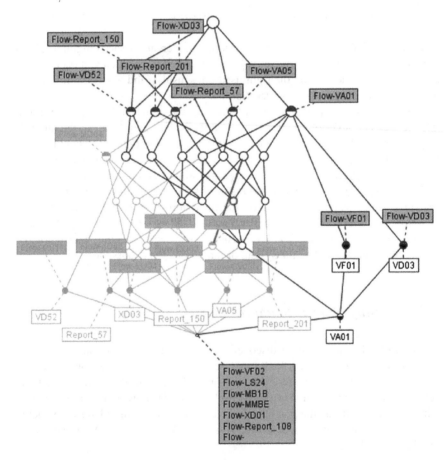

Fig. 11 Lattice for transaction flow for an individual user

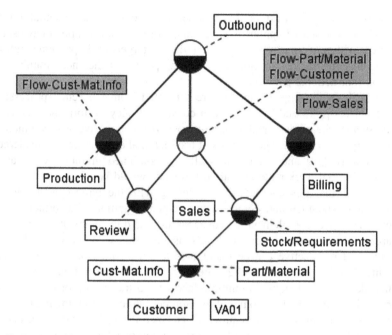

Fig. 12 Lattice with VA01 and related transactional areas

investigate the consolidation of these report. For the purposes of this explanation 'XD03—display customer' has been included in the report count, 'VD52' is the only transaction with a maintenance function. It should be also be noted that this only addresses the data available within the ERP system, it is likely that there are even more sources that should be considered.

From a BPM perspective it should be questioned why so much non value adding time is being incurred, an obvious question to ask is what areas are the reports reflecting? Using the ability to add multiple levels Fig. 12 has been produced. This focusses on a specific transaction 'VA01—create sales order' and applies the AREA attributes for all other transactions. From an analysis viewpoint this provides an easier conceptual model for understanding and reduces the need for expert knowledge. This example indicates that reports are being run against a range of other departments when creating a sales order. The salient point here is that these are separate transactions being used to collect information beyond the minimum that is required to process the transaction.

3 Evaluation

Many useful features have been demonstrated, more importantly the potential for including this type of analysis as part of an organisation's Business Intelligence capability has been highlighted. Two points of view have considered; the results of the analysis in the context of the data set and secondly the potential for this analysis method as a BI application. The range of techniques applied have shown how areas within an Enterprise can be identified as targets for improvement efforts. Specifically this relates to interface design, process adherence, training and common transactional patterns.

The analysis techniques applied are separated from the actual process; the unique objects processed have not been considered. Key Performance Indicators (KPIs) such as the product quality or delivery performance have not been included in the analysis at present. The analysis of transactional data has been considered in previous work [22] and could potentially be combined in future work as the analysis of actual work patterns and performance would be of interest.

It is clear that this approach is in keeping with the philosophical ideas of context and semantics and complements human cognition. The principles and structure appear to complement human capabilities and in combination could push towards the attainment of expert levels as defined by Dreyfus and Dreyfus. A contextual perspective can be maintained and a detailed or targeted analysis combined within a view containing more descriptive and general information. The method demonstrated supports multiple labels for a transaction or event and has the potential to include data from multiple sources while maintaining relationships. There is still significant work required to enhance the users experience, particularly around data preparation and lattice navigation or interrogation.

From an analysis viewpoint, it would indicate that the sales department is querying information from many areas including production, inventory and billing during the order capture process. This in itself is not unusual when accepting customers orders, using separate transactions to access this information suggest a non optimal system or a process that is out of control, certainly highlighting an area for further investigation.

Applications of this method could indicate how resources such as reports and transactions should be unified or combined to support specific job functions, there is a potential for automated mash-up reports in the context of the task being performed. Another move forward would be to associate the success of objects against the patterns associated with the specific job function. The individual manner by which a user approaches a tasks could be attributable to the outcome.

In memory techniques such as Hana could enable this analysis directly from the data source without intermediary data preparation steps. The data preparation steps will not differ significantly from those describe. The concept of holding all ERP data in memory coupled with significant performance gains in database access speeds as described by Plattner [7] have the potential to make this a real time analysis technique.

Process and interface design could benefit from the sequence and overlapping use of transactions. For example, transactions that are used concurrently to perform a task could be candidates for process and or interface redesign. Integrated interfaces such as SAP Business Suite 7 are based on an open SOA platform and aimed at enhancing integration [23], incorporating this analysis and evolving alongside the users actual job function is a possibility.

User management could include grouping users based on actual use and comparing commonalities within hierarchies of groups. The result may be useful in defining authorisations that are tailored to the users job function and far less generic than is common in organisations today.

3.1 Framework

Finally the rudimentary capabilities required in a framework for discovering semantics from ERP system data are listed below. These have been discussed in the main text but are presented in a concise list below.

Data Requirements

- Hierarchical levels of granularity in the source data
- Single or standardised data repository
- Simple data format for representing facts

Semantics and Context Requirements

- Traverse through data in multiple dimensions and maintaining focus
- Prominently retain and promote context
- Support for combined human and system discovery

Graphical Interaction Requirements

- Promote or eliminate common features between lattices
- Graphical filtering to remove or replace attributes
- Control of layer transparency
- Ability to record analysis at key points

It is suggested that ERP is positioned as the temporal reference for an organisation, a reference for the orchestration and choreography involved. The captured knowledge using this approach does not mandate a direct reference to the transactional system providing the data and models have been constructed correctly. The data sources may be external to this system and a standard reference is necessary particular when considering sequence and usage. ERP typically represents the core of the business functions and is central to an organisations operation; this makes it a logical choice as the central reference of interactions and sequencing.

Post analysis the reference maybe redundant, particularly if the analysis if focussed on relationships, classification, sequence or usage; however the ability to construct these consistently requires a frame of reference. With the growing trend towards ubiquitous computing the ability to align processes accurately will become even more fundamental, it may not be possible to maintain a direct link and navigate between the results and source data. It is highly likely that as data volumes continue to increase much of the source will be lost or loosely coupled data. After the completion of the analysis and storage of results, this repository may effectively become the storage and application for contextual knowledge.

4 Conclusion

This use case has proven that FCA has a practical application for the development of ERP systems and is a useful exemplar for the Internet of Things. Capturing transaction activity related to the core functions of an Enterprise coupled with the capability of representing multiple levels and data sources makes the method and use case a complementary pair.

A emergent theme is the agility that this method supports, a view supported generally of Semantic Technologies [24]. This analysis has demonstrated that it is possible to discover useful knowledge and semantics through applying FCA, lattice diagrams and graphical exploration. Important capabilities demonstrated include the interactive interrogation and discovery enabled by traversing levels and maintaining context with a focus.

The data used for the analysis could be extracted and stored in a triple store and combined with data from other sources. Extraction and storage of user data into a triple store would also negate the need to permanently store the source log files within the ERP system. The data set analysed contained approximately 35,000 records, a fraction of the four million records for the whole enterprise over the same period. In the context of big data, the method described is highly suitable.

The knowledge represented in the lattices could be used to build a useful ontology for logical deduction and analysis. The examples constructed have demonstrated that it is possible to extract and build knowledge that represents the actual use of systems from large data sets relatively easily and with differing levels of granularity. The hierarchical capability provides this from a time frame and sequential perspective. For specific examples, as demonstrated, the use of lattices to display results has enabled meaningful graphical analysis. Lattice diagrams can add value in this environment, the limiting factor is the graphical complexity.

Data preparation at this point in time is manual, however, everything demonstrated has only used database queries and data available from the source SAP ECC 6.0 system.

A number of distinct applications have been highlighted in BPM and system interaction. These include process design and monitoring based on actual use and the individuality of users. Applying the same method to data captured from the Internet of Things could equally result in useful semantics and knowledge about Network Horizons [25] which is the interaction between Enterprises and also towards a Supply Chain viewpoint focussed on the actual objects handled or traded.

4.1 Further Research

There is clearly development work required in order to make this a technique suitable for implementation in real organisations. Data preparation and interaction with the graphical interface are two areas where considerable improvements could be made.

Integrating the techniques described with object based transactional data may produce even more insight into the semantics discoverable from ERP data. Differences between users or systems could be directly related to operational performance highlighting useful knowledge and semantics at various stages of the process.

References

1. T. Debevoise, R. Geneva, *The Microguide to Process Modeling in BPMN 2.0*, 2 edn. (Advanced Component Research, Lexington, 2008)
2. R. Wille, *Conceptual Graphs and Formal Concept Analysis*. (Technische Hochschule Darmstadt, Fachbereich Mathematik, 1997)
3. D. McComb, *Semantics in Business Systems: The Savvy Managers Guide* (US, Elsevier, San Francisco, 2004)
4. B. Ganter, R. Wille, *Formal Concept Analysis: Mathematical Foundations*. (Springer, Berlin, 1998) (Translated by C, Franzke, 1998)
5. K. Devlin, *Goodbye Descartes—The End of Logic and the Search for a New Cosmology of the Mind*. (Wiley, NY, 1997)
6. R. Sabherwal, Succeeding with business intelligence: Some insights and recommendations. Cutter Benchmark Rev. **7**(9), 5–15 (2007)
7. H. Plattner, A. Zeier, *In Memory Data Management: An Inflection Point for Enterprise Applications*, 1st edn. (Springer, Berlin, 2011)
8. H.L. Dreyfus, S.E. Dreyfus, *Mind Over Machine: The Power of Human Intuition and Expertise in the Age of the Computer*. (Basil, Blackwell, Oxford, 1986)
9. S. Andrews, C. Orphanides: FcaBedrock, a Formal Context Creator. In: 18th International Conference on Conceptual Structures (Kuching, Malaysia) 26–31 July, (2010)
10. S. Andrews, in Conceptual Structures for Discovering Knowledge, eds. by S. Andrews, S. Polovina, R. Hill, B. Akhgar. In-Close2, a High Performance Formal Concept Miner, vol. 6828 (Springer, Berlin, 2011), pp. 50–62
11. ConExp: Concept Explorer http://sourceforge.net/projects/conexp/. Accessed 01 April 2012
12. S. Andrews, C. Orphanides, S. Polovina, Visualising computational intelligence through converting data into formal concepts. In: F. Xhafa, L. Barolli, H. Nisino, M. Aleksy, (eds.) Proceedings of the 2010 International Conference on P2P, Parallel, Grid, Cloud and Internet Computing (3GPCIC). IEEE Computer Society, pp. 302–307 (2010)
13. Oxford (2013): Oxford Dictionary: Isomorphic http://oxforddictionaries.com/definition/english/isomorphic. Accessed 01 Dec 2012
14. S. Andrews, K. McLeod, Gene co-expression in mouse embryo tissues. CEUR Workshop Proc. **753**, 1–10 (2011)
15. J. Poelmans, G. Dedene, G. Verheyden, H. Van der Mussele, S. Viaene, E. Peters, *Combining Business Process and Data Discovery Techniques for Analyzing and Improving integrated Care Pathways* (Springer, Berlin, 2010)

16. J. Poelmans, P. Elzinga, G. Dedene, S. Viaene, S.O. Kuznetsov, A concept discovery approach for fighting human trafficking and forced prostitution. in: S. Andrews, S. Polovina, B. Akhgar, R. Hill (eds.) Proceedings of the 19th International Conference on Conceptual Structures for Discovering Knowledge, (ICCS'11), (Springer, Berlin, 2011), pp. 201–214
17. P. Hitzler, M. Krtzsch, S. Rudolph, *Foundations of Semantic Web Technologies*. (CRC Press, Boca Raton, 2009)
18. J.F. Sowa, *Knowledge Representation Logical, Philosophical, and Computational Foundations* (Course Technology, Cengage Learning, 2000)
19. S. Poonen, SAP Solutions Portfolio—"The Big Picture" http://www.sap-tv.com/video/7344/sap-solutions-portfolio-the-big-picture-presented-by-sanjay-poonen. Accessed 01 Apr 2012
20. Forbes: It Doesn't Take Two Years to Create a Good Strategy http://www.forbes.com/sites/sap/2011/10/07/saps-bill-mcdermott-it-doesnt-take-two-years-to-create-a-good-strategy. Accessed 01 Apr 2012
21. SAP: SAP Help—STAD http://help.sap.com/saphelp_nwpi711/helpdata/en/ec/af4ddc0a1a46 39a037f35c4228362d/-content.html. Accessed 10 Mar 2012
22. M. Watmough, Evaluation of an Approach for Teaching Formal Concept Analysis. In: CEUR Workshop Proceedings, Proceedings of the 1st CUBIST. Workshop, Vol. 753, pp. 57–67 (2011)
23. N. Muir, I. Kimbell, *Discover SAP*, 2nd edn. (Galileo Press, Bonn, 2010)
24. F. Dau, Semantic Technologies in Enterprises. In: S. Andrews, et al. (eds.) ICCS 2011, LNAI 6828, pp. 1–18. (Springer, Berlin, 2011)
25. D.W. Liere, *Network Horizon and the Dynamics of Network Positions* (Erasmus University Rotterdam, ERIM Electronic Series Portal, 2007)

High Quality of Service and Energy Efficient MAC Protocols for Wireless Sensor Networks

Bilal Muhammad Khan and Rabia Bilal

Abstract Wireless sensor networks (WSNs) are increasingly gaining impact in our day to day lives. They are finding a wide range of applications in various domains, including health care, assisted and enhanced-living scenarios, industrial and production monitoring, control networks, and many other fields. In future, WSNs are expected to be integrated into the "Internet of Things", where sensor nodes join the Internet dynamically, and use it tocollaborate and accomplish their tasks. As wireless sensor networks being used in many emerging applications the requirement of providing high quality of service (QoS) is becoming ever more necessary. This highlights major issues like collision, scalability, latency, throughput and energy consumption. In addition mobile sensor network faces further challenges like link failure, neighbourhood information, association, scheduling, synchronisation and collision. Medium Access Control (MAC) protocols play vital role in solving these key issues. This chapter presents the fundamentals of MAC protocols and explains the specific requirements and problems these protocols have to withstand for WSN. The QoS is addressed for both static and mobile sensor networks with detailed case study of the IEEE 802.15.4 WPAN standard. Research challenges with literature survey and further directions are also discussed. The chapter ends with conclusions and references.

B. M. Khan (✉)
Department of Electrical and Power Engineering, National University of Science and Technology (NUST-PNEC), Karachi, Pakistan
e-mail: bmkhan@pnec.edu.pk

R. Bilal
Department of Electronic, School of Engineering and Informatics, University of Sussex, Brighton, UK
e-mail: rb212@sussex.ac.uk

F. Xhafa and N. Bessis (eds.), *Inter-cooperative Collective Intelligence: Techniques and Applications*, Studies in Computational Intelligence 495, DOI: 10.1007/978-3-642-35016-0_12, © Springer-Verlag Berlin Heidelberg 2014

1 Fundamentals of Wireless MAC Protocols

MAC layer is a part of the Data link layer in the OSI reference model. The task of the MAC layer is clear and well defined; it determines the access mechanism and time for a node in order to try to transmit data, control or management packets to another node in the case of unicast scenario or to a set of nodes as in the multicast scenario. The two important responsibilities of the remaining part of the Data Link Layer are Error Control and Flow Control. Error Control techniques are used to ensure that transmission incurs no errors and to take corrective measures if there are errors in transmission, whereas flow control regulates the rate of transmission which is crucial in the case of slow receivers to prevent them from being overloaded with data.

1.1 Requirements and Design Constraints for Wireless MAC Protocols

The most important performance measurements for MAC protocols are throughput, fairness, stability, low access delay, low transmission delay, low overhead and energy consumption. There are several causes for overhead in MAC protocols; it can be the result of length per packet like MAC header and trailer, by collisions or by exchange of extra control packets. Packet collision tends to occur if the MAC protocol allows two or more nodes to access the shared communication medium at the same time. Collision is the result of inability of the receiver to decode the packets correctly this triggers the mechanism of retransmission of data packets. For time critical applications this delay and retransmission process is unacceptable; in some applications the transmission of important packets are preferred over unimportant ones which generates another class of MAC protocols which are based on the concept of priorities.

The operation and performance of MAC protocols is heavily influenced by the properties of the physical layer. As the signal is transmitted over the wireless link its strength decreases with the increase in the distance between transmitting and receiving nodes. This loss in signal strength is a very important factor in terms of successfully demodulating the signal at the receiving node since transceivers need minimum signal strength to carry out this task. This leads to a maximum range that a sensor node can reach with a given transmit power. However if the two nodes are out of reach from each other they are unable to receive each other's on-going communication which gives rise to one of the challenges which MAC protocols have to overcome in the wireless domain of hidden and exposed nodes [1].

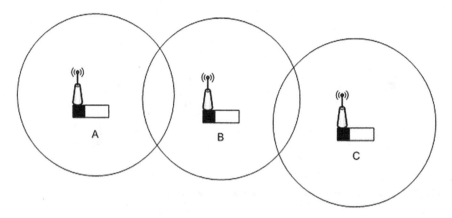

Fig. 1 Hidden node problem in wireless sensor networks

1.1.1 Hidden Node Problem

The hidden node problem occurs in Carrier Sense Multiple Access (CSMA) protocols in which a node senses the medium before the start of transmission. If the medium is found to be busy the node defers its transmission to avoid collision and retransmission due to the result of collision. Figure 1 gives a typical scenario for hidden node problem. There are three nodes A, B and C in the network. The arrangement of the nodes is such that node A and node B are in communication range as well as node C and node B; however node A and node C are out of communication range from each other. Assuming that node A starts to communicate data packets towards node B , after some time node C senses the channel and find it idle as it is out of the communication range of node A it also starts to transmit packets towards node B resulting in collision due to hidden node. Using simple CSMA MAC protocol in such a scenario leads to collision and needless retransmission of data packets.

1.1.2 Exposed Node Problem

Consider Fig. 2 in which there are four nodes A, B, C and D. Node C and node D are out of communication range from node A and node B respectively. If node B starts to transmit data to node A and after some time node C attempts to transmit data towards node D, in normal circumstances this communication is possible however in the case of wireless communication and especially using simple CSMA protocol node C transmission will be suppressed, this is due to the fact that before transmission node C according to the CSMA procedure tends to sense the channel and although the on-going transmission is between node B and node A the node C considers that the medium is busy and defers its transmission causing waste of bandwidth.

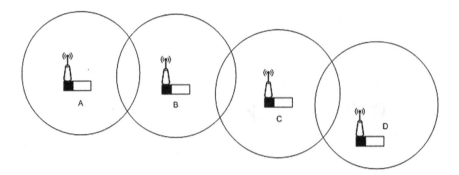

Fig. 2 Exposed node problem in wireless sensor networks

The solutions for hidden and exposed node terminals are proposed in the shape of busy tone [1] and RTS/CTS handshaking signals used in IEEE 802.11 [2] and are implemented in MACA [3] and MACAW [4] MAC protocols.

2 Classification of QoS Wireless MAC Protocols

There are several wireless MAC protocols available in literature. These protocols can be broadly classified into; Fixed assignment protocols, Demand assignment protocols and Random access protocols.

2.1 Fixed Assignment Protocols

The fixed assignment class of protocols are collision free and all the available resources are divided between the nodes, such that each node can use its resources exclusively. In case of topology changes due to mobility, joining of new nodes or malfunction of nodes or changes in the load pattern different signalling is used in order to renegotiate the resources to the remaining nodes in the network. Hence such type of protocols may not be viable in terms of scalable networks.

Typical protocols that fall in this category are TDMA, FDMA, CDMA and SDMA. Time Division Multiple Access [5] divides the time into several super-frames depending upon the size of network and these superframes are further subdivided into time slots. These time slots are assigned to the nodes in the network which transmit data in the specified slots periodically in every super-frame. In order to avoid overlapping of different time slots the TDMA system requires synchronisation. In the case of Frequency Division Multiple Access the frequency band is divided into a number of sub-channels and the nodes are individually assigned these sub-channels in order to transmit the data. In such protocols frequency synchronisation is required and the transceiver of FDMA is more complex than that of TDMA. In Code Division Multiple Access [6–8] the nodes spread the signal over a much larger bandwidth using different codes to separate their transmissions. The receiving node must know the codes used by each transmitter. However in Space Division Multiple Access spatial separations are used by nodes to separate their transmissions. The SDMA protocols requires an array of antenna and complex signal processing algorithms [9] hence it is not a suitable candidate to be used in resource constrained wireless sensor networks.

2.2 Demand Assignment Protocols

In this class of protocols the resources are allocated to the nodes on a temporary short term basis upon the request and usually last long enough to support the data burst. This class of protocol is divided into centralized and distributed protocols. The examples of centralized protocols include HIPERLAN/2 protocol [10–14], DQRUMA [15] and MASCARA [16] as well as polling schemes [17–19]. In such protocols nodes send out requests for the allocation of resources to the central coordinator which accepts or rejects the request depending upon the availability of resources. In the case of successful allocation the node is sent acknowledgement with the information of the allocated resources i.e the time slot in TDMA and the duration in which the node has to complete its transmission. The request from the nodes to the coordinator is usually contention based. Alternatively the nodes combine the requests with the data packets transmitted in their specified time slot avoiding transmission of a separate request. Another approach is that the coordinator polls all the joining nodes for their request to reserve or allocate the resources. In such protocols the central node or the coordinator needs to be awake all the time in order to receive any incoming request and to allocate the resources to the requesting node. The de-allocation of resources is performed by the central coordinator if the node is not using its allocated slot. Thus the node doesn't have to send an extra de-allocation packet to the coordinator. This class of protocol requires the coordinator performing several tasks which are energy exhaustive. Thus the protocols are suitable for the networks which have no constraints over energy, or more than one node having no constraints of energy.

Token Bus [20] is an example of a distributed demand assignment protocol. The node capturing the token has the right to initiate the transmission. Token frame is rotated among the nodes which are in logical ring topology. Special management procedures are devised in order to include or exclude a node from the ring as well as to handle failure like the token being lost. Token passing protocols have also been considered for wireless applications [21–23] but they suffer greatly due to the dynamic changing channel conditions and struggle to maintain logical ring topology. Due to the adhoc nature of the network maintaining ring topology involves a significant amount of signalling on top of token passing. Moreover the node transceiver has to be in the ON state as the timing of token passing is variable in order to avoid token loss and breaking of the ring.

2.3 Random Access Protocols

The first random access protocol was ALOHA or slotted ALOHA [24]. In the case of ALOHA the node which wants to transmit the packet starts the transmission immediately without coordinating with other nodes thus increasing the chances of collisions at the receiver. In order to find out about the successful transmission, the receiver sends an acknowledgement signal to the transmitter, in the absence of the acknowledgement the transmitter assumes that a collision occurred at the receiver. In this condition the transmitter backs off for a random amount of time and starts the process again. Thus ALOHA provides short access and under low traffic load, however under high traffic and dense network conditions the number of collisions increases and the network efficiency and throughput is degraded significantly. In the case of slotted ALOHA, the superframe is subdivided into time slots of equal lengths. The nodes have to synchronize in order to transmit the data in these time slots as only those nodes will be allowed to transmit which begin transmission at the start of the time slot, during a specific time slot any other node that wants to transmit data has to wait for the starting of a new time slot thus by synchronization number of collision is minimised and the throughput efficiency of the slotted ALOHA is better than the pure ALOHA.

In order to improve the efficiency of ALOHA protocols and maintain an acceptable quality of service CSMA protocols [25] are proposed. In these protocols the transmitting node senses the channel this procedure is known as carrier sensing, if the channel is available or in other terms there is no ongoing transmission over the channel, the node starts transmitting. If the node finds the channel busy during the sensing procedure then it defers its transmission by using several possible algorithms. In case of nonpersistent CSMA the node waits for a random time after it senses the channel. Before this random waiting time the node does not care about the state of the channel. In persistent CSMA the node waits for the ongoing transmission and then follows the backoff algorithm. In case of p-persistent the CSMA transmitting node starts the transmission in a time slot with some probability p and waits for the next slot for transmission with probability 1-p. If some other node in the network starts to transmit the node defers and repeats the entire procedure. Selection of smaller values of p makes collisions highly unlikely but at the cost of large access delays, conversely larger values of p yields higher number of collisions.

In the case of the backoff algorithm used in Distributed Coordination Function(DCF) the transmitting node chooses a random value and starts its timer. The timer is decremented after each time slot, if another node starts to transmit in these time slots the node freezes its timer and resumes after the transmission is over. If the timer of the node reaches zero the node starts transmitting its data packet over the channel. If the node transmission confronts an error such as non-reception of acknowledgement frame from the intended receiver, then it doubles the value of the contention window and the whole procedure of contention continues again. The CSMA algorithm is susceptible to hidden terminal problems.

3 QoS Challenges in Wireless Sensor Networks

In the case of WSN the requirement is different from traditional wireless MAC protocols. The major concern is to conserve energy. The significance of energy efficiency in the design of MAC protocols is relatively new and several of the traditional protocols such as ALOHA and CSMA have no provision to accommodate this feature. In [26–28] energy conservation accompanied by typical performance factors such as fairness, throughput and latency is discussed. Moreover scalability and robustness also emerges as prime and significant requirements in the design of MAC protocol for wireless sensor networks. These QoS challenges for WSNs are explained in this section.

- Resource Constraints: WSNs typically lack in bandwidth, memory, energy and processing capabilities. However energy is by far the most crucial factor as in many scenarios it is almost impractical to replace or recharge batteries of sensor nodes. Therefore it is of paramount importance that any proposed QoS MAC should be simple in order to operate on a highly resource constrained sensor node.
- Collisions: Collision causes unnecessary waste of energy in the retransmission of the collided packet at the transmitting node as well as receiving the same packet again at the receiving node. Collision incurs high latencies and affects the throughput of the network. Tackling the problem of collision is the prime goal in designing any QoS MAC protocol.
- Node deployment: Node deployment in WSNs can be deterministic or random. In case of deterministic deployment the nodes are deployed in fixed known topology and routing can be performed through pre-schedule paths. Whereas in random deployment the nodes organize them self randomly.
- Topology Changes: Due to node mobility, link failure, energy depletion or security attacks the topology is changed. One more factor which is inherent to WSN is that in order to save energy most MAC protocol use sleep-listen mode and turn off the radio completely for energy saving purpose, this result in topology change as well. Hence the dynamic nature of topology change in WSN introduces an extra challenge for QoS support.
- Data redundancy: Most of the applications for WSNs comprises of homogenous nodes, therefore an event can be detected by several nodes. This redundancy is helpful in reliable data transmission; it also causes unnecessary data delivery which results in congestion. There are data aggregation protocols [29, 30] which deal with such type of problems but they may also introduce additional delays and complexity in the system. Hence efficient and effective QoS mechanisms are required to deal with data redundancy issue.
- Real time traffic: There are certain applications in which data is valid for a certain amount of time frame. For such kind of application QoS requirements are very high and a suitable protocol is required which meets this criteria without compromising other attributes of MAC.

- Overhearing: Wireless medium is broadcast in nature and all the nodes which are within the radio range of the source node overhear the transmitted packet. The node to which the packet is not destined drops the packet on receiving it [31, 32]. In cases of dense networks avoidance of overhearing can contribute in substantial savings of energy. However completely avoiding overhearing is also not an efficient scheme as nodes also want to be aware of channel state as well as the state of the neighbouring nodes.
- Protocol Overheads: In order to tackle different problems such as collision MAC introduce protocol overhead related to control frames. RTS/CTS control packets are overhead and occur on a per packet transmission basis. In one way these overhead tend to improve the overall quality of service of the network by addressing some of the major hurdles faced by MAC protocols but on the other hand excessive use of these overheads can cause loss of energy, increase in latency and make the network less efficient.
- Idle Listening: A node in the state of readiness to receive a data packet but not receiving it is said to be in the state of idle listening. Idle listening also contributes to the energy loss of the network.

Above are most of the challenges related to WSN especially in order to maintain QoS. These challenges make it difficult for providing deterministic QoS guarantees such as hard time bound for data arrival, packet loss or guaranteed bandwidth. However different applications have different QoS requirements and it is feasible to provide an acceptable QoS for these various applications as discussed in the rest of this chapter.

4 QoS Aware MAC Protocols for WSNs

Most of the MAC protocols designed for WSN try to tackle one of the above mentioned problems in order to save energy and to improve quality of service of the network. These protocols can be broadly classified into Low duty cycle protocols, Contention based protocols and Schedule based protocols. In the following section these classes and major protocols within them are discussed along with how much impact they made on achieving the challenge of maintaining quality of service and energy conservation for wireless sensor networks.

4.1 Low Duty Cycle Protocols

In order to tackle the problem of idle listening low duty cycle protocols are proposed. The main theme behind these protocols is to avoid spending valuable energy in idle state and reduce the communication activities of a sensor node. In such schemes the node spends most of the time in sleep state and only wakes up when it is about to either transmit or receive data packets.

In order to implement such protocols several different approaches are used, in [33] cycled receiver approach is used. In this approach a node spends most of the time sleeping while it wakes up periodically in order to receive the packets from the neighbouring nodes. The scheme is divided into listen, wakeup and sleep periods as illustrated in Fig. 3. In such schemes the node listens to the channel during the listening period and goes back to sleep mode if there is no activity over the channel or in other words no node tries to direct packets towards the listening node. In order to communicate the transmitting node must acquire knowledge of the listening period for its receiving node. This can be achieved by letting the node send a short beacon at the start of its listening period [33], another approach is the transmitting node continuously sends request packets towards the destination until one hits the destination listen period. There are several problems in such schemes and algorithms, by choosing small duty cycle the traffic from the neighbouring nodes towards the destination node causes congestion especially in high traffic load scenarios this causes severe degradation in network performance, however by selecting low duty cycle a considerable amount of energy can be conserved. On the other hand a long sleep period causes high latency; in the case of multi-hop scenarios the per-hop latency significantly increases causing high end to end latency.

There are several variants of Low duty cycle protocols available some of the major protocols are discussed in the following section.

4.1.1 Sparse Topology and Energy Management

The STEM protocol [34] was developed to provide a solution for the idle listening problem. STEM is suitable for those networks which occasionally transmit data or that have no hard latency requirements an example of such networks could be a habitat monitoring system in which the node has to wait a long time before it transmits any considerable change in the environment, thus in order to enhance battery life and reduce the amount of energy consumed STEM protocol can provide a potential solution for idle listening problems in such types of networks. Figure 4 illustrates the STEM protocol duty cycle for a single node. It can be seen from the figure that STEM uses two channels which requires two transceivers in each node one for the wakeup channel and the other for the data channel. The data channel is always in sleep mode except when transmitting or receiving the data packets. The wakeup channel is subdivided into listen period, wakeup period and

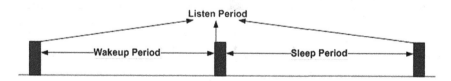

Fig. 3 Periodic wakeup scheme

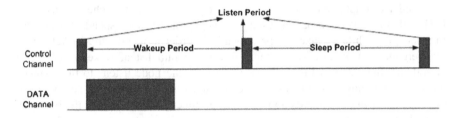

Fig. 4 STEM duty cycle

sleep period. If the node does not receive any signal or request from the neighbouring node it goes directly to the sleep period switching its transceivers to off state as well. However upon detection of any request the node starts packet transfer on the data channel. STEM protocol has two different variants on the basis of methodology for the transmitter to acquire receiver acknowledgement, these variants are STEM-B and STEM-T.

4.1.2 Sensor MAC

S-MAC protocol proposed in [32, 35] tends to give a possible solution for idle listening, collisions and overhearing problems. Unlike STEM protocols S-MAC doesn't have two channels. S-MAC adopts a periodic wake up scheme subdivided into wakeup and sleep periods. However unlike STEM protocol during the listen period in S-MAC a node can receive and transmit packets. Figure 5 illustrates the duty cycle adopted by S-MAC protocol. S-MAC tries to coordinate with other neighbouring nodes such that their listen period starts at the same time. In S-MAC protocol listen period of the node is further subdivided into three states.

The first state is also known as the synchronisation state. In this state a node accepts its neighbouring nodes synchronisation packet which contains the neighbouring node schedule; the receiving node stores this information in a table also known as a schedule table. The synchronisation phase is subdivided into time slots

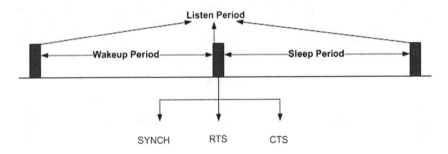

Fig. 5 SMAC duty cycle

and all the neighbouring nodes of the intended receiver contend using CSMA protocol with backoff algorithm. In this case each neighbouring node of the intended receiver chooses a random time slot and starts to transmit its synchronisation packet if no signal or activity is detected in that particular slot otherwise the node will go to sleep mode and wait for the next wakeup cycle of the intended receiving node. On the other hand in order to maintain synchronisation in the network and to allow new nodes to be aware of their surrounding topology, the node sends a synchronisation packet in the broadcast to all the neighbouring nodes, in this way all the nodes which are initially part of the system can synchronise again and the new node can be accommodated in the network. This period is known as synchronisation period.

After the synchronisation state the next state is RTS state. In this state the node listens for any RTS packet from the neighbouring nodes. S-MAC uses RTS/CTS handshake in order to minimise collision due to hidden node problem. The interested nodes having data to transmit contend using CSMA with backoff to send RTS packet towards the intended receiver.

Finally the node after receiving the RTS packet sends out a CTS packet to the transmitting node, after this data packet exchange starts between these two nodes.

Hence by using a combination of RTS/CTS handshake, virtual carrier sensing and network allocation vector tables (NAV) S-MAC protocol tries to tackle the problem of overhearing, idle listening and collision . Hence a virtual cluster is formed in S-MAC protocol if all the nodes within the network know about their neighbouring node schedule they all wake up at the same time to perform data or control packet exchange activity. In general the working of S-MAC protocol uses the following steps. A new node switches on its transceiver during the listen period for a time equal to synchronisation period. If the node receives any synchronisation packet from any of the neighbouring nodes, it adapts to this schedule and broadcasts it again in the next listen period of its neighbouring node. If this is not to be the case then the new node randomly picks up its own schedule and broadcasts it. However if the node receives another schedule during the broadcast synchronisation period it drops its own schedule and adapts to the new one. Moreover there can be a scenario in which the node receives a new schedule after it has chosen itself and its schedule has been adopted by some of the other neighbouring nodes. In this particular condition the node will retain both of the schedules and transmit synchronisation packets in the two different schedules which it receives. On the other-hand if the node knows that there is no other neighbouring node sharing its previously adopted schedule then the node drops its own schedule and adopts the new one.

In this case S-MAC makes virtual clusters of different schedules; the border nodes of these clusters have to adopt both of the schedules and are most likely to lose more energy in order to work like a bridging node between the two virtual clusters having different synchronisation schedules. S-MAC in terms of latency pays a high price as most of the time the nodes are in sleep mode. A variant to S-MAC is proposed in [32] which introduce the concept of adaptive listening to reduce the latency of the S-MAC protocol itself. In such schemes a node listens to the RTS/CTS handshake between the two nodes. These control packets also carries the duration of the transmission between the two nodes. The overhearing node for these packets knows the fact that it is also in the neighbourhood of the intended receiving node and hence it increases its duration of listening just in case the packet is destined to some other node which is one hop away from the actual destination node and the over-hearing node can provide this hop. In this way the adaptive listening procedure actually reduces the per-hop latency of the packet.

S-MAC protocol has one major drawback it is hard to adopt the wakeup and listen period according to the changing traffic load. Another variation of S-MAC is T-MAC [33]. This protocol enables nodes to shorten the time of listen period if the node doesn't receive any activity on the channel. If a node doesn't receive any signal during a set defined duration within its listen period it is allowed to go in to sleep mode whereas in the case of S-MAC the node has to keep listening the whole time of listen period before it makes this decision.

4.1.3 Mediation Device Protocol

Mediation device protocol [34] can work concurrently with wireless sensor network industrial standard IEEE 802.15.4 especially in peer to peer mode [35–37]. As all other preceding protocols the prime objective of this protocol is also to conserve energy and hence it allows nodes in wireless sensor networks to periodically go to sleep mode and wake up only for a short duration of time in order to exchange the data. The protocol doesn't involve any complex arrangements between the nodes; each node has its own individual sleep schedule and is not required to gain information about schedule for other nodes.

In general when a node wakes up it transmits a query beacon, this beacon contains the address of the node. After transmitting the query beacon the node waits for some time according to the protocol in order to receive any packet or signal. If the node during this wait time does not receive any packet it goes to sleep mode. For a transmitting node when it transmits the packet it has to synchronise with the schedule of the neighbouring receiving node, in order to achieve this the transmitter would stay awake in order to receive a query beacon, this whole exercise consumes a lot of energy of the transmitter which is undesirable especially in the case of wireless sensor networks. The problem is tackled by using dynamic synchronisation in the protocol, in order to achieve such synchronisation mediation device is used (MD). The protocol assumes that the mediation device is not energy constrained and has a full duty cycle. Because the device has a full duty cycle it will remain active all the time and is able to receive packets at any time from all the nodes within its radio range and will be aware of the schedules of all the nodes in the surrounding area. The scenario is depicted in Fig. 6. In the figure there are three nodes X, Y and Z. The node Z acts as mediation device for node X and Y. Node X wakes up and sends its query beacon towards node Z, since the node X has data to transmit it sends RTS signal towards node Z, after transmitting the RTS packet the node X waits for some time but as it doesn't receive any signal

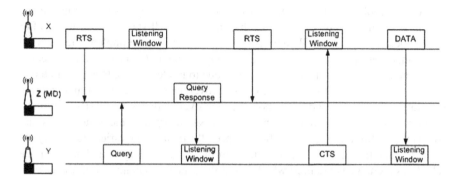

Fig. 6 llustration of mediation device protocol

from any other node it goes to sleep mode. As node Y wakes up which is the intended receiving node in this scenario, it sends a query beacon towards node Z. The mediation device Z sends query response packet towards node Y which has the wakeup detail and address of node X. In this way node Y knows about the wakeup cycle for node X more specifically the receiving window timing just after the query beacon from node X. Hence in the next wakeup cycle when node X sends its RTS packet node Y responds with a CTS packet and the data packet transmission occurs between the two nodes, after the successful transmission the node Y sends acknowledgement for the data reception, dislocates itself from node X wakeup schedule and follows its own previous wakeup cycle.

Under the main assumption that the mediation device is not energy constrained and can have a full duty cycle the protocol has some advantages. First it does not require any global time synchronisation between the nodes, all the synchronisation is done through the mediation device. Secondly the burden of energy is only shifted to mediation device and not the transmitting or receiving nodes. This enables them to have a low duty cycle and conserve as much energy as they possibly can. However the idea of having a mediation device not affected by energy especially in the case of wireless sensor networks is not practical. Moreover the nodes as they wake up send query messages, in the case of a dense network where nodes can have their duty cycle overlap there can be a collision of query beacons at the start of the wakeup cycle. The collisions will increase as more and more active nodes enter the network. Secondly in order to cover the entire network the number of mediation devices should be more than one, again making it difficult to support such devices in energy constrained wireless sensor networks.

4.2 Schedule Based Protocols

Many MAC protocol designs for wireless sensor networks fall in the category of schedule based protocols. The advantage of schedule based protocols is that they tackle the problem of idle listening by allocating or assigning time slots for the transmission and reception nodes so that the nodes will stay in sleep mode after their allocated time. Secondly a transmission schedule can be implemented in such a way that there is no collision at the receiving node and hence avoiding specific and complex algorithms to avoid collisions due to hidden nodes. However in terms of drawback such protocols use extensive overhead signalling for maintaining scheduling in cases of variable topology, such overheads are too costly in term of resource constrained wireless sensor networks. Moreover the schedule is not adaptable to different load conditions especially if the load varies frequently. In order to maintain a schedule and the information regarding neighbouring node schedules memory is required which is not that much in abundance in the case of tiny sensor networks. Following are some of the major schedule based protocols designed for wireless sensor networks to handle such problems and difficulties.

4.2.1 Low Energy Adaptive Clustering Hierarchy Protocol

The LEACH protocol is presented in [38] it uses TDMA based MAC protocol with clustering and routing technique. The protocol targets dense homogeneous wireless sensor networks with energy constrained nodes. The protocol divides the network into several clusters and each cluster has a cluster head which is responsible for creating and maintaining the TDMA schedule among the nodes and the rest of the nodes are the member nodes. The member nodes use these assigned TDMA slots to exchange data among themselves and to the cluster heads. The cluster heads then aggregate all the data and transmit it to the sink node or to another node for relaying purposes. Since most of the time the sink node is far away in large networks the cluster head uses more power to transmit the data towards the sink, whereas the rest of the nodes use the same low power in order to reach the cluster head. The cluster heads are the nodes which consume a lot of energy since they are responsible for communicating data towards the sink and they have a full duty cycle as well. Hence if a single node is chosen to be the cluster head for a long duration it will deplete all its energy resources in a quick time which results in a communication breakdown between the cluster and the rest of the network including the network sink. In order to overcome this problem the cluster head responsibility is rotated within the cluster so that no one node exhausts its resources. The process of becoming cluster head is independent and each node decides on its own that it has to act as a cluster head without any kind of signalling and election process. The decision to become cluster head depends on when the node became the cluster head for the last time and if the time duration is longer than the chances of a node opting to become cluster head become higher, the process is also described in [39]. Moreover the nodes which choose to remain non cluster head chose their cluster head on the basis of received signal strength from the potential cluster heads.

After the formation of a cluster each cluster head randomly selects a CDMA code for its cluster which it broadcasts to all its member nodes. The reason behind this is to stop the interference from the boarder nodes sitting at the edges of each cluster. The protocol comprises of different phases. The initial phase is known as the setup phase in which the node self elects itself as the cluster head. The next phase is termed as the advertisement phase in which the cluster head informs all neighbours with an advertisement packet . The cluster head for this phase uses the CSMA protocol. The non-cluster head member nodes pick up the advertisement packet of the strongest signal and send their information to the cluster head in the joining phase. The nodes use the CSMA protocol in the join phase as well. After the join phase the cluster head knows the number of members in its cluster and constructs a TDMA schedule for the neighbour node alongside picking up a random CDMA code for the cluster communication. The cluster heads then broadcast the schedule along with the code to the respective nodes. After this phase TDMA steady state phase starts in which nodes exchange data packets with the cluster heads.

LEACH protocol suffers from collision as well; due to the reason of collisions between advertisement or join packets the protocol doesn't guarantee that all the non-cluster nodes will be part of a cluster. The cluster head has to be switched on all the time whereas the non-cluster head should be awake during setup phase. This creates an extra burden in terms of loss of energy and the entire setup time causes an increase in the latency of data transmission. Moreover for large areas LEACH protocol is not suitable as the cluster head has a finite amount of range beyond which it cannot communicate with the sink directly, this in terms of scalability is the major disadvantage of the protocol as no two cluster heads communicate with each other.

4.2.2 Self Organizing Medium Access Control for Sensor Networks Protocol

The SMACS presented in [40, 41] is a part of wireless sensor network protocol suite that covers MAC, mobile nodes, multihop routing protocol, neighbour discovery and local routing protocol. SMACS assume that the available spectrum is divided into several channels and each node can tune its transceiver to any one of them as well as the nodes have several CDMA codes at their disposal. As far as topology is concerned the network is static for a long time. Each node has fixed time slots and these are divided into superframes. All the nodes within the network have the same superframe length; the superframes are also divided into time slots. The general working principal of this protocol is illustrated in Fig. 7. Assume that there are two nodes in the network X and Y as shown in Fig. 7. In the start of SMACS protocol the nodes perform neighbourhood discovery. Consider node X wakes up and listens to the frequency channels, if the node does not find any activity or any message over the link it transmits an invitation message with its own address and number of neighbours attached to it, assume that the total number of nodes attached to node X is zero. When node Y receives a message from node X it waits for some duration of time and then replies with a message containing node

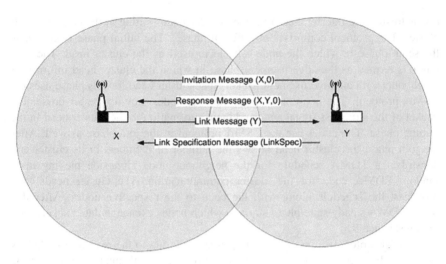

Fig. 7 Neighbourhood discovery mechanism in SMACS

X and Y addresses and the amount of neighbours attached with it, so far assuming that the total number of neighbours with node Y is also zero. As node X receives this message it invites Y to construct a link by sending another message to Y containing Y's address only. This message indicates to node Y that it has been selected and that it can select any time slot for the transmission and reception of data as no other nodes are attached with node X. In response node Y establishes a link with node X by transmitting a link specification in which it gives the information of the time slots and code over which these two nodes will communicate to each other.

In a scenario when both the nodes have neighbour nodes attached to them the protocol runs by starting with the transmission of an invitation message from node X with the address of itself and information regarding the attached nodes. Node Y also responds by sending the address of it and node X with all the information regarding neighbouring nodes. In response to that node X sends it's entire schedule to node Y which has all the timing details and slots which are reserved for communication with other nodes. Node Y after carefully selecting mutually common time slots which are not overlapping with the existing neighbour nodes of either of nodes sends back the link specification to node X. In this manner the entire protocol works in each cycle.

The major drawback of this protocol is the length of superframe. If the superframe is too short then all the neighbour nodes will not be visible to the node. Another drawback will be in terms of network load and number of nodes. If the network load is low but the number of nodes in the neighbourhood is high the node will be awake on every set schedule just to find no data transfer. The protocol incurs high overheads and can consume energy.

4.3 Contention Based Protocols

In contention based protocol the nodes in a network have to compete in order to transmit their data from one node to another and eventually to the sink node. The quality of the network and performance of the MAC protocols significantly depends upon the number of active nodes, traffic load and topology of network. If in a given network the number of active nodes are increased this would lead to an increase in the chances of collision due to nodes that select the same slot to transmit but are also in the communication range of each other as well those nodes which are hidden to the transmitting node. This causes a significant loss of transmitter and receiving node's energy as well as adding to network latency and overall throughput degradation of the entire network. The two most important and common contention based protocols which are already described in the earlier section are ALOHA and CSMA. In the following section variants of these protocols are discussed which are especially designed to fulfil the needs of wireless sensor networks and tackle the problems which are unique to such networks.

4.3.1 Carrier Sense Multiple Access Protocols

In [42] several CSMA variants are discussed and their performance measures are presented. The study includes fairness and energy consumption in these protocols. The author considers multihop networks with single and multiple sinks. The network is considered as being in the idle state for a long duration and becomes active after sensing something substantial. Upon triggering all the nodes within the network become active and try to communicate the data this results in collisions and degradation of quality of service of the network. The state diagram used to represent CSMA in [42] is presented in Fig. 8 for clarity of the protocol.

It is shown in the figure that a node undergoes several stages before transmitting data using The CSMA protocol. As the data packet reaches the buffer of the node it starts a random delay and initialises its trial counter. During this random delay period the node transceivers are in sleep mode, when the delay period is over the node goes to the listening state, if there is an on going activity over the channel the node goes to the backoff stage once again provided its number of trials is smaller than the maximum number of allowable trials. During this backoff the node once again can go in to sleep mode. After the backoff delay period is over the node once again enters to listening state, if the medium is found busy again the node either drops the packet if the number of the retrial counter exceeds its maximum limit or it goes to backoff delay once again. However if the medium is found to be idle the node sends an RTS signal towards the intended receiver, this handshaking signalling can be dropped if the load over the network is small. After a RTS the node waits for a CTS signal for a bounded duration of time, if it receives a CTS then it starts the data transmission and waits for acknowledgement of data reception otherwise the node goes to the backoff state once again if the number of retrials is

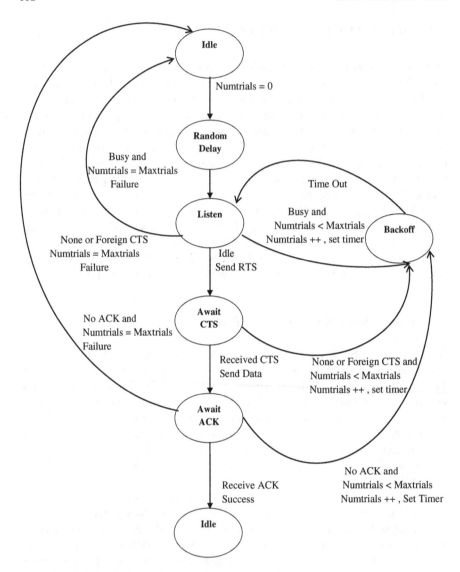

Fig. 8 Block diagram of CSMA protocol

still smaller than the maximum limit or drops the packet. Several variants of this protocol with random delay, non-random delay, fixed window backoff, exponentially increasing backoff, exponentially decreasing backoff and no backoff have been proposed and investigated in [42], the author suggested that a protocol with random delay, backoff algorithm and fixed listening time performs better in single hop scenarios.

4.3.2 Power Aware Multi-Access with Signalling

PAMAS protocol is presented in [43], the protocol provides a mechanism against overhearing by combining RTS/CTS handshaking and the busy tone used by the MACA protocol in [3]. PAMAS uses two channels one for control signalling and the other for data packet transmission. If a node becomes activated and has some data packets in its buffer it starts by sending a RTS packet over the control channel towards the intended receiving node without performing any carrier sensing. The RTS packet carries both the addresses of transmitting and receiving nodes including the length of transmission. As the intended receiver recycles the RTS it responds by sending a CTS packet towards the transmitter provided there is no ongoing transmission near the receiving node going on. As the transmitting nodes receive CTS packets they start sending data packets over the data channel towards the receiver, however the receiver receiving the data packet also sends a busy tone on the control channel. If the transmitting node doesn't receive any CTS packet from the intended receiver it waits for a bounded time interval and then goes into an exponential backoff state.

As it is clear from the discussion that protocol requires some sort of synchronisation or awareness to the node so that when they wake up from the sleep mode they should realise that the channel is available for any ongoing transmission taking place in its neighbourhood, moreover a mechanism is needed if CTS and RTS packets are corrupted and the node doesn't know exactly how long the transmission which is taking place in the neighbourhood will last for. For this purpose the protocol supports a probing procedure. The node runs the probing protocol in order to find out how long the data transmission which is ongoing will last so that it can start its data transmission setup and most importantly during the time of ongoing data transmission can switch its transceiver off and goes into sleep mode. In probing protocol the node using a query sends out probing packets on the control channel, and any transmitter which finishes its transmission in the time interval sends to the query node the query response signal on the control channel in this way the node knows how long the ongoing transmission will take. Most importantly in dense wireless sensor networks this process requires more computation and retransmission of the probing packet until the node finds out the actual duration of the ongoing transmission.

The major drawback of the protocol is it requires heavy computation in terms of probing and in the case of wireless sensor networks the resources are limited moreover the control packets are transmitted without any carrier sensing this could also lead to high latency before the transmission of data packets and degrade the quality of service of the network.

5 IEEE 802.15.4 MAC Protocol

The industrial standard for wireless sensor networks is the IEEE 802.15.4 developed in October 2003 [44, 45]. The standard covers the physical layer and the MAC layer for low rate wireless personal area networks. The targeted application for the IEEE 802.15.4 standard requires low bit rate and energy conservation. The physical layer provides a bit rate of 20 Kbps for a single channel in the frequency range of 868 – 868.6 MHz, 40 Kbps for ten channels in the range of 905 – 928 MHz and 250 Kbps for 16 channels in 2.4 GHz ISM band. There are a total of 27 channels available but the MAC protocol used in this standard uses only one channel at a time as it is not multichannel. The MAC protocol uses both schedule based as well as contention based periods and the nodes can be homogenous as well as heterogeneous within the same network.

5.1 Network Architecture

The MAC protocol uses two types of nodes Full function devices and reduced function devices. The Full function device nodes are divided into three operational categories Personal Area Network Coordinator or PAN, a simple coordinator or just a simple device. The reduce function node works only as a simple device for transmission and reception. A device is associated with a coordinator node which must be a FFD and only communicate with it, thus in principle forming a star network with the coordinator. However the FFD nodes can communicate in a peer to peer fashion and multiple coordinator nodes form a PAN and have a central coordinator known as a PAN coordinator which also serves as the central command of the network as well as the sink. The role of coordinator node in IEEE 802.15.4 standard is defined as a node which manages a list of all the associated devices. Devices can associate, disassociate and re associate with the coordinator. The coordinator is responsible for the allocating of short addresses to all the devices associated with it. In order to maintain synchronisation the coordinator node in beacon enabled mode sends a continuous beacon which has information about the start of the next superframe, outstanding requests and other parameters; moreover the coordinator upon the request of the node can also allocate a fixed slot for the transmission of data.

5.2 Superframe

In beacon enable mode the coordinator organises channel access and data transmission using the superframes presented in Fig. 9. All the superframes are equal in lengths, the coordinator starts the superframe by sending a beacon packet which

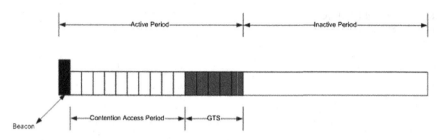

Fig. 9 Superframe structure of IEEE 802.15.4

marks the starting of the frame and also contains information regarding the lengths of various components of the superframe. The superframe of IEEE 802.15.4 beacon enable mode is subdivided into an active period and inactive period. During the inactive period all the nodes including the coordinator switch off their radios and go to sleep mode. All nodes become active just before the inactive period of superframe ends in order to receive the beacon. The active period is divided into 16 time slots, the first time slot is occupied by the beacon frame whereas the remaining is portioned as contention access period and guaranteed time slot period which comprises of a maximum seven time slots and it is solely at the discretion of the coordinator to allocate to the requesting node. Only the node which has been allocated the GTS slot is activated during its specific time slot whereas the rest of the nodes in the network goes to sleep mode.

5.3 Slotted CSMA/CA Protocol

The node in CAP of the superframe uses the slotted CSMA/CA protocol for data transmission. There is a difference in CSMA/CA protocol use in wireless LAN IEEE 802.11 and wireless Sensor network IEEE 802.15.4, in the latter the protocol has no defence mechanism against the hidden nodes i.e the protocol doesn't supports the RTS/CTS handshake algorithm. The operation of this protocol is described in Fig. 10. The 16 time slots in the superframe are further subdivided into smaller slots known as the backoff period. One backoff period has a length of 20 channel symbols, the CSMA/CA considers each backoff period as a single slot. The protocol contains three variables NB (Number of Backoff), CW (Contention Window) and BE (Backoff Exponent). When a node has packets in the buffer it initialises these parameters to the values of NB = 0, CW = 2 and BE = macMinBE respectively. The node waits for the next backoff period and computes a random number in the range of [0, 2BE -1]. This random number is the backoff delay and the device waits for this period before attempting any further steps towards the transmission of data. After the delay period is over the node with the next backoff period performs clear channel access (CCA), if the node finds the channel to be

Fig. 10 Slotted CSMA/CA
in IEEE 802.15.4

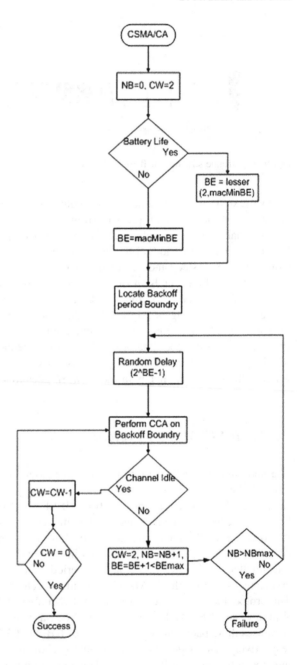

idle then it decrements the value of CW and with the next backoff period boundary performs CCA again, if the channel is found to be idle again the node assumes that it has won the contention and starts data transmission. If during any one of the CCA processes the channel is found to be busy the node increments the value of

NB as well as BE, if the number of retrials exceeds the allowable NB limits the packet is dropped otherwise the node selects once again a random backoff delay in the range of [0, 2BE -1]; Moreover for each retrial the value of CW is also initialised to its original value of 2.

5.4 Un-Slotted CSMA/CA Protocol

As discussed earlier IEEE 802.15.4 works in beacon enable and non-beacon enable mode. In non-beacon enable mode the protocol uses un-slotted CSMA/CA MAC. Figure 11 shows the operation of un-slotted CSMA/CA. In this mode the coordinator node doesn't send any beacon and there are no GTS slots hence in this mode the nodes are not synchronised. The node performs clear channel access

Fig. 11 Un-slotted CSMA/
CA in IEEE 802.15.4

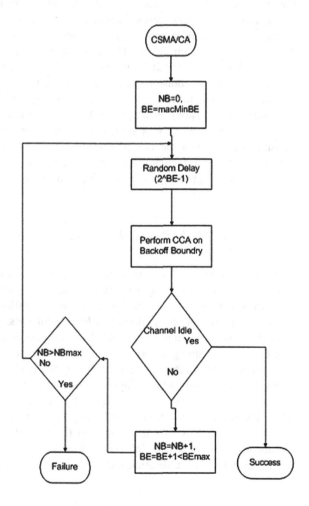

once and then transmits the data packet. In this protocol the coordinator is always in active state where as all other nodes in the network can go to sleep mode. The major drawback of this protocol is as the network is scaled up more and more nodes will face the problem of collision as there is no set universally defined super frame available which all the nodes follow. Moreover the latency of the network will also increase as the sleep and wakeup schedule of every single node in the network will differ from the other nodes because of lack of synchronisation.

6 Mobile WSN Key Problems

Mobility in wireless sensor networks raises new challenges and problems that occur at MAC level due to the existing link failure and new link formation. This in turn introduces delays which are associated to establish connectivity. Moreover due to mobility neighbourhood information is variable which causes problems in scheduling. The mobile nodes may experience loss of connectivity in a network using schedule based protocols. The reason is as the node enters in the radio range of a new cluster has to wait for the next schedule synchronisation point, this leads to temporary loss of connectivity and an increased latency for the packet which the mobile node wants to transmit. Moreover mobility may also increase the chances of number of collisions in the network, as the mobile node has no information about the neighbouring nodes in the new cluster, thus can interfere in the on-going transmission between the two nodes and causes collision in the network. This collision will lead to increase in number of retransmissions, resulting in increase in packet latency, high energy consumption and often resulting in packet drop.

6.1 Effect of Mobility on the Quality of Service

Node mobility in wireless sensor network under certain scenarios can be beneficial, as in case of increasing network lifetime and network coverage [46, 47], and [48]. However as discussed in earlier section mobility also imposes new challenges and problems on the design of MAC protocols. As the node moves away from the communicating node frame loss and packet drop can occur due to the significant signal strength variations [49]. Moreover due to hidden node and non-synchronisation with the new cluster the mobile nodes contribute in the increase of number of collision which plays a significant role in the degradation of Quality of Service in wireless network. Especially in case of wireless sensor network where the resources are limited and energy conservation is significant, collision causes excessive loss of energy. Due to the loss of packet number of retransmission increases resulting in sever degradation in throughput, loss of energy, inefficient bandwidth utilisation and higher latency for the network.

6.2 Challenges and Motivation for Mobile WSN

Designing MAC protocol for mobile scenario is a challenging task; especially in case of WSNs which are inherent resource constrained. The prime objective of such type of protocols in mobile scenario is to maintain connectivity and acceptable quality of service while incurring less collision in the network. The protocols should be less complex and conserve energy. Moreover the time associated for connectivity, neighbourhood discovery and synchronisation should be minimised as these factors contributes significantly in the increase of latency.

6.3 Classification of Mobility in Wireless Sensor Network

The breaking of existing link and formation of new due to one hop neighbours moving out of radio range due to mobility causes packet loss, increase in latency, increase in number of collision and high energy consumption [50–52]. Mobility causes rapid changes in topology of the network [53]; hence become increasingly challenging especially for MAC protocols to adapt to such fast changes.

Mobility in wireless sensor network can be broadly classified into the following three main categories:

- Nodes are mobile while the Sinks are stationary.
- Sinks are Mobile and Nodes are stationary.
- Nodes and Sinks both are mobile.

In a typical WSN the sink is located centrally and stationary, however the area around the sink experience high level of communication activities, due to this reason the energy supplies of the node near the vicinity of sink depletes considerably fast. This may result in the disconnection of the sink with the entire network. In order to mitigate this problem Mobile Sink concept is introduced. [54] proposes sink relocation methods on the basis of energy consumption of individual sensors are evenly distributed and overall network energy is minimised. Load balancing and data transmission is investigated in [55], the authors in order to maximise the network life time proposes joint mobility and routing scheme. In this scheme the mobile sink moves on a circular trajectory inside the area where the nodes are deployed. The sensing nodes which are inside this trajectory send the data to the mobile sink using shortest possible path to conserve energy.

In order to address the problems of connectivity in sparse network where the numbers of sensing nodes are low mobile data collector scheme is introduced [56]. In this scheme the mobile node gathers all the buffer information stored in an individual node. Investigation over different mobility patterns is presented in [57]. A random mobility pattern is introduced for mobile data collector in [58]. In this scheme a mobile data collector randomly moves to collect the data buffered in the nodes of a sparse network. Since the movement of the data collector is random therefore the network experiences latency. In [59] knowledge based mobility or predictable mobility for mobile data collector is introduced to meet some latency bounds as well as save energy of the individual nodes. The nodes in such scenario are aware of data transfer time and based on this go in sleep mode to save energy, hence increasing network life time [60]. In majority of WSNs sensor nodes generate data at different rates. Data losses in sensor node occur if the buffer if the node is full and the contents of the buffer are not being transferred to the mobile data collector. In order to resolve such issues a scheme referred to as controlled mobility [61, 62] is proposed. To achieve controlled mobility Mobile Element Scheduling is proposed in [63]. The mobile element (ME) is schedule in real time to visit the nodes before the buffers are full. In order to improve upon the protocol two more proposals Earliest Deadline First (EDF) and Minimum Weight Sum First (MWSF) are proposed in [63].

In EDF the decision of the next node to be visited by ME depends upon the node having its buffer overflow deadline first. EDF scheme leads to significant amount of data loss due to the fact that nodes are far away as the network is sparse and two consecutive nodes can have the same deadline for buffer overflow. In MWSF this problem is tackled by introducing distance factor. The ME scheduling not only depends upon the earliest deadline of the buffer overflow it also take into account the fact of how far the nodes are within the network. Even MWSF introduces the distance factor before deciding which node to visit but the back and forth movement in order to reach far away nodes not completely avoided. In [64] multiple mobile data collector with relay data collection algorithm is proposed to increase the scalability of the sparse network and to curb the problem of distance between the nodes. In this algorithm nodes which are out of range from mobile data collector finds the nearest sensor nodes to the MDC and relay their data to them using shortest possible route.

6.4 Case Study for IEEE 802.15.4 MAC Protocol for Mobile Sensor Network

In [65] IEEE 802.15.4 industrial standard CSMA/CA MAC protocol is presented. CSMA/CA uses random backoff values as collision resolution algorithm. Since WSN is resource constrained therefore the entire backoff phenomenon is performed blindly without the knowledge of the channel condition, this factor contributes in high number of collision especially when the number of active nodes are high. The IEEE 802.15.4 MAC protocol was mainly used in static scenarios and initial research work and performance studies are done while considering static applications [37, 66–68]. Recently the use of mobile nodes in WSN becomes a very hot topic [62, 69], and [70]. Introducing node mobility in WSN raises many challenges and problems, especially at MAC level. IEEE 802.15.4 MAC CSMA/CA is also used in mobile applications however the QoS is very poor

[71–73]. The protocol incurs high latency over the mobile node as it requires long association process for the node leaving from one cluster to another cluster. If the node miss the beacon from the new cluster head consecutively for four times then the node will go in orphan realignment process hence inducing more delays. In order for the node to join the new cluster and not to miss the beacon from the new cluster head it tends to wake up for long duration which in turns significantly increases the energy consumption of the individual node. Moreover as the mobile node succeeds in migrating from one cluster to another it has to compete with the already available active nodes in the cluster which contributes in significant number of collision and packet drop, resulting in overall degradation of network QoS.

6.4.1 Association, Synchronisation and Orphan Scan in IEEE 802.15.4

The process of association and synchronization in IEEE 802.15.4 CSMA/CA MAC protocol is presented in Fig. 12. The node association starts with an active scan procedure that scans all listed channels by sending beacon requests to all nearby coordinators. All the information received in a beacon frame will be recorded in a PAN descriptor. The results of the channel scan will be used to choose a suitable PAN. The node then sends a request to associate with the chosen coordinator. The node updates its current channel and PAN id while waiting for an acknowledgement from the coordinator. Upon receiving an acknowledgement, the node then waits for the association results. The coordinator will take aResponse-WaitTimesymbols (32*aBaseSuperframeDuration, about 0.49 s) to determine whether the current resources are available on the PAN in order to allow the node to associate. If sufficient resources are available, the coordinator then allocates a short address to the node and sends an association response command containing a new address and a status indicating a successful association. If there are not sufficient resources, the node will receive an association response command with a failure status.

After the node associates with its coordinator, it will send a request to synchronize and start tracking the beacons regularly. If the node fails to receive a beacon aMaxLostBeaconstimes (equal to 4 times), it may conclude that it has been orphaned. The node then has the option either to perform the orphan device realignment procedure or perform the association procedure. If the node chooses to perform an orphan device realignment, it will do the orphan scanning by sending an orphan notification command to relocate its coordinator. The node waits for aResponseWaitTimesymbols to receive a coordinator realignment command. The coordinator that receives the orphan notification command will search its list looking for the record of that node. If the coordinator finds the record, it will send a coordinator realignment command to the orphaned node together with its current PAN id, MAC PAN id, logical channel and the orphaned node's short address. The process of searching the record and sending the coordinator realignment command takes within aResponseWaitTimesymbols.

Fig. 12 Association and
synchronization process in
IEEE 802.15.4 CSMA/CA
MAC protocol

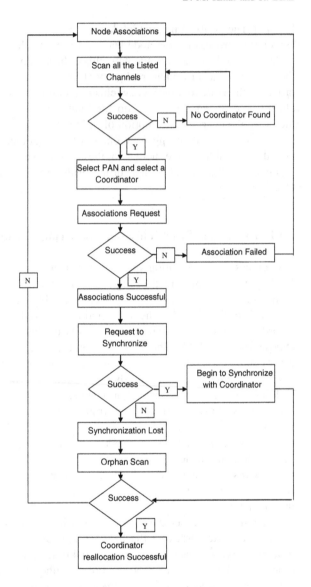

6.5 Other Mobile MAC Protocol for Wireless Sensor Networks

In [49] MOB-MAC is presented for mobile wireless sensor network scenario. MOB-MAC uses an adaptive frame size predictor to significantly reduce the energy consumption. A smaller frame size is predicted when the signal characteristics becomes poor. However the protocol incurs heavy delays due to the variable size in the frame and is not suitable for mobile real time applications.

AM-MAC is presented in [74]; it is the modification of S-MAC to make it more useful in mobile applications. In AM-MAC as the mobile node reaches the border node of the second cluster copies and hold the schedule of the approaching virtual cluster as well as the current virtual cluster. By adopting this phenomenon the protocol provides fast connection for the mobile node moving from one cluster and entering the other. The main drawback is the node has to wakeup according to both the schedule but cannot transmit neither receive data packet during the wakeup schedule other than the current cluster. This contributes to significant delay and loss of energy due to idle wakeup. In [75] another variation of S-MAC is presented by the name of MS-MAC. It uses signal strength mechanism to facilitate fast connection between the mobile node and new cluster. If the node experience change in received signal strength then it assumes that the transmitting node is mobile. In this case the sender node not only sends the schedule but also the mobility information in the synchronous message. This information is used by the neighbouring node to form an active zone around the mobile node so that whenever the node reaches this active zone it may be able to update the schedule according to the new cluster. The down side of this protocol is idle listening and loss of energy. Moreover nodes in the so called active zone spend most of the time receiving synchronous messages rather than actual data packets thus resulting in low throughput and increase latency. In [76] another real time mobile sensor protocol DS-MAC is presented. In this protocol according to the energy consumption and latency requirement the duty cycle is doubled or half. If the value of energy consumption is lower the threshold value (Te) than the protocol double its wakeup duty cycle to transfer more data, thus increasing the throughput and decreasing the latency of the network. However the protocol requires overheads and during the process of doubling of wakeup duty cycle if the value crosses Te the protocol continues to transmit data using double duty cycle resulting in loss of energy. MD-MAC is proposed in [77] which is the extension of DS-MAC and MS-MAC. The protocol enforces Te value and during any time if the value of energy consumption is doubled then it halves the duty cycle. Moreover the mobile node only undergoes neighbourhood discovery rather than other neighbouring node which forms an active zone as in the case of MS-MAC. MD-MAC is complex and requires high overheads.

7 Open Issues and Future Research Directions

WSNs are being rapidly used in various applications; these applications have different QoS requirements. This expansion in the utilisation of WSN creates new future directions and generates various challenges to overcome in order to meet and satisfy the demands of various applications. Following are some of the major issues and research directions which are open for investigation due to the increase in applications for WSNs.

- Heterogeneous networks: Traditionally WSNs comprises of homogenous nodes, but as the WSNs are being used in diverse applications the networks are now comprises of mostly heterogeneous nodes. Hence one research direction can be to develop novel MAC protocols which have the ability to fulfil the diverse QoS requirements of heterogeneous sensor networks.
- Multimedia applications: Due to the size and ease of deployment WSNs are very strong candidates for multimedia applications. As it is known that the delivery requirements of multimedia data is different than that of scalar data. Such type of data requires higher throughput, bounded delay, low congestion and reliability. Therefore another potential area for research is developing novel MAC protocols that meet the requirements of multimedia WSNs.
- Multi applications over single platform: With the advent new powerful operating system for WSN, it is now possible to run multiple applications over a single node. This on the other hand required large amount of data flow over the network and becomes even more challenging if this data has different priority levels. Development of protocols that support QoS for multiple applications having diverse priority levels is another research direction that should be further investigated.
- Coexisting networks: As WSN is being used in multiple application scenarios, therefore more than one WSN can be in the neighbourhood of other sensor networks. This causes sharing of resources which are already limited. Moreover different networks in the neighbourhood need to collaborate to achieve fairness in sharing the wireless resources. Hence collaborative QoS providing MAC protocols can be another topic of research.
- Mobile sensor networks: Traditionally majority of the applications in which WSN is used were static, as new applications emerges WSNs are also being used in mobile applications. This brings up new challenges and problems to tackle. Mobile WSN can be one of the existing and challenging research directions which need further investigation in order to meet QoS requirements.
- Energy efficiency: Mostly QoS and energy conservation are reciprocal to each other. As the applications are varied the requirement of QoS along with energy conservation is increased. This provides a unique challenge and need further research and analysis to develop such a novel MAC protocol that can satisfy both energy as well as acceptable QoS for modern applications.
- Hardware implementation: Majority of the protocol available in the literature shows the performance on the basis of simulations results. However there is a strong need to implement these protocols on hardware platforms and verify their performance in practical scenario. Implementation of new and old protocol over test bed and relaxing the assumptions over the protocols which are set due to simulation environment is another open research area which has a lot of potential.

Cross layer approach: Providing QoS solution for WSNs is not an easy task; since these networks are resource constrained. Therefore it is worth investigating cross layer approach to relax the limitation of MAC layer alone in order to achieve the required QoS for emerging applications.

8 Conclusions

At present WSNs are not only used for low data rate applications but also for more complex applications which requires reliability, time bounded and efficient transmission and reception of data. Moreover the WSNs are comprises of heterogeneous nodes that range from simple data collecting sensor nodes to video and microphones. Such increase in utilisation of sensor nodes in variety of applications requires high QoS providing protocols for WSNs. Keeping in view of such requirements this chapter focus on challenges, metrics, parameters and requirements of QoS aware MAC protocols for WSNs. The chapter provide in-depth and detail analysis and comparisons between the developed protocols and their achievements. Moreover mobile sensor networks are also presented and a case study of IEEE 802.15.4 in mobile scenario is discussed. The chapter also provides designing tradeoffs and open research issues in the field of QoS providing MAC protocols for WSNs to contribute to further research efforts in the fields of WSNs.

References

1. F.A. Tobagi, L. Kleinrock Packet Switching in Radio Terminals Part II The Hidden Terminal Problem in CSMA and Busy Tone Solution, IEEE Transaction on, Communications, 23(12), pp. 1417–1433, 1975
2. The Editors of IEEE 802.11; IEEE standard for Wireless LAN Medium Access Control (MAC) and Physical Layer (PHY) specifications, Nov 1997
3. P. Karn, A New Channel Access Method for Packet Radio, Proceedings of ARRL/CRRL Amateur Radio, 9th Computer Network Conference, pp. 134–140, Sep 1990
4. V. Bharghavan, *MACAW: A Medium Access for Wireless LANS, Proceedings of ACM SIGCOMM 94* (UK, London, 1994)
5. I. Rubin, Access Control Disciplines for Multi-Access Communication Channels reservation and TDMA Schemes. IEEE Trans. Inf. Theo. 25(5), 516–536 (1979)
6. S. Glisic, B. Vucetic, *Spread Spectrums CDMA Syst. Wireless Commun.* (Artech House, Boston, MA, 1997)
7. A.M.J. Goiser, *Spread Spectrum Techniques* (Springer Verlag, New York, 1998)
8. A. H. M. Ross, K. S. Gilhausen, CDMA Technology and IS-95 North American Standard, IEEE Press, the Communications Handbook, pp. 199–212, 1996
9. P. H. Lehne, M. Petersen, An overview of smart antenna technology for mobile communication systems, IEEE Communication Surveys and Tutorials, vol. 2, no. 4, 1999
10. A. Doufexi, S. Armour, M. Butler, A. Nix, D. Bull and J. McGehan A Comparison of Hyper LAN/2 and IEEE 802.11 a Wireless LAN standard, IEEE Communication Magazine, vol. 40, no.5, pp. 172–180, 2002
11. ETSI. TR 101 683, HYPERLAN Type 2: System Overview, Feb 2000
12. ETSI. TR 101 475, BRAN HYPERLAN Type 2: Physical Layer, March 2000
13. ETSI. TR 101 761-1, BRAN HYPERLAN Type 2: Data Link Control Layer Part1 Basic Data Transport Function, March 2000
14. ETSI. TR 101 761-2, BRAN HYPERLAN Type 2: Data Link Control Layer Part2, Radio Link Control Protocol Basic Functions, March 2000
15. M.J. Karol, Z. Liu, K.Y. Eng, An efficient demand-assignment multiple access protocol for wireless (ATM) networks. Wireless Netw. 1(3), 269–279 (1995)

16. N. Passas, S. Paskalis, D. Vali, L. Merakos, Quality of service oriented medium access control for wireless ATM networks. IEEE Commun. Mag. **35**(11), 42–50 (1997)

17. O. Sharon, E. Altman, An efficient polling MAC for wireless LAN. IEEE/ACM Trans. Netw. **9**(4), 439–451 (2001)

18. H. Takagi, *Analysis of polling systems* (MIT Press, Cambridge, MA, 1986)

19. F.A. Tabagi, L. Kleinrock, Packet switching in radio channels, Part III polling and dynamic split channels reservation multiple access. IEEE Trans. Commun. **24**(8), 832–845 (1976)

20. IEEE 802.4 Token Passing Buss Access Method, 1985

21. H.J. Moon, H.S. Park, S.C. Ahn, W.H. Kwon, Performance degradation of IEEE 802.4 Token buss network in noisy environment. Comput. Commun. **21**, 547–557 (1998)

22. N. Malpani, Y. Chen, N. Vadiya, J. Welch Distributed Token Circulation on mobile Adhoc Networks, IEEE Trans. Mobile. Comput. **4**(2), pp. 154–165, 2004

23. A. Willig, A. Wolisz, Ring stability of the PROFIBUS token passing protocol over error prone links. IEEE Trans. Ind. Electro. **48**(5), 1025–1033 (2001)

24. N. Abramson, Development of the ALOHANT. IEEE Trans. Inf. Theory **31**(2), 119–123 (1985)

25. L. Kelinrok, F.A. Tobagi, Packet Switching in radio channels Part1, Carrier Sense Multiple Access Models and their Throughput/Delay Characteristics. IEEE Trans. commun. **23**(12), 1400–1416 (1975)

26. W. Ye, J. Heidemann, and D. Estrin, Medium Access Protocol with coordinated adaptive sleeping for Wireless Sensor Networks, IEEE/ACM Transaction on Networking, 2004

27. E.Y.A. Lin, J.M. Rabaey, A. Wolisz, *Power Efficient Rendez-vous schemes for dense wireless sensor networks, IEEE ICC04* (June, Paris, 2004)

28. C. Schurgers, V. Tsiatsis, S. Ganeriwal, M. Srivastava, Optimizing sensor network in the energy latency density design space. IEEE Trans. Mobile Comput. **1**(1), 70–80 (2002)

29. J. Taneja, J. Jeong, and D. Culler,Design, modeling and capacity planning for micro-solar power wireless sensor networks, IPSN08, 7th International conference on information processing in sensor networks, IEEE Computer society, Washington DC, USA, pp. 407–418, 2008

30. C. Intanagonwiwat, R. Govindan, D. Estrin, J. Heidemann, F. Silva, Directed diffusion for wireless sensor networking. IEEE/ACM Trans. Netw. **11**(1), 2–16 (2003)

31. A. J. Goldsmith and S. B. Wicker Design Challenges for Energy Constrained Ad-hoc Wireless Networks, IEEE Wireless Communications, vol. **9**, no. 4, pp. 8–27, 2002

32. W. Ye, J. Heidemann, D. Estrin", An energy-efficient MAC protocol for wireless sensor networks. IEEE INFOCOM 2002, 1567–1576 (2002)

33. T. V. Dam and K. Langendoen, An Adaptive Energy Efficient MAC Protocol for Wireless Sensor Networks, Ist International Conference on Embedded Networked Sensor Systems, pp. 171–180, Nov 2003

34. E.H. Callaway, *Wireless Sensor Networks Architecture and Protocols* (Florida, Boca Raton, 2003)

35. E. Callaway, P. Gorday, L. Hester, J.A. Gutierrez, M. Naeve, B. Heile, V. Bahl, Home network with IEEE 802.15.4: A developing standard for Low rate wireless personal area network. IEEE Commun. Mag. **40**(8), 70–77 (2002)

36. LAN/MAN standard committee of the IEEE Computer Society. IEEE standard for information technology, Telecommunications and information exchange between systems, Local and Metropolitan area network specific requirements Part 15.4, Wireless Medium Access Control and Physical Layer specifications for low rate wireless personal area, network, Oct 2003.

37. G. Lu, B. Krishnamachari and C. S. Raghavendra, Performance Evaluation of the IEEE 802.15.4 MAC for low rate low power wireless networks, IEEE International conference on performance computing and communications, Phoenix, pp. 701–706, April 2004

38. W.B. Heinzelman, A.P. Chandrakasan, H. Balakrishnan, Adaptive protocol for information dissemination in wireless sensor networks. IEEE Trans. Wireless Netw **1**(4), 660–670 (2002)

39. W. B. Heinzelman, A. P. Chandrakasan and H. Balakrishnan, Energy Efficient Communication protocol for wireless microsensor networks, 3rd Hawaii international conference on system services, Hawaii, pp. 174–185, Jan 2000

40. K. Sohrabi, J. Gao, V. Ailawadhi, G.J. Pottie, Protocol for self organization of a wireless sensor network. IEEE Personal Commun. **7**(5), 16–27 (2000)

41. K. Sohrabi, G. J. Pottie Performance of a Novalself organize protocol for wireless ad-hoc sensor networks, IEEE 5th Vehicular technology conference, pp. 1222–1226, 1999

42. A. Woo, D. Culler, A transmission control scheme for media access in sensor networks. ACM/IEEE Int. Conf. Mobile Comput. Netw. (Mobicom) 2001, 221–235 (2001)

43. C.S. Raghavendra, S. Singh, PAMAS power aware multi access protocol with signaling for Ad-hoc networks. ACM Comput. Commun. **27**, 5–26 (1998)

44. J. Zheng, M. J. Lee, Will IEEE 802.15.4 Make Ubiquitous Networking a Reality?, IEEE Commun. Mag. vol. **42**, no.6, pp. 140–146, 2004

45. J.A. Gutierrez, M. Naeve, E. Callaway, V. Mitter, B. Heile, IEEE 802.15.4 A developing standard for low power low cost wireless personal area network. IEEE Netw. Mag. **15**(2), 12–19 (2001)

46. M. Grossglauser, D. Tse, Mobility increases the capacity of Adhoc wireless networks. IEEE Infocom 2001: The Conf. Comput. Commun. **1**(3), 1360–1369 (2001)

47. J. Luo, J. Panchard, M. Piorkowski, M. Grossglauser, J. P. Hubaux, MobiRoute: Routing towards a mobile sink for improving lifetime in sensor networks, 2nd IEEE/ACM International Conference on Distributed Computing in Sensor Systems, San Francisco, pp. 480–497, Jun 2006

48. Z. Vincze, R. Vida, *Multi-hop Wireless Sensor Networks with Mobile Sink* (ACM Conference on Emerging Network Experiment and Technology, Toulouse, France, Oct, 2005)

49. Prasad Raviraj, Hamid Sharif, Michael Hempel, Song Ci, MOBMAC- an energy efficient and low latency MAC for mobile wireless sensor networks. IEEE Syst. 370–375, 14–17 (Aug 2005)

50. S. A. Munir, B. Ren, W. Jiao, B. Wang, D. Xie, J. Ma, Mobile wireless sensor network architecture and enabling technologies for ubiquitous, Conference on Advanced Infonnation Networking and Applications Workshops (AINAW '07), May 2007, pp. 113–120

51. M. Rahimi, H. Shah, G.S. Sukhatme, J. Heideman, D. Estrin, Studying the feasibility of energy harvesting in a mobile sensor networks. Proc. IEEE Int. Conf. Robotics Automation, Taipai **1**, 19–24 (May 2003)

52. A. Chakrabarti, A. Sabharwal, B. Aazhang, Using predictable observer mobility for power efficient design of sensor networks, 2nd International Workshop on Infonnation Processing in Sensor. Networks **2634**, 129–145 (Apr 2003)

53. M. Ali, T. Suleman, Z. A. Uzmi, MMAC: a mobility-adaptive, collision-free mac protocol for wireless sensor networks, Proceedings of the 24th IEEE IPCCC'05, Phoenix, pp 401–407, 2005

54. S. R. Gandhamet al., Energy Efficient Schemes for Wireless Sensor Networks With Multiple Mobile Base Stations, Proc. IEEE GLOBECOM, 2003

55. J. Luo, J.-P. Hubaux, *Joint Mobility and Routing for Lifetime Elongation in Wireless Sensor Networks* (Proc, IEEE INFOCOM, 2005)

56. E. Ekici, Y. Gu, D. Bozdag, Mobility-based communication in wireless sensor networks. IEEE Commun. Mag. **44**(7), 56–62 (Jul 2006)

57. A. Kansalet al., Intelligent Fluid Infrastructure for Embedded Networks, Proc. 2nd Int'l. Conf. Mobile Systems Applications and Services, 2004

58. R. Shah et al., *Data mules: Modelling a Three-Tier Architecture for Sparse Sensor Networks* (Proc. IEEE Wksp, Sensor Network Protocols and Apps, 2003)

59. M. Ghassemian, H. Aghvami, An investigation of the impact of mobility on the protocol performance in wireless sensor networks, 24th Biennial Symposium on, Communications, pp 310–315, Jun 2008

60. S. Narwaz, M. Hussain, S. Watson, N. Trigoni, P.N. Green, *An Underwater Robotic Network for Monitoring Nuclear Waste Storage Pools* (Sensors and Software Systems, Springer, In, 2009)

61. A. Pandya, A. Kansal, G. Pottie, Goodput and delay in networks with controlled mobility 2008 IEEE Aerospace Conference, pp 1323–1330, Mar. 2008

62. K. Dantu, M. Rahimi, H. Shah, S. Babel, A. Dhariwal, G.S. Sukhatme Robomote: Enabling Mobility in Sensor Networks, IEEE/ACM, 4th International Conference on Information Processing in Sensor Networks, (IPSN/SPOTS), pp 404–409, April 2005

63. A. A. Somasundara, A. Ramamoorthy, M. B. Srivastava, Mobile Element Scheduling for Efficient Data Collection in Wireless Sensor Networks with Dynamic Deadlines Proc. 25th IEEE Int'l. Real-Time Systems, Symposium, 2004

64. D. Jea, A. A. Somasundara, M. B. Srivastava, Multiple Controlled Mobile Elements (Data Mules) for Data Collection In Sensor Networks Proc. IEEE/ACM Int'l. Conf. Distrib. Comp, in Sensor Systems, 2005

65. S. Lam, A carrier sense multiple access protocol for local networks. Comput. Netw. **4**, 21–32 (1980)

66. W.T.H. Woon, TCWan, Performance evaluation of IEEE 802.15.4 wireless multi-hop networks: simulation and testbed approach International Journal of Ad-Hoc and Ubiquitous. Computing **3**(1), 57–66 (2008)

67. J. Zheng, Myung J. Lee, A comprehensive performance study of IEEE 802.15.4 Sensor Network Operations, IEEE Press, Wiley Interscience, Chapter 4, pp 218–237, 2006

68. A. Koubaa, M. Alves, E. Tovar, YQ Song, On The Performance Limits of Slotted CSMA/CA in IEEE 802.15.4 for Broadcast Transmissions in Wireless Sensor Networks, IPP-HURRAY Technical, Report, TR-060401, April 2006

69. M. Laibowitz, J.A. Paradiso, Parasitic Mobility for pervasive Networks, 3rd International Conference on Pervasive Computing, (PERVASIVE 2005) (Munich, Germany, May, 2005)

70. L. Hu, D. Evans, Localization for Mobile Sensor Networks, ACM, Mobi-Com 2004, Sep, 2004

71. C. Chen, J. Ma, Simulation Study of AODV performance over IEEE 802.15.4 MAC in WSN with Mobile Sinks, Proc of Advanced Information Networking and Applications, Workshop 2007, (AINAW'07), pp. 159–163, 2007

72. S.B. Attia, A. Cunha, A. Koubaa, M. Alves, Fault Tolerance Mechanism for Zigbee Wireless Sensor Networks, 19^{th}Euromicro Conference on Real Time Systems (ECRTS'07) (Pisa, Italy, July, 2007)

73. K. Zen, D. Habibi, A. Rassau, I. Ahmed, Performance Evaluation of IEEE 802.15.4 for Mobile Sensor Networks, 5^{th}International Conference on Wireless and Optical Communications Networks, Surabaya, Indonesia, 2008

74. Sung-Chan Choi, Jang-Won Lee and Yeonsoo Kim, An Adaptive Mobility-Supporting MAC protocol for Mobile Sensor Networks, IEEE Vehicular Technology Conference, pp 168–172, 2008

75. H. Pham, S. Jha, An adaptive mobility-aware MAC protocol for sensor networks (MS-MAC), in: Proceedings of the IEEE International Conference on Mobile Ad-hoc and Sensor Systems (MASS), pp 214–226, 2004

76. P. Lin, C. Qiao, X. Wang, Medium access control with a dynamic duty cycle for sensor networks. Proc. IEEE Wireless Commun. Netw. Conf. (WCNC) **3**, 1534–1539 (2004) ·

77. S.A.Hameed, E.M.Shaaban, H.M.Faheem, M.S.Ghoniemy, Mobility-Aware MAC protocol for Delay Sensitive Wireless Sensor Networks IEEE Ultra Modern Telecommunications & Workshops, pp 1–8, Oct 2009

ZERO: An Efficient Ethernet-Over-IP Tunneling Protocol

Inara Opmane, Leo Truksans, Guntis Barzdins, Adrians Heidens,
Rihards Balodis and Pavel Merzlyakov

Abstract An Ethernet over IPv4 tunneling protocol is proposed, which catego-
rizes all Ethernet frames to be tunneled into NICE and UGLY frames. The UGLY
frames are tunneled by traditional methods, such as UDP or GRE encapsulation,
resulting in substantial overhead due to additional headers and fragmentation
usually required to transport long Ethernet frames over IP network typically
limited to MTU=1,500 bytes. Meanwhile the NICE Ethernet frames are tunneled
without any overhead as plain IPv4 packets due to non-traditional reuse of
"fragment offset" or "identification" field in the IP header. It is shown that for
typical Internet traffic transported over Ethernet, the proposed ZERO tunneling
protocol classifies 99.94 % of Ethernet frames as NICE and thus results in nearly
zero-overhead, non-fragmented Ethernet over IP tunneling. The proposed tunneling method extends also to the Ethernet frames containing VLAN and MPLS
tags, as well as IPv6 packets—also these can be classified as NICE and transported
with zero-overhead over Internet or private IPv4 transport network. Unprecedented

I. Opmane · L. Truksans · G. Barzdins (✉) · A. Heidens · R. Balodis
Institute of Mathematics and Computer Science, University of Latvia, Riga, Latvia
e-mail: guntis.barzdins@lumii.lv

I. Opmane
e-mail: inara.opmane@lumii.lv

L. Truksans
e-mail: leo.truksans@lumii.lv

A. Heidens
e-mail: adrians.heidens@lumii.lv

R. Balodis
e-mail: rihards.balodis@lumii.lv

P. Merzlyakov
Institute of Solid State Physics, University of Latvia, Riga, Latvia
e-mail: pavel.merzlyakov@gmail.com

F. Xhafa and N. Bessis (eds.), *Inter-cooperative Collective Intelligence:*
Techniques and Applications, Studies in Computational Intelligence 495,
DOI: 10.1007/978-3-642-35016-0_13, © Springer-Verlag Berlin Heidelberg 2014

efficiency of the proposed tunneling protocol enables wide use of OSI Layer 2 transparency across existing Layer 3 infrastructures thus enabling new network design patterns essential for novel applications such as Internet of things.

1 Introduction

Internet of things might fail to take off in near future for low-level technical constraints, such as IP addressing and routing inflexibility of current Internet protocols.

Therefore we propose an OSI Layer 2 tunneling method, which in terms of robustness is comparable to NAT (which is a staple of current IPv4 Internet overcoming the addressing limitations of the early Internet), yet provides a remedy to a different Internet limitation. The problem addressed by our approach is that current Internet is a Layer 3 network, while many "private network" or "partially private network" applications (such as VPN, tunneling, distributed server farms, cloud computing, Internet of things, etc.) would benefit, if Internet would have been able to provide robust Layer 2 connectivity transparent to dynamic routing, auto discovery and multicast within overlaid "private" Layer 2 clouds.

This problem has partially been addressed by MPLS technology widely deployed in carrier networks and providing Ethernet-over-MPLS (EoMPLS) Layer 2 VPN service [1, 2] in combination with VLAN [3] technology. But due to its cost, complexity, and reliance on extended MTU Carrier Ethernet [4], MPLS is not practical for use in existing access networks.

The current solutions [5, 6, 13, 14, 16–18] for tunneling Layer 2 traffic over access networks and Internet are highly inefficient, because they all relay on extensive encapsulation, packet fragmentation, and reassembly. This might be justified in cases where VPN encryption of traffic is additionally applied, but for mere transport of Layer 2 traffic over Layer 3 Internet here we propose an original and highly efficient tunneling method with nearly zero overhead.

We have not found any other non-fragmenting Ethernet tunneling protocol that would work through the public Internet—all the mentioned Ethernet-over-IP protocols are encapsulating tunneled Layer 2 frames and thus trigger fragmentation. The proposed method shares some similarity with ROHC [7] header compression algorithm, but is different in that Layer 3 packets rather than Layer 2 frames used for tunnel transport.

This chapter describes our proposal—the ZERO tunneling protocol[1] that is simplex, in most cases does not fragment Ethernet frames carrying even maximum size IP packets, works over public Internet, and its synchronization is resilient to packet loss.

---

[1] Patent application P-12–89 (LV) [31.05.2012]

2 The ZERO Protocol Concept and Design

By observing a typical Internet user traffic one may notice a set of fields that usually have identical content in a burst of Ethernet frames. Also, some fields are rarely used at all. Say, a computer is using SMTP protocol to send a large e-mail to a remote server. Every outbound frame during the whole session will have the same source and destination MAC addresses. Also, the *Fragment Offset* (FO) field will probably be 0 in all packets. The idea behind ZERO protocol is to distinguish the often unused fields in the packet's Layer 3–4 PDUs (we call them X-fields as in "Index fields") and fill those with indexed Layer 2–3 information that needs to be preserved (we call them S-fields as in "Saved fields") during tunneling over the Internet. Presumably, no Ethernet or IP fields are universally unused—any field may contain a non-null value in some scenarios. To cope with such rare cases, we will sort Ethernet frames into NICE and UGLY frames and tunnel them by different means.

For the rest of this chapter we will use term *Entrance* for the tunnel entrance entity (device or software) and term *Exit* for the tunnel exit entity. We also emphasize the simplex nature of this protocol and with *Entrance* and *Exit* we really mean only one direction. The reverse direction (if necessary) is identical in functionality but uses fully separate data structure, really a different tunnel. A bidirectional pair of such simplex tunnels forms an Ethernet pseudowire over IP network, such as Internet.

We use the term *Transported* network for the network to be transported into the tunnel. An example would be a company private network. The network carrying the tunneled traffic is referred to as *Transport* network. Internet is an example of *Transport* network. Figure 1 shows the terms associated with a simplex tunnel (in one direction).

The ZERO protocol relies on categorizing all transported Ethernet frames into NICE and UGLY frames as defined below, and transporting each category by different means.

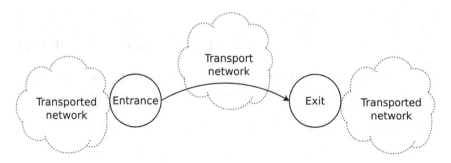

Fig. 1 Tunneling terms

Fig. 2 Structure of a NICE
Ethernet frame

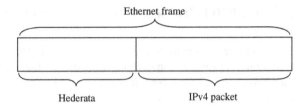

Definition 1 Ethernet frame is said to be NICE if it fulfills the following criteria:

- it includes a legal IPv4 packet as the last part of Ethernet frame data field (see Fig. 2.);
- the header of the included IPv4 packet is 20 bytes long (has no options field, see Fig. 4.);
- the included IPv4 packet is not fragmented and its total length does not exceed MTU 1500 bytes.

All these criteria are typically met by majority of Internet traffic packets sent over Ethernet. All fields preceding an IP packet (included as the last part of the Ethernet frame) will be referred to collectively as *hederata* (see Fig. 2.). Besides native Ethernet header, *hederata* may include VLAN tags, MPLS headers, etc. NICE frames in most cases will be tunneled with zero-overhead.

Definition 2 Ethernet frame is said to be UGLY if it is not NICE.

UGLY frames will be tunneled in less optimal fashion using traditional full encapsulation.

2.1 Selection of S-Fields and X-Fields

Figure 3 shows the roles of IP header fields in the NICE packet transformation process during the ZERO tunneling:

- contents of S-fields (along with contents of *hederata*, see Fig. 2) will be saved in the *Channel table* shared by both tunnel *Entrance* and *Exit;* the corresponding table raw INDEX will be communicated from *Entrance* to *Exit* via ZERO SYN packets and X-fields;
- in the transformed packet S-fields are overwritten by the X-field and T-field values (such as tunnel *Exit* IP address) necessary for tunneling the packet over transport network towards the *Exit*;
- the C-fields are copied to the transformed packet without change.

The IPv4 [19] header structure (see Fig. 4) imposes some restrictions on the choice of S, X, T, C field roles.

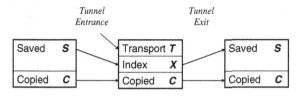

Fig. 3 IP header field roles

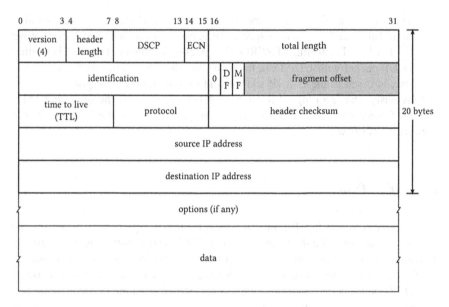

Fig. 4 IPv4 packet header format

We propose the *Fragment Offset* (FO, 13 bits) field of IP header to be used as a X-field. As T-fields we will use *Evil bit* (EB, 1 bit) [9] and *Destination IP address* (DA, 32 bits) fields of IP header, because DA needs to be overwritten by the IP address of the tunnel *Exit* device IP address. All other IP header fields are C-fields and may go into *Transport* infrastructure unmodified and will remain so during the transportation (except for *TTL* and *header checksum* fields changing on every hop). Alternate choices for X-fields and T-fields will be discussed later in this chapter.

Note that also the whole frame header (*hederata*, see Fig. 2) is treated as part of the S-field since it has meaning only to the local link and will inevitably be lost during NICE transit through Internet. We leave outside *hederata* the frame Preamble field which is a constant label for marking frame start and also the frame Checksum field which will be regenerated on every link anyway.

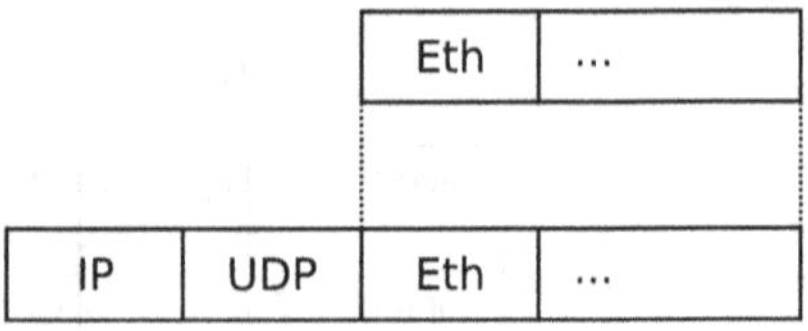

Fig. 5 ZERO ENC
transformation

	Eth	...	
IP	UDP	Eth	...

2.2 UGLY Tunneling

The transformation of UGLY frame on *Entrance* is performed by encapsulating the whole Ethernet frame into a special IP packet with a UDP datagram that we call ZERO ENC. The format of ZERO ENC packet is shown in Fig. 5. It is a rather usual UDP packet with UDP Data field carrying the whole original frame. ZERO ENC is addressed to the *Exit* IP address and to a dedicated UDP port.

Obviously, the resulting ZERO ENC packet may exceed the outbound MTU and get fragmented. That will not disrupt the protocol since the X-fields are not used in ZERO ENC.

2.3 NICE Tunneling

The transformation of NICE frame on tunnel *Entrance* is performed by storing the original values of S-fields in the *Channel Table* and substituting the them by T-fields (EB and destination IP address of the tunnel *Exit*) and X-fields (*Channel Table* INDEX) as described earlier in Sect. 2.1. We call the transformed packet ZERO IP. On the tunnel *Exit* the S-fields are restored from the *Channel Table* and the original Ethernet frame is thus recreated.

EB is a kind of labeling mechanism necessary to mark the ZERO IP packets on the *Transport* network so that *Exit* will recognize those from the other traffic. The *evil bit* (EB, 1 bit) IP field is unused in the public Internet as described in [9].

Replacing a destination IP address in an otherwise unmodified packet would void it's integrity in regards to its header checksum. So, during both *Entrance* and *Exit* transformations the packet *header checksum* gets recalculated.

We call *Channel* an established transformation associated with a unique S-field value. Thus for multiple S-field values a tunnel will contain multiple *Channels*.

The proposed S-field list (*headerata*, EB, FO, DstIP) is at least 158 bits large, hence larger than the proposed X-field (FO) of only 13 bits. Therefore a *Channel Table* is used here to enumerate the *Channels*. *Channel Table* has 3 columns: *s_fields* (S-field values), INDEX, and *last_synced* (to be discussed in Sect. 2.5). The INDEX is the same size as the X-field. During *Entrance* transformation the associated INDEX is written into the X-field to represent the S-field values. See Fig. 6.

Fig. 6 Transformation of
NICE packets on *Entrance*
using *Channel Table*

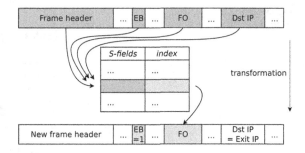

If the S-field values for a new frame on *Entrance* are already present in the *Channel Table*, the *Channel* is reused; otherwise the new S-field values are stored in a new *Channel Table* entry with an unused INDEX. The *Channel Table* synchronization is described further.

2.4 Synchronizing the Channel Table

When a new *Channel Table* entry is created on the *Entrance*, the NICE frame (first in it's *Channel*) is tunneled in the ZERO SYN format. Figure 7 illustrates the ZERO SYN format which is similar to ZERO ENC format in that it encapsulates the original frame allowing *Exit* to extract S-fields for the new *Channel*. However, ZERO SYN has one extra field immediately following the UDP header—the INDEX field. It's length is equal to X-field length rounded up to byte boundaries. The field contains the *Channel Table* INDEX to be used with the S-field data extracted from the encapsulated frame.

Upon receiving ZERO SYN the *Exit* saves the S-field contents into it's *Channel Table* under the given INDEX. This allows *Exit* to simultaneously deliver the first frame and to create a new *Channel Table* entry. The *Exit* must update the appropriate *Channel Table* entry upon receiving any ZERO SYN packet. This convention allows same format to be used for establishing new *Channels* as well as updating existing ones.

Since the proposed S-fields (their *hederata* part) are not fixed in length, different *Channels* may have different bit patterns to check in frames. We choose not to synchronize the patterns explicitly—instead, the same S-field recognition algorithm is used at *Entrance*, as well as *Exit* to identify the S-fields directly from the transported Ethernet frame. That way we guarantee that whatever S-field

Fig. 7 ZERO SYN
transformation

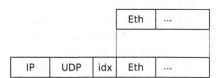

Fig. 8 ZERO SYN update
interval

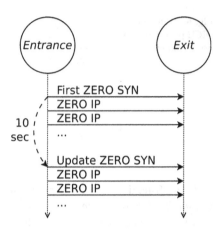

pattern will be determined for the frame at *Entrance*, the same S-field pattern will
be determined for the same frame at *Exit*.

Since the ZERO protocol is simplex, the tunnel endpoints do not use any
confirmations or requests for the synchronization. The ZERO SYN packet for a
new *Channel* may be lost resulting in *Exit* being unable to reconstruct S-fields for
following ZERO IP packets with the same S-field values. A simple redundancy
technique is used to remedy this problem. Every 10 s *Entrance* will send a NICE
frame again in the ZERO SYN form, allowing the *Exit* to update the entry in it's
Channel Table. If the *Exit* has missed a ZERO SYN packet it will have a chance to
acquire that information in 10 s. See Fig. 8.

It should be noted that update ZERO SYN is sent in no less than 10 s, but only
when a frame of that *Channel* is being tunneled—no updates are sent for inactive
Channels. The length of 10 s between ZERO SYN updates is chosen to result in
3 ZERO SYN packets to be transferred inside the TCP timeout window. The
proportion of 3 updates per timeout is often used in routing protocols.

Two different UDP destination ports are used for ZERO ENC and ZERO SYN
packets to make them distinguishable on *Exit*.

2.5 Managing the Channel Table

When not explicitly stated, this section talks about the processes to manage the
Channel Table only at the *Entrance*. It is the *Entrance* that does the logic of
building and updating the table. The *Exit* is just following the updates brought by
ZERO SYN packets.

In case the *Channel Table* is full and a new *Channel* needs to be added, a Least
Recently Used (LRU) mechanism is used to replace the oldest record with a new
one. To make looking for the least recently used entry in *Channel Table* efficient
we propose a linked list with pointers to *Channel Table* entries. For every NICE

frame on *Entrance* whenever an existing *Channel* is found or new one created the pointer to this *Channel* entry is moved to the head of the list. By doing so the other list entries are falling towards the end of list. The list's head would represent most recently used *Channels*, the tail—least recently (or never) used *Channels*. Thus, looking for the least recently used *Channel* becomes trivial—looking at *LRU list*'s tail.

We also have a *last_synced* column to the *Channel Table*. It shows the time when the last ZERO SYN packet was sent for the *Channel*. It is used to manage *Channel* updates in the following way. For every NICE frame on *Entrance* whenever an existing *Channel* is found in the *Channel Table* the *last_synced* time is checked. If it is older or equal to 10 s, the necessity to encapsulate the frame into ZERO SYN is triggered. Also, the trigger updates *Channel*'s *last_synced* field with current time. See Fig. 8.

So far we have not detailed the S-fields lookup algorithm in *Channel Table*. For the sake of simplicity we assume full seek of the table looking for INDEX of given S-field values. A more efficient approach might be to store S-fields in a hash-table that could be looked up in constant time.

Note that the presented algorithm has no locking issues because each frame to be tunneled is completely processed by a single thread without any helper or background processes.

The *Exit* has significantly simpler *Channel Table* management functions: lookup of *Channel* in the *Channel Table* by a given INDEX; or updating a given *Channel* with a given S-fields data at the given INDEX. Both are rather trivial.

2.6 TTL Compensation

Most applications default to rather large TTL for new IP packets disregarding if the target is in local or corporate network or in the public Internet. Often used values are in range between 64 and 255. We propose the default behavior to leave the TTL for NICE packet as is. The value will decrease during propagation over the *Transport* network, the target will receive a frame with a lower IP TTL value.

Two problems emerge:

- Some applications depend on TTL correctness in their logic. Traceroute is an example.
- If a packet will have lower TTL value than necessary to traverse the *Transport* network, it will be lost in transit.

To solve these problems the following TTL compensation mechanism is used. The idea is to simply add a certain number *ttl_delta* to the TTL field. *ttl_delta* would represent the number of hops between the tunnel endpoints. However, there are three issues with this idea:

- If *ttl_delta* is added at *Entrance*, for NICE packets with TTL already at the maximum 255 or close (TTL + *ttl_delta* > 255) the method would lose correctness.
- If *ttl_delta* is added at *Exit*, for NICE packets with low TTL (TTL < *ttl_delta*) the packet would be lost in transit.
- The path from *Entrance* to *Exit* may change at some point in time due to routing change in Internet. In that case a constant *ttl_delta* would not represent the actual hop count between tunnel ends and the method would lose correctness.

We chose not to fight the second issue and tunnel all NICE frames with low TTL in the ZERO ENC form. We believe such frames to be rare for typical Internet applications. Both other issues are solved by two procedures:

- Compensating at the *Exit*, meaning the *Exit* always increases TTL of the received ZERO IP packet by *ttl_delta* value.
- Using a dynamic *ttl_delta* update process.

The *ttl_delta* update process is a part of the earlier described *Channel* synchronization process. The TTL of any ZERO SYN packet will always be set to 255 at *Entrance*. The original TTL altogether with the original frame is encapsulated inside the ZERO SYN packet. Thus, upon receiving a ZERO SYN the *Exit* is synchronizing not only S-fields but also the *ttl_delta*. It is looking for the incoming ZERO SYN packet TTL (*syn_ttl*) and calculating the new *ttl_delta*:

$$ttl\_delta = 255 - syn\_ttl \qquad (1)$$

If the new *ttl_delta* differs from previously used, it is updated. This simple check is done for every received ZERO SYN. But the update of *ttl_delta* will be done in those rare situations when it actually changes. See Fig. 9.

Note that the proposed TTL compensation mechanism ensures correct *ttl_delta* only at *Exit*. The *Entrance* does not know that number and so it can not correctly evaluate the "too low TTL" criteria. If the *Entrance* also has the reverse tunnel from the *Exit* side, it could look for detected *ttl_delta* in reverse direction and hope

Fig. 9 ttl_delta update example

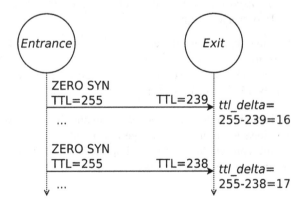

the path in both directions is the same or at least with the same hop count. We propose a different approach that suits the asymmetric infrastructures like satellite communications. The delta numbers will have different semantics on each tunnel end. The *Exit* calculates the *ttl_delta*, as described. But the *Entrance* has a different number—*ttl_min*. It will be a configuration parameter and will serve as the "too low TTL" criteria. Network administrators may choose to set that number little lower than the lowest default TTL in the tunneled infrastructure, 63, for instance.

2.7 Tunneling the Internet

Since Destination IP address is one of the S-fields the ZERO protocol mandates to create a new *Channel* for any IP host connected to. That would be acceptable for communication among two parts of a small to medium sized network (more on this further). But it does not scale well if the number of IP destinations is large or unpredictable. It would be so if a branch office is going to access Internet through a ZERO tunnel to central office, for instance.

Note that the returning traffic in the opposite tunnel would scale well because the packets in there would have limited set of Destination IP addresses (back to branch office) and the highly variable Source IP addresses would be carried along unchanged in ZERO IP packet. In other words, the scalability problem is with "local to global" *Channels*, not with "global to local" or "local to local" *Channels*. By "local" we mean branch IP networks that have limited inside subnets and reach out to Internet through one gateway.

To address this scalability issue we propose the following terminology and solution. The tunnels that are directed towards branch networks are called *Direct Tunnels* (*DT*) and it's *Channels* are called *Direct Channels* (*DC*). The tunnels that are directed towards central and global networks are called *Indirect Tunnels* (*IT*) and it's *Channels* are called *Indirect Channels* (*IC*). Both *Entrance* and *Exit* will have a configuration parameter for a mutual tunnel that determines tunnel's mode: *DT* or *IT*. Figure 10 shows an example topology with asymmetric tunnel types.

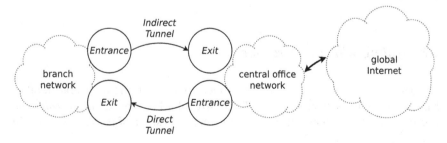

Fig. 10 Direct and *Indirect Tunnels* topology example

And the solution is to add one step before packet transformation for *ICs*: we swap Source IP and Destination IP addresses. Further, Source IP becomes S-field instead of Destination IP and the former Destination IP address is carried in the transformed packet as Source IP. This solves the scalability problem, assuming the branch networks are not subject to scalability issues themselves.

Theorem 1 Every branch computer requires only one Channel in each direction through ZERO tunnel to communicate with all global Internet hosts connected behind the central office network.

Assumptions: the MAC and IP addresses of computer and gateway are fixed and there are no other variable fields in *hederata*. The tunnel from branch is configured *Indirect* and the tunnel from central network is configured *Direct*.

proof.

(1) The *Entrance* in branch network will create *Indirect Channel* for traffic from the computer. A priori, the *IC* will be represented by Source MAC of the computer, Destination MAC of the gateway, and Source IP of the computer. Since all these three parameters are fixed, the ZERO tunnel will use and reuse single *IC* for all traffic from branch computer to global Internet hosts.
(2) The *Entrance* in central network will create *Direct Channel* for traffic to the computer. A priori, the *DC* will be represented by Source MAC of the gateway, Destination MAC of the computer, and Destination IP of the computer. Since all these three parameters are fixed, the ZERO tunnel will use and reuse single *DC* for all traffic from global Internet hosts to the local computer.

From (1) and (2) Theorem 1 is proved.

It should be noted that *ICs* are equally well suited for traffic through a computer chosen gateway, as well as through a Proxy ARP gateway [8]. No matter how delusional ARP table would form in local computers, they would generate same frames for Proxy ARP gateway as for a computer chosen gateway: computer's Source MAC and Source IP, Destination MAC of gateway (Proxy ARP) and Destination IP of global host. When a tunnel is configured *Indirect,* then the *Entrance* would again create an *IC* and reuse it for all traffic from a computer to Internet via Proxy ARP gateway.

2.8 Dealing with Convergence

As long as the *Channel Table* is not full (according to Theorem 1—branch office has less computers than *Channel Table* size of 8192) the *Channel Tables* at *Entrance* and *Exit* will eventually converge in the presence of packet loss and will not change any more.

Fig. 11 Entrance flow chart

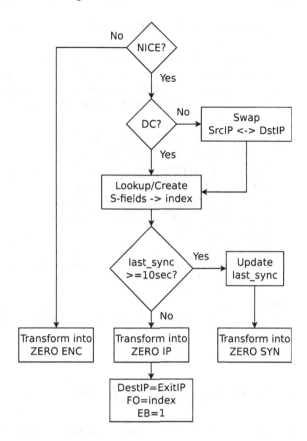

Once the *Channel Tables* at *Entrance* and *Exit* have converged the tunneling is always correct even in the presence of packet loss in *Transport* network.

To minimize corruption of tunneled packets during convergence the Exit discards any packet for unestablished *Channel* (Fig. 11).

Theorem 2 In converged state ZERO protocol will correctly tunnel NICE frames with zero-overhead through Transport infrastructures that: (T1) do not fragment IPv4 packets with size not exceeding 1,500 bytes; (T2) do not filter IPv4 packets by Source IP address; (T3) do not alter IP packet contents besides normal TTL and IP header checksum modification during forwarding.

Proof.

(1) All frame fields that are lost during tunneling are saved in *Channel Table* as S-fields. As long as the *index* value saved in NICE IP FO field is preserved during travel in *Transport* network, all S-fields will be correctly reconstructed on tunnel *Exit*. For this to happen the NICE IP packets may not be fragmented during travel in *Transport* network. That would ruin the *index*. Given (T1) we

can state that FO field will stay unchanged in *Transport* network and the correct *index* value will be delivered to *Exit*.

(2) Besides the unused EB and FO fields of the IPv4 packet embedded in NICE frame (Definition 1) Destination IP field is also modified for travel in *Transport* network. That field is also an S-field and will be reconstructed on *Exit*, if 1) holds. The originally empty EB and FO fields will also be restored to null value on *Exit*.

(3) If EB field is not cleared during travel in *Transport* network (T3) the Ext will correctly recognize NICE IP packets.

(4) The only remaining critical concern is that Source IP address of original source is carried along into the *Transport* network and there is a possibility an intermediary service provider may filter out packets with unexpected Source IP addresses or set EB. Given (T2) this is not an issue and NICE IP packets will not be filtered.

(5) The only other IP field that will change during travel in *Transport* network is TTL. That correctly restored by previously described TTL compensation method.

Using (1), (2), (3), (4) we can state that all NICE frame fields that will or may lose original content during tunneling will be restored on *Exit*. With that and (5) Theorem 2 is proved.

3 The Prototype Implementation

Here we describe a ZERO protocol prototype[2] that we implemented and tested in GNU/Linux operating system. We believe the implementation does not have any obstacles for porting it to other operating systems.

3.1 Zero Server

We implemented the ZERO protocol prototype logic in a userland process named *Zero server*. It attaches to the networks via Linux sockets. The Zero server on *Entrance* reads frames from the transported network, processes and then sends them via transport network to *Exit* Zero server according to protocol logic. Figure 12 shows data input/output points of the Zero server. From here on in this chapter the Zero server is illustrated functioning in both directions (full-duplex pseudowire). According to the protocol design terminology we illustrate Zero

---

[2] Available at http://zero.lumii.lv

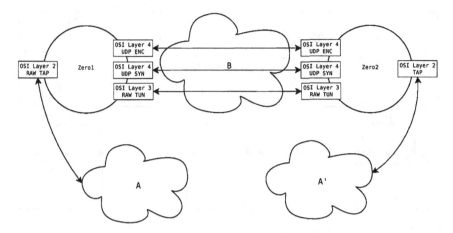

Fig. 12 Zero server bidirectional data paths

server here as both *Entrance* and *Exit*, establishing outbound tunnel and accepting inbound tunnel, and building the appropriate data structures.

Zero server has four different input/output points:

- *RAW TAP* uses raw socket (OSI Layer 2) to establish connection to the *Transported* network. It works with Ethernet frames.
- Second point *RAW TUN* uses raw socket (OSI Layer 3). It allows sending modified ZERO IP packets into *Transport* network and intercepting all ZERO IP packets from the *Transport* network.
- *UDP ENC* and *UDP SYN* points use standard UDP sockets to send and/or receive the ZERO ENC and ZERO SYN packets respectively.

For ZERO protocol to work in both directions between two sites a Zero server needs to be configured and functional at both tunnel ends according to Fig. 12. The A and A′ clouds represent two parts of a single *Transported* network. B represents a *Transport* network.

3.2 Testing Environment

Although ZERO protocol has been tested to work well also over public Internet (see Sect. 4), for the controlled testing of the prototype we created an environment that resembles a typical OSI Layer 2 tunnel usage between two sites. Conceptually the testing environment is as seen previously in Fig. 12. The IP addresses and networks of the testing environment are shown in Fig. 13.

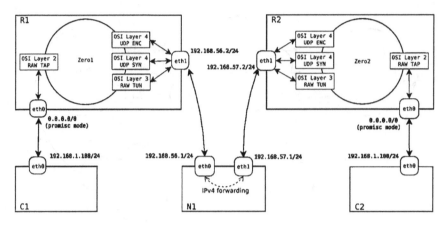

Fig. 13 Zero server prototype testing environment

Five hosts were used in the test:

- C1 and C2 are the end-hosts each residing on it's own side of the *Transported* network (192.168.1.0/24). The end-hosts have no knowledge of the tunneling between them.
- R1 and R2 are the ZERO tunneling end systems, each connected with one interface to the *Transported* network and another interface to the *Transport* network. To test bidirectional communication each of the tunnel end systems acts as *Entrance* for traffic to the other part of the *Transported* network and *Exit* for traffic from the other part.
- N1 is a router simulating internetwork between the R1 and R2 systems.

R1 has the following configuration: *eth0* interface is configured in the promiscuous mode and allows R1 to accept any frames coming from C1 *Transported* network. The *eth0* interface is left without IP address which is acceptable for a transparent OSI Layer 2 device the R1 system pretends to be. The raw socket is bound to *eth0* device that is used by the Zero server process to send and receive Ethernet frames.

On the *Transport* network side Zero server uses: raw socket to send and receive NICE IP packets; UDP sockets with two different ports numbers send and receive ZERO ENC and ZERO SYN packets respectively. The R1 host routing table is adjusted so that: packets destined to the other end of tunnel (R2) be properly routed to the next gateway (N1) out via *eth1* interface.

Same configuration principles are applied to the system R2.

Table 1 Protocol counters for both tunnel ends

Protocol	R1	R2
ARP	5	5
IPv4	57753	9936
LLC	3	3

Table 2 NICE and UGLY counters

Status	R1	R2
NICE	57728	9914
UGLY	33	30

Table 3 R1 and R2 entering frame counters

Frame size	Count (R1)	Count (R2)
1514	57743	0
74–1081	11	8
66	2	9931
60	5	5

3.3 ZERO Protocol Test

The above described testing environment was used for actual testing. An http server was set up on the C1 computer. During the test the C2 computer downloaded ~80 MB file at bit rate 400 kB/s from C1 computer. Packet analyzer tcpdump was used for collecting Zero protocol traffic on computer N1.

Table 1 shows the protocol counters for both tunnel ends. The numbers represent packets coming into the tunnel from *Transported* network side.

Table 2 shows the counters of NICE and UGLY frames detected by the tunnel ends. From Tables 1 and 2 we can deduce:

- Almost all IP packets coming into R1 and R2 tunnel are NICE (respectively 99.94 % and 99.69 %) and eligible for optimal tunneling.
- More then 1/4 of UGLY frames are non-IPv4.

Table 3 shows the most popular frame sizes coming into R1 and R2. This confirms the asymmetric traffic nature of this experiment. 99.96 % of the frames coming into R1 were maximum size frames—1,514 bytes (1,500 MTU + frame header). And almost all those frames were NICE and were transported in the optimal NICE IP format. The total percentage of frames that were transported in NICE IP format was 99.94 %. The rest were either ZERO ENC (for the UGLY frames) or ZERO SYN (for syncing NICE frame *Channels*).

Table 4 R1 generated transport packet overhead

(R1)	Zero (%)	EoIP (GRE) (%)
Packet count overhead	+0.06	+102.47
Volume overhead	+0.01	+6.33

Table 5 R2 generated transport packet overhead

(R2)	Zero (%)	EoIP (GRE) (%)
Packet count overhead	+0.19	+0.00
Volume overhead	+0.32	+60.95

Table 6 R1 and R2 generated transport packet overhead

(R1 + R2)	Zero (%)	EoIP (GRE) (%)
Packet count overhead	+0.08	+79.57
Volume overhead	+0.01	+7.04

3.4 Comparison to Legacy GRE Based EoIP Tunnel

The test setup from previous Sect. 3.3 was repeated to compare ZERO protocol behavior to a typical legacy Ethernet over IP tunneling protocol (based on unconditional full encapsulation). The comparison was made towards GRE based EoIP implementation in MicroTik RouterOS version 5.22 [20, 21]—EoIP tunnel adds 42 byte encapsulation overhead (8 byte GRE + 14 byte Ethernet + 20 byte IP).

Tables 4, 5 and 6 illustrate Zero server and EoIP (GRE) comparison for both tunnel directions. The numbers show packet count and volume overhead for packets forwarded into the *Transport* network.

The only parameter where Zero server does not improve performance is packet count on R2 because that system mostly tunnels small request and acknowledgement packets that don't get fragmented with EoIP. The volume, however, differs greatly because Zero server tunnels NICE packets optimally as opposed to EoIP that encapsulates unconditionally. In this aspect R2's Zero server uses only ~ 60 % of the volume that EoIP uses.

On R1 the gain is different, because there Zero server tunnels mostly large packets. Because of optimal tunneling Zero server needs only 1/2 of the packet count required by EoIP, which, on the other hand, would encapsulate and fragment all maximum sized frames. The relative reduction in total volume is less significant since the encapsulation overhead for long frames is relatively smaller.

As shown in Table 6 above, in general Zero server is considerably more efficient than EoIP.

4 Discussion

The core ZERO protocol described in Sect. 2 is intended primarily for the controlled service provider networks adhering to all conditions stated in Theorems 1 and 2—we cannot guarantee that it will work over public Internet due to various anti-spoofing or connection-tracking filters employed by some ISPs. Therefore in this Section we discuss several extensions enabling ZERO protocol operation over public Internet at the expense of non-essential tunneled frame modification or occasional integrity violation. Our tests show that the proposed extensions work well both over national and international public Internet.

4.1 Identification Field Feasibility as X-Field

The often enabled TCP segmentation offload [15] functionality in modern Ethernet NIC (Network interface cards) and connection tracking option in Linux kernel Netfilter module tend to defragment IP packets in transit and thus interfere with the FO field use in core ZERO protocol as described in Sect. 2. For transport networks where FO field cannot be used as X-field, there is an option to use *identification* field instead. Although use of *identification* field formally violates Theorem 2., it usually does not cause any problems in practice as long as ZERO is the only protocol manipulating the *identification* field.

The intended use of *identification* field in IPv4 is described in the RFC 791 [19] that defines fundamentals of IPv4 protocol:

> The identification field is used to distinguish the fragments of one datagram from those of another. The originating protocol module of an internet datagram sets the identification field to a value that must be unique for that source-destination pair and protocol for the time the datagram will be active in the internet system.

Therefore it is rather safe to assume that the only requirement for the content of *identification* field is that it needs to be unique for any (Source, Destination, Protocol) triplets for the time packets are active and could be fragmented.

Since ZERO IP packets are not expected to be fragmented in the transport network, the content of identification field becomes irrelevant. Thus *identification* field can serve as alternative to FO field for the role of X-field. In that case the 16 bits of *identification* field can be used as INDEX, provided that on tunnel *Exit* *identification* field is filled with a pseudo-random or incremental values.

4.2 Alternative Treatment of TTL Field

Besides the TTL compensation method described earlier another and rather obvious method to recreate the original TTL value at the tunnel Exit would be to include TTL field in the S-field list. Then, there would be no minimum TTL requirement for NICE classification, any TTL value would be allowed for NICE packets. Also, the earlier TTL compensation method may recreate incorrect TTL value if a load sharing with oscillating hop count appears on the path in the transport network. Adding TTL to S-field would guaranty correct TTL recreation. And since the original TTL value would be saved in the Channel table, tunnel TTL field may be set to value 255 or any large number to make sure the packet does not expire in the transport network.

A drawback of adding TTL field to S-fields is that variations in this field would produce separate Channels for otherwise equal S-field lists.

4.3 ZERO NAPT Traversal

ZERO protocol as described so far is not designed to traverse Network address translator (NAT) or Network address port translator (NAPT) [22]. Nevertheless such functionality would be of great interest due to massive and increasing use of NAPT devices for Internet access. ZERO protocol can be made to operate via NAPT gateway by forcing the ZERO tunnel endpoints to behave as regular UDP client and server; additionally also two more NICE frame criteria should be introduced: *the packet needs to include a TCP or UDP segment and for TCP segment Urgent pointer field must be 0.*

Under these conditions all data exchange between ZERO tunnel server and client can be carried in UDP packets. ZERO ENC and ZERO SYN are UDP packets by definition and thus do not require any modification apart from UDP port selection in line with client-server model to enable NAPT traversal. Meanwhile ZERO IP packets can be converted into UDP packets (without increasing their length) using the same field substitution principle used earlier on IP header fields, only this time applied to UDP and TCP header fields:

- Original source and destination ports of TCP or UDP header are added to the S-fields list along with source and destination IP addresses and IP protocol number of IP header; afterwards source and destination IP addresses and port numbers are overwritten by new source and destination IP addresses and UDP port numbers in line with client-server model to enable NAPT traversal between Zero client and server.
- The IP packets containing TCP segment are transformed into IP packets containing UDP segment (without increasing their length, see Fig. 14) by changing IP protocol field value from 6 to 17 to make it appear as UDP segment. Additionally restorable on *Exit* TCP header fields *Checksum* and *Urgent Pointer* are deleted (4 bytes total) to provide room for UDP header fields *Length* and *Checksum* (also 4 bytes total).

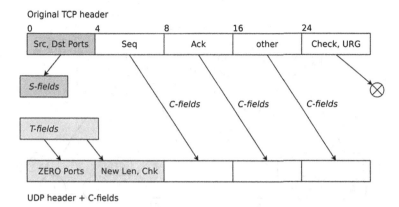

Fig. 14 TCP to UDP segment header transformation

- IP *identification* field must be used as X-field (see previous section), because FO field cannot be used with NAPT, which by definition performs packet defragmentation.

The reverse UDP-to-TCP transformation performed on *Exit* can be easily deduced.

The proposed NAPT solution provides ZERO tunneling efficiency gain for small or lightly used networks. Even for large networks where the *Channel Table* might get exhausted often, the ZERO protocol would still maintain communications integrity by frequently replacing old *Channel* entries with new ones. This fully UDP-based ZERO protocol version is compatible with Hole punching techniques [23] popular in P2P networks for one or both ZERO tunnel endpoints behind 3rd party NAT.

4.4 Multi-Point Tunneling Topology

So far the ZERO protocol is described as operating from one *Entrance* host to one *Exit* host. The following two data structures are associated with a single tunnel, one—for each end.

The *Entrance* data structure:

- configuration parameters: *Entrance* IP, *Exit* IP, *ttl_min*, implementation specific (UDP port, attached interfaces, etc.);
- *Channel Table*—built and updated during operation.

The *Exit* data structure:

- configuration parameters: *Entrance* IP, *Exit* IP, implementation specific (UDP port, attached interfaces, etc.);
- *ttl_delta*—learned during operation from ZERO SYN packets;

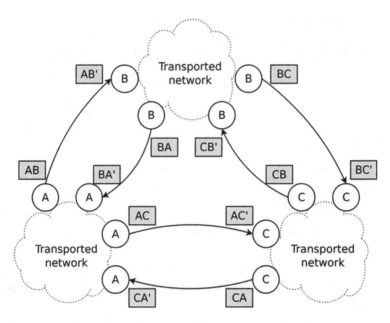

Fig. 15 Mesh ZERO tunneling topology example

- shadow *Channel Table*—a copy of the *Entrance Channel Table* that is learned
  during *End* operation via ZERO SYN packets;

On both ends the data structures are identified by the $<$ *Entrance* IP, *Exit* IP $>$
pair. To create a tunnel to another tunneling host, a new data structure is created
identified by the $<$ *Entrance* IP, *Another Exit* IP $>$ pair.

With these definitions it becomes possible to build multi-point tunneling
topologies where any tunneling host can create a tunnel with any other, thus
building full or partial mesh tunneling topologies similar to MPLS VPN.

Figure 15 shows an example of 3 tunneling hosts configuration. The data
structures are depicted as gray boxes and identified with host letters. For instance,
AB identifies an *Entrance* data structure for tunnel from A host to B host. The
shadow tables have the same notion, except marked with asterisk, for instance,
AB' identifies an *Exit* data structure for tunnel from A host to B host.

4.5 On Security Implications

ZERO protocol, like any unencrypted tunneling protocol (such as GRE [13]) is
prone to third party injecting spoofed tunnel packets. The described use of Evil bit
(EB) is a weak authentication mechanism to minimise spoofing. A stronger
mechanism to minimize consequences of spoofing would be to scramble data of

tunneled packets. A simple scrambling method would be to XOR the data of tunneled packets with a random shared bit sequence—this would corrupt the spoofed packets during descrambling. An even stronger protection method would be to use DES as the scrambling mechanism—DES encryption does not change data length, yet potentionally makes ZERO protocol VPN grade secure.

4.6 IPv6 Handling

According to Definition 1, NICE can only be frames containing IPv4 packets. Actually, Ethernet frames containing IPv6 packets can also be transformed into IPv4 packets and tunneled with zero-overhead. Since the *Channel Table* stores Destination IP address of IP packets contained in NICE frames, the said address is not included in ZERO IP tunnel packets. In case of IPv6 this allows to reduce 16 bytes in ZERO IP tunnel packets. Four more bytes can be reduced by mapping three IPv6 header fields *Payload length*, *Next header*, *Hop limit* into three IPv4 fields *Total length*, *Protocol*, *Time to live*, respectively. This results in 20 bytes reduced for any Ethernet frame containing IPv6 packet sufficient to add whole IPv4 header (*sans*, the three fields mapped from IPv6 header).

5 Conclusions

The core Zero protocol for efficient Ethernet-over-IP tunneling has been presented in Sect. 2 along with formal proof of its transparency, efficiency, and convergence. The core Zero protocol is suitable for controlled service-provider networks where guaranteed transparency and efficiency is required. The core protocol has been developed with satellite service-provider networks in mind, but it could equally benefit also other infrastructures where true OSI Layer 2 transparency is required for the Internet of things or other purposes.

The overhead-less nature of Zero tunneling enables new IP network design patterns, where user IP addressing and routing is fully isolated from the service provider IP addressing and routing through the L2 abstraction. This design principle extends also to the Zero protocol capability of tunneling IPv6 without any overhead over IPv4 legacy infrastructure thus providing an easy migration path.

The Zero protocol extensions discussed in Sect. 4 disrupt full transparency and efficiency guarantee, but enable Zero protocol use over un-controlled public Internet, including support for NAPT gateway traversal. The extended Zero protocol is aimed at end-users ready to tolerate non-essential frame modification to achieve overhead-less L2 connectivity through public Internet.

The extended Zero protocol can operate on top of service-provider core Zero protocol—the overhead-less operation is preserved for both thanks to their reliance on modifying different (*identification/port* or FO/EB respectively) header fields.

Zero protocol prototype implementation has been demonstrated and tested both in lab and across public Internet. The test results confirm nearly zero overhead efficiency of the ZERO protocol.

Acknowledgments This work has been partially supported by the Latvian National Research Program Nr. 2, Development of Innovative Multifunctional Materials, Signal Processing and Information Technologies for Competitive Science Intensive Products" within the project Nr. 5, New Information Technologies Based on Ontologies and Model Transformations.

References

1. RFC 4448, Encapsulation Methods for Transport of Ethernet over MPLS Networks, http://www.ietf.org/rfc/rfc4448.txt 15 Oct 2012
2. RFC 4447, Pseudowire Setup and Maintenance—Using the Label Distribution Protocol (LDP), http://www.ietf.org/rfc/rfc4447.txt 15 Oct 2012
3. IEEE Std. 802.1Q-2005, Virtual Bridged Local Area Networks
4. Carrier Ethernet, http://en.wikipedia.org/wiki/Carrier_Ethernet 15 Oct 2012
5. OpenVPN, http://openvpn.net/ 15 Oct 2012
6. RFC2661, Layer Two Tunnelling Protocol "L2TP", http://www.ietf.org/rfc/rfc2661.txt 15 Oct 2012
7. RFC 3095, RObust Header Compression (ROHC): Framework and four profiles: RTP, UDP, ESP, and uncompressed, http://www.ietf.org/rfc/rfc3095.txt 15 Oct 2012
8. RFC 1027, Using ARP to Implement Transparent Subnet Gateways, http://www.ietf.org/rfc/rfc1027.txt 15 Oct 2012
9. RFC3514, The Security Flag in the IPv4 Header, http://www.ietf.org/rfc/rfc3514.txt 15 Oct 2012
10. Linux man-pages project, release 3.35, packet (7) function manual, http://man7.org/linux/man-pages/man7/packet.7.html 15 Oct 2012
11. W.R. Stevens, B. Fenner, A.M. Rudoff, Unix Network Programming, Vol 1: The Sockets Networking API (3rd Edition), Addison Wesley, 2003
12. The Linux Kernel Module Programming Guide, http://tldp.org/LDP/lkmpg/2.6/html/lkmpg.html 15 Oct 2012
13. RFC1702, Generic Routing Encapsulation over IPv4 networks, http://www.ietf.org/rfc/rfc1702.txt 15 Oct 2012
14. RFC3378, EtherIP: Tunneling Ethernet Frames in IP Datagrams, http://tools.ietf.org/html/rfc3378 15 Oct 2012
15. Large Segment Offload, Wikipedia, http://en.wikipedia.org/wiki/Large_segment_offload 15 Oct 2012
16. VXLAN: A Framework for Overlaying Virtualized Layer 2 Networks over Layer 3 Networks, http://tools.ietf.org/html/draft-mahalingam-dutt-dcops-vxlan-02 15 Oct 2012
17. NVGRE: Network Virtualization using Generic Routing Encapsulation, http://tools.ietf.org/html/draft-sridharan-virtualization-nvgre-00 15 Oct 2012
18. A Stateless Transport Tunneling Protocol for Network Virtualization (STT), http://tools.ietf.org/html/draft-davie-stt-01 15 Oct 2012
19. RFC 791, INTERNET PROTOCOL, http://tools.ietf.org/html/rfc791 15 Oct 2012
20. Mikrotik RouterOS wiki, http://wiki.mikrotik.com/wiki/MikroTik_RouterOS wiki 15 Oct 2012

21. RFC1701, Generic Routing Encapsulation (GRE), http://www.ietf.org/rfc/rfc1701.txt 15 Oct 2012
22. RFC3022, Traditional IP Network Address Translator (Traditional NAT), http://www.ietf.org/rfc/rfc3022.txt 15 Oct 2012
23. RFC5128, State of Peer-to-Peer (P2P) Communication across Network Address Translators (NATs), http://www.ietf.org/rfc/rfc5128.txt 15 Oct 2012

Energy-Efficient Indoor Spaces Through Building Automation

José Santa, Miguel A. Zamora-Izquierdo, María V. Moreno-Cano, Antonio J. Jara and Antonio F. Skarmeta

Abstract Building intelligence and, more recently, energy efficiency are key concepts to bear in mind when future smart spaces are considered. Common automation capabilities in the field of domotics only presented the first building blocks for the indoor spaces of the future. In this framework, energy consumption requires a special treatment, due to the strident environment preservation issues that the society is facing nowadays. For this reason, new solutions are needed to deal with the increasing power requirements of indoor spaces. In this line, an intelligent building platform which embraces not only well known automation necessities of indoor spaces, such as automatic lighting, security, remote access, etc., but also novel concepts in the fields of context-awareness, resident tracking and profiling, and net-zero/positive energy building are considered in a new proposal: the PLATERO platform. This chapter provides a detailed background on building automation/intelligence and energy efficiency and then details the novel architecture of PLATERO, its main elements and energy efficiency subsystems, and describes the different prototypes developed and deployed in a reference energy-efficiency building.

J. Santa (✉) · M. A. Zamora-Izquierdo · M. V. Moreno-Cano · A. J. Jara · A. F. Skarmeta
Department of Information and Communications Engineering, Computer Science Faculty,
University of Murcia, 30100 Murcia, Spain
e-mail: josesanta@um.es

M. A. Zamora-Izquierdo
e-mail: mzamora@um.es

M. V. Moreno-Cano
e-mail: mvmoreno@um.es

A. J. Jara
e-mail: jara@um.es

A. F. Skarmeta
e-mail: skarmeta@um.es

F. Xhafa and N. Bessis (eds.), *Inter-cooperative Collective Intelligence:*   375
Techniques and Applications, Studies in Computational Intelligence 495,
DOI: 10.1007/978-3-642-35016-0_14, © Springer-Verlag Berlin Heidelberg 2014

1 Introduction

The integration of sensors, actuators, control processes, among others, but more recently, information and communication technologies have implied that the wide and challenging terms *Intelligent Building* and *Building Automation* have become a reality in the recent years. According to experts in the field [1], an intelligent building is one that provides a productive and cost-effective environment through optimizations based on its three basic elements: people, considering owners, tenants, visitors, etc; products, standing for materials, fabrication, structure, facilities, equipment and services; and processes, composed of automation, control, systems, maintenance and performance evaluation. Based on most advanced sensing technologies, intelligent buildings must be able to monitor status parameters, analyze these data and, finally, actuate to reach the previous objectives, but considering transversal and essential goals such as cost and, more recently, energy efficiency. The latter is obviously interrelated to cost, but, in the last years, energy efficiency trend to be linked to environment preservation.

1.1 The Starting Point: Domotics

There is limited knowledge about what home automation (or domotics) exactly is. People usually understand this concept as "a set of expensive gadgets which make the house smart", and a large part thinks that domotics is not essential. Probably they were right some years ago, but perhaps an up-to-date and whole explanation about what current technologies and home automation systems can offer could persuade them. Initial solutions which switched on the lights when the inhabitants were present, have given way to automated systems which are able to control the operation of most appliances, windows, lighting, blinds, locks, etc., and, what is more and more demanded, monitor the house state.

The main fields where home automation can be applied are security, entertainment, labour-saving white goods, environmental control, eHealth and remote control [2]. The potential customers of such systems are working adults who need to save time, an aging population that needs assistance and users wanting a remote control of the house. Thus, the number of services offered and the wide variety of clients make the adaptation of commercial solutions a challenge for companies involved in the sector. Such adaptation must also take into account the usefulness of the system in the target environment. The boundary between a system which helps inhabitants with daily tasks, and a system which performs undesired automatic actions, is sometimes narrow [3]. And this is the reason why it is necessary to identify user requirements of home automation solutions. These generally fall into the following groups:

- Efficient automation of routine tasks.
- Security of automation systems.

- Easy to use.
- Local and remote access.
- Tele-monitoring.
- System cost and flexibility.

Probably obviating the previous requirements, the domotics world has been immersed in a competition of communication standards and specifications since early 90's [4]. Nowadays the state of wired domotics protocols, in Europe at least, is more established, and EIB (European Installation Bus) appears to be the most used specification; however, a similar problem has recently arisen in the field of wireless communication technologies, due to continuous advances in the area of in-home networking. Apart from initial radio-based solutions and the more recent Bluetooth, new wireless communication technologies are suitable for home automation [5]. ZigBee and Z-Wave are currently competing to become the in-home networking reference of future houses.

As a way to deal with this diversity of protocols in indoor automation, later on this chapter, a real solution is presented using EIB and proposing a low-power wireless solution based on 6LowPAN (IPv6 over Low Powered Wireless Personal Area Networks), as an alternative to ZigBee. Moreover, an IP-based communication protocol between the main controller of the indoor space (home automation module) and the rest of local embedded computers and remote equipment is proposed.

1.2 Energy Efficiency in Buildings

World energy requirements are continuously growing [6], in spite of the imminent exhaustion of natural resources and the serious environmental impacts that involve both the production and consumption of energy by traditional processes. Contrarily to many people think, one of the most important energy consumption fields are buildings, both residential and commercial. This tendency has become more pronounced in the last years, above all in developed countries, where energy consumption by buildings implies nowadays among 20 and 40 % of the total. This is the reason why the research community in building automation is currently interested in energy efficiency. Intelligent buildings must be capable not only of improving inhabitant experience and productivity, but also they must provide mechanisms to make the most of the purchased energy and be able to integrate their own energy sources.

Since the reduction of the carbon footprint at a global scale is a need, energy efficiency in buildings become a key issue as new countries such as China continue growing. International measures are necessary to improve energy efficiency of buildings. The European Commission, for example, is aware of the problem and has recently issued a recast of the Directive about Energy Performance of Buildings (2010/31/EU) [7], which pushes the adoption of measures to improve the performance of the energy used in buildings appliances, lighting and, above all, boilers and ventilation systems. It is important to remark the great impact of Heating, Ventilation and Air Conditioning (HVAC), comprising 50 % of the building energy consumption, and, in many developed countries 20 % of the total energy consumption [8]. This is the reason why automation systems must save their use. Standardization organizations are also aware of the problem [9], such as the International Organization for Standardization (ISO), which has set up the technical committees ISO/TC 163, Thermal Performance and Energy Use in the Built Environment, and ISO/TC 205, Building Environment Design. Across their works, these groups recognize that, apart from the physical building architecture, intelligent and automation systems are needed to improve comfort and energy efficiency, as it is considered, for example, in the ISO 16484 proposal, Building Automation and Control Systems.

Of course, intelligent automation systems can help to improve energy efficiency, although the delimitation of the problem and, above all, the way to resolve it is not direct. First of all, it must be understood the context of the problem. Although the technological plane is essential as an enabler, it is not the only need to improve indoor energy efficiency. According to [10], reducing energy consumption on these environments comply a combination of providing consumption feedback, automation, economic strategies and social factors. A concrete proposal presented later on this chapter focuses on the first two fields, as a way to exemplify how information and communication technologies can support the reduction of power consumption.

1.3 Requirements of a Power-Aware Intelligent Building

An indoor intelligent management system must provide proper monitoring and automation capabilities to cope with most important comfort and energy efficiency requirements [11]. On one hand, a high comfort level is desired, by guaranteeing thermal, air quality and luminance requirements of inhabitants. On the other hand, energy savings should be addressed, by establishing a tradeoff between comfort measures and the energy resources required. As can be deducted, apart from collecting and providing consumption information in an effective way, a set of subsystems is subject to be automated to accomplish these objectives:

- Electric lighting.
- Shading systems, to take advantage of the solar light and reduce solar radiation and glare.
- Windows and door opening and ventilation systems, to take advantage of the natural ventilation/air and assure the air quality.
- Boilers.
- Heating/cooling systems.

In this line, the solution presented later on this chapter offers a real designed and developed solution to: monitor energy consumption of the most important subsystems of a reference building, i.e. lighting, HVAC and most energy consuming devices, and able to manage a wide range of automated devices in order to save energy.

The rest of the chapter describes in detail the architecture of a system able to provide scalable building automation and efficient energy monitoring. Nevertheless, a review of current works on indoor automation and energy efficiency in buildings is first presented next.

2 First Steps in Small-Scale Indoor Automation

The most common design approximation in home automation until now has been considering user's needs and current technologies to deploy ad-hoc solutions. This methodology has led to works which are too technology-dependant, even present in the research world. In [12] authors describe a home automation system based on an Internet-accessible server. This hosts a Java application that manages a set of digital I/O lines connected to some appliances. A similar solution is described in [13]. However, communication with appliances is now carried out using a radio frequency (RF) link and a non-standard management protocol. This requires slave nodes that supports the protocols over the RF link to be deployed in the house and wired to appliances. The solution described in this chapter, apart from supporting digital I/O (and other inherited communication technologies), is compliant with standardized protocols in domotics. The work presented in [14] is focused on the specification of the logical model of the house and appliances, in order to enable the implementation of IF/THEN sentences to manage the devices. Although the gateway has been developed, the work lacks integrating it in a whole automation solution. Furthermore, another issue that clearly differentiates that work of the solution presented in this chapter is the integration of critical automation tasks in a distributed management system set-up remotely. As it is later explained, we claim on a high reliability solution based on an embedded home automation unit instead.

In these previous works researchers try to solve scalability issues and also offer different remote management capabilities through Internet. The previous works are mainly centred on the home automation unit, highly linked to a Web-based gateway. However, there are more elements to be considered in an integral home automation solution. A suitable HMI (human-machine interface), not only by means of a gateway, but also using local control panels, have also been developed in the solution presented in this chapter. Likewise, a complete security system for the house involves the communication with more entities, such as the local security staff and the security company. Moreover, an insufficient security treatment for IP-based communications in current solutions is also noticeable. The platform presented in this work covers this lack by means of secure communication channels using symmetric cryptography.

A concept that is gaining a gradual importance these days is ambient intelligence. The penetration of this idea in home automation is more and more evident in the literature, and proposals which offer context-awareness and ubiquity capabilities are being introduced on the automation basis in order to provide houses with real intelligence [15]. The proposal presented in this chapter adapts ambient intelligence concepts to the special case of assisted living systems, field in which the platform has also been applied successfully [16]. In [17] authors present a research project where a house has been automated to offer pervasive services to inhabitants. Although interesting ubiquitous services are proposed in that work, thanks to the wide deployment of sensors, actuators and a pervasive middleware, the reliability of the system is not properly considered. Thus, services are offered by means of a software gateway, which communicates with deployed devices by means of RF. Also, the house security and the remote management and monitoring are not considered in the work. These services are supported by the networked home platform presented in [18]. Although the work only includes the logical model of the architecture, and it is specially focused on UPnP (Universal Plug and Play), the functionalities pretended in the solution are quite similar to the ones included in the architecture presented in the current chapter.

3 When Energy Efficiency Comes into Play

Achieving energy efficiency in buildings is considered an integrated task that comprises the whole lifecycle of the building. According to the literature [19], the main stages in this context are:

- Design, by using simulations to predict the energy performance.
- Construction, by testing individual subsystems.
- Operation, by monitoring the building and controlling actuators.
- Maintenance, by solving infrastructure problems due to energy deficiencies.
- Demolition, by recycling materials and usable elements.

During these phases it is necessary to continuously reengineering the indices that measure energy efficiency, to adapt the model to the building conditions. However, the optimization of these parameters comprises a complex task, full of variables and restrictions. In [20], for instance, a multicriteria decision model to evaluate the whole lifecycle of a building is presented. In [21] this problem is tackled from a multi-objective optimization point of view, and the authors conclude that finding an optimal solution is really complex, being only feasible to approximate it.

The long-term goal of all these tasks about energy efficiency in the literature is reaching a net-zero/positive energy building [22], where the power consumption will be null or even negative, thus generating energy that can be stored o sold to energy providers. In this kind of system the availability of alternative (and green) energy sources is essential. In [23] a building automation system which integrates a dynamic energy management system is presented. Here the building is provided with photovoltaic, wind and biomass generators, together with a oil-based one which works as an alternative energy source. As it is explained later on the chapter, the real proposal presented follows this line, complementing the common external provision of energy with solar panels, including photovoltaic solar units on the whole roof of the building, and integrating a diesel energy generator in the cellar.

The way to measure energy efficiency in smart buildings is usually done in terms of energy consumption, energy costs and carbon emissions. In [22], for instance, three main indices are computed in quasi-real time: the Generation-Consumption Effectiveness Index (GCEI), the Cost and, more innovative, the Comfort Index (CI), which considers thermal, visual and air quality measurements. Some of these parameters are supposed to be directly calculated with monitoring data collected from sensors, while others are obtained from surveys. The proposal given by [24] pays a especial attention to the building usage, as compared with the energy consumed. This work first describes the input data used to compute the efficiency indices, considering static information, such as dimensional character- istics of the building or activities performed inside, and dynamic data obtained from the monitoring subsystem, such as consumption information or external temperature. Then the work proposes to measure six partial indices, based on energy, hours of use and volume of conditioning areas, and a global one computed on the basis on them.

Some of the previous works present partial developments regarding monitoring and data collection from sensors. In [19], for instance, a reference implementation of an energy consumption framework is given, to analyze the efficiency of a ventilation unit. The real deployment of the system presented in [24] is based on a common client/server architecture focused on monitoring energy consumption, which has been analyzed on real buildings. A similar proposal is given in [25], but this work is less focused on efficiency indexes and more on a cheaper an practical device to cope with a broad pilot deployment, to collect the feedback from users and address next improvements for the system. The proposal given in [26] describes an interesting approach to model energy consumption information with ontologies, presenting a good solution to avoid compatibility problems between different systems, however, only a reference prototype to collect energy con- sumption information is presented. In [27], the authors describe an automation system for smart homes on top of a sensor network. Although the goal pretended in this work is similar to the one aimed in the current chapter, the system lacks on automation flexibility, since each node of the network only offers limited I/O capabilities through digital lines, there is not a local friendly interface for users in the house, and, what is most important, the integration of energy efficiency capabilities is in fact weak. The work presented in [28] is also based on a sensor network to cope with the building automation problem, but this time the messages of the routing protocol include monitoring information of the building.

In contrast to previous works, the current chapter presents a real experience on a general purpose platform for building automation, which addresses the problem of energy efficiency, dealing with monitoring issues, by means of a flexible sensor architecture which allows the collection of energy efficiency parameters, and able to control a wide range of automated parts of the building.

4 A Real Proposal: PLATERO

The PLATERO platform gives a building automation solution that covers current and future requirements in the indoor smart building field. It was first designed in its early stages as DOMOSEC (Domotics and Security) [29], and now it has been improved with new subsystems (such as the SCADA and energy efficiency monitor) and a deep refactor of the hardware base. PLATERO avoids a tight dependence on technologies, considering successful experiences, and proposing innovative subsystems. All of this thanks to the analysis, design, implementation and deployment of an entire building automation, management and monitoring architecture. This underlying architecture is used to provide common domotic capabilities, but taking into account the design of an integral platform to offer novel pervasive services in smart indoor spaces. The **main features** of PLATERO are:

- A variety of in-building communication technologies are supported to control a wide range of devices.
- Integral control and data fusion from sensors using a home automation module.
- In-building management through distributed and adapted control panels.
- Fault-tolerant design by means of database replication and a security subsystem developed on alternative communication technologies.
- Value-added monitoring and control services available in a distributed management logic.
- Remote management and monitoring through a 3D interface.
- Remote re-programming of automation module.
- Flexibility, since the system can be adapted to specific solutions, through the use of a base platform complemented by optional components.
- Integrated solution for saving on energy consumption.

The whole architecture of the PLATERO system is showed in Fig. 1. As can be seen, the platform is divided into the in-home system, the supporting security infrastructure, the system set-up access, the tele-assisstance feature for technical support, the remote access subsystem, and the energy efficiency module. External entities communicate with in-home subsystems by means of Internet. The SCADA (supervisory control and data acquisition) acts as a gateway for most users and external systems (e.g. the energy efficiency logic) to monitor/manage the building by means of a secure access later explained.

Although the diagram showed in Fig. 1 considers all the possible elements of a complete configuration for automating an individual building, the system is

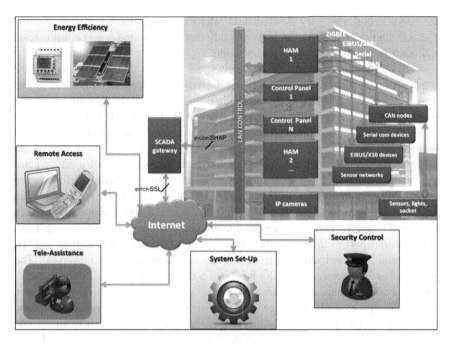

Fig. 1 Overall building automation system provided by PLATERO

completely modular and it is designed to cover the management of a network of buildings. Moreover, the generic nature of the system enables us to apply the automation architecture in houses, offices, schools, shopping centres, hospitals, resorts and, in general, any other domain where indoor automation and intelligence takes place.

4.1 Building Monitoring and Control

The main element of the architecture is the **Home Automation Module** (HAM). This comprises an embedded computer that is connected with all appliances, sensors and actuators. In this way, the HAM centralises the "intelligence" of an administrative domain of the building (e.g. a floor or a laboratory), since it contains the configuration used to control all installed devices.

In order to provide a local human-machine interface of the system, several **Control Panels** can be spread in the building to manage automated spaces. These comprise an embedded solution with an HMI adapted to the controlled devices. For example, in a three-storey office building, each floor could have an HAM and a pair of control panels, in order to set the automatic opening of windows in the floor, switch on the air conditioning to set the desired temperature, or close/open the blinds according to the desired light intensity before using artificial lighting. These examples are developed cases of study that diminish the power consumption and contribute to environment preservation.

A **SCADA** platform acts a as gateway to offers value-added services for management and monitoring, but it is not in charge of performing any control over appliances or actuators directly. Instead, this gateway communicates with the HAM using a UDP-based protocol later explained. This SCADA will be usually provided remotely in a high-end server. Some other solutions leave these control actions to a local PC-based gateway, which is understood as a not appropriate strategy. In [30] a software implementation of a gateway is also used as auto-mation station, which is executed over a common PC/Java platform. The embedded solution used in our HAM proposal offers a fault-tolerant architecture and assures the correct operation of devices. A SCADA-based solution is used in PLATERO to give extra services to inhabitants, and perform networking tasks from the transport to the application layer in the OSI stack, as described in [31].

A remote data and request processing management system is proposed. The SCADA has been designed on the basics of a distributed data collection logic. It collects building data, sensor measurements and energy efficiency information from buildings in a reliable way, and provides processed information to users/ administrators through a SCADA access. Its architecture is shown in Fig. 2. As can be seen, data from HAMs is collected by a set of Data Collection Points (DCPs) by means of the SHAP protocol. HAMs choose one of these DCPs according to the observed performance and an initial priority list. All data col-lected by DCPs is then sent to Data Base Proxies, in charge of turning HAM measurements into data records. Several Data Base Proxies provide reliability to the system for accessing the database. Finally, an intermediate stage for providing a buffered and synchronized access to the database is provided by DB Writer. All this information flow provides a fault-tolerant design against eventual problems in the different modules.

As can be seen in Fig. 2, two management modules have been included in the data collection system: HAM Manager and System Manager. HAM Manager is used to keep track of all building connections, and it enables administrators to check the HAM firmware. System Manager is an always-on service that monitors the operation of all modules. It periodically reads status information of all modules (DCPs, DB Proxies, etc.) from the database, since each new record also includes status stamps of each system module.

All collected information from HAMs is finally provided to users/administra-tors (and the rest of building-external entities) through a SCADA access (called SCADA-Web). This is also illustrated in Fig. 2. By means of a PC platform, a common Web Browser can be used to access a URL of the system. After the user is authenticated, a secure HTTPS link is established. At this moment, a JNLP (Java Networking Launching Protocol) application is automatically downloaded. This software operates at the client side, and provides a graphical front-end to access building information of all the monitored HAMs. This information is available for the JNLP application by means of an SSL (Secure Socket Layer) link with a Data

Distributed Data Collection Logic

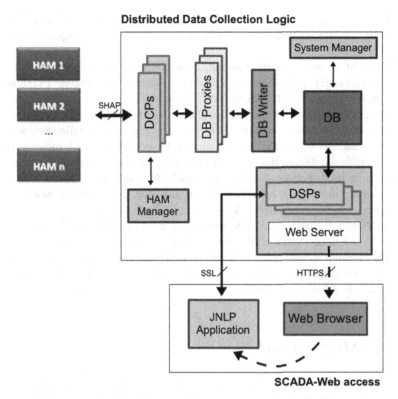

Fig. 2 Architecture of the SCADA distributed logic

Server Point, which access the database. This feature also improves the system reliability when the building information is accessed. Moreover, the JNLP technology offers flexibility to the system, since external entities dynamically download the Java (platform-independent) application from a Web server, but only when it is accessed for the first time or a newer version is available at the SCADA server.

4.2 In-Building Networking

To date, most efforts in designing novel communication protocols in the home/building automation field have been focused on communicating a home automation controller with appliances. In [32], for instance, a bus-based protocol is defined over the power line. The PLATERO system, on the contrary, bets on current specifications to connect the HAM with indoor devices, and it proposes a novel communication protocol that connects IP-based elements of the architecture through UDP. As IP-based elements are considered the HAM, the SCADA, control panels, IP cameras, and the architectural elements outside the house: energy efficiency module, remote access, tele-assisstance, system set-up and security control.

The HAM supports several communication controllers in order to connect with many devices. By complementing the direct digital and analog I/O through common wiring, a CAN (Controller Area Network) bus can be used to extend the operation range or provide a more distributed wiring solution. X-10 connections over the power line are also available for low-cost domotics installations, whereas the EIB controller offers a powerful solution for connecting with more complex appliances. Serial-485 devices can be connected and the powerful Modbus protocol can be used. Finally, ZigBee (or 6LowPAN) and Bluetooth can be used to avoid wiring in already built houses, for instance, and connect new devices through a wireless sensor network.

A LAN installation is used in the building to connect all IP-based elements with the HAM, whereas a changeable communication technology can be used to connect the in-building network with Internet. Fibre optic, common ADSL, ISDN or cable-modem connections could be enough to offer remote monitoring/management and a basic security system.

4.3 The SHAP Protocol

The Superior Home Automation Protocol (SHAP) is used to connect the IP-based components of PLATERO. In our architecture it connects each HAM with the in-home and remote IP-based entities, following a sliding window strategy with UDP packets to assure the data flow control.

Management messages are sent from control panels or the SCADA to the HAM by using the SHAP protocol. Moreover, the SHAP protocol includes a set of messages that are used to flash the HAM microcontroller code memory remotely. A feature which performs this task (**System Set-Up**) has been especially developed and integrated in the SCADA-Web access. In this way, it is possible to reduce maintenance costs. This software is later explained in the chapter.

Due to the security hazards that imply Internet communications in a building automation system, secure communication channels between IP-based elements of the architecture are used. Building networks accessed from outside imply a number of security issues focused on confidentiality, authenticity, authorisation and Integrity, as it is described in [33]. The SHAP protocol implements an approach based on symmetric cryptography to deal with these questions. SHA-2 (Secure Hash Algorithm) is used to calculate a hash of the packet payload and then the resultant value is encrypted, jointly with the heading, with AES (Advanced Encryption Standard), by means of a symmetric key shared among the terminals of the system. The message is completed with a CRC of the whole packet. The SHA hash assures the integrity of the payload, and the AES encryption applies authenticity to the information transmitted, since the synchronization between the sender and the receiver (included in the heading) is hidden. Confidentiality is not directly provided for the packet payload since encryption is supposed to be included only in desired messages. For example, applying encryption to the payload is unnecessary for memory map messages, when the HAM memory is flashed, because decoding all packets would delay the process too much. Alarm messages, on the other hand, can offer encryption by themselves. At the service level (SCADA-Web access), confidentially is offered by means of a secure HTTP access in the JNLP application. Remote clients are authorised to access the graphical monitoring application by an authentication stage executed at the beginning of the session.

4.4 Technical Tele-Assistance

One of the main functionalities of the SCADA, which is only available for authorized personnel, is the technical **Tele-Assistance**. This allows technicians to remotely diagnostic the operation of the various devices and subsystems installed in buildings. Apart from a common monitoring access, where the technician can check the operation parameters of devices, it is possible to receive alarms that indicate anomalies in the system. Unattended ones are stacked and notified until the devices in question are checked.

When accessing the status of subsystems or individual devices, the tele-assistance support of PLATERO is able to provide the manufacture-recommended operation parameters and compare them with the current ones. This makes easier the work of technicians, who can avoid traveling for in-situ assistance when the problem can be solved remotely. This way the maintenance costs during the system exploitation are reduced.

4.5 Securing the Building

Due to the relevance of safety services in current building automation systems, the PLATERO architecture includes an integrated security system. Local sensors connected to the HAM, such as presence, noise and door opening detectors, are used as inputs for the security system. As it can be seen in Fig. 1, a **Security Control** entity is in charge of receiving security events from the building. PLATERO interfaces with the software installed at the security company using standardized alarm messages. More details about the security provisioning in buildings can be found in [29].

Sometimes, automated buildings are included in an administrative domain (e.g. housing developments) and monitoring/securing tasks can be applied by local staff. The SCADA system can be used in these cases to receive security events from buildings and provide notifications to local security staff, apart from the security notifications sent to external security companies if desired.

4.6 Energy Efficiency Subsystem

One of the most important capabilities of the PLATERO platform is that the system gathers information from all devices about the energy production and consumption, to propose new settings for HVAC appliances. This process is carried out by the **Energy Efficiency** module depicted in Fig. 1, with the aim of minimizing the energy consumption while maintaining the desirable comfort conditions.

The energy efficiency module monitor the building spaces through the SCADA access and, as it has been explained above, is in charge of proposing the needed actuation commands over key devices with the aim of saving energy (adapts lighting and HVAC systems, switch on/off appliances, etc).

4.6.1 Maintaining Comfort Conditions

The CEN standard EN 15251 [34] specifies the design criteria (thermal and visual comfort, and indoor air quality) to be used as input for calculating the energy performance of buildings, as well as the long-term evaluation of the indoor comfort. Our intelligent building system takes into account this standard of quality measurements, the user comfort requirements, gathered from the residents activity, and the generated energy in the building (i.e. by using solar panels, as it is the case in the use case later explained), in order to define the best environmental comfort while energy efficiency is maintained.

Monitoring variables are measurable thanks to SCADA-Web access in PLA-TERO, therefore, the developed models to control energy efficiency and comfort conditions in our test building are fully integrated into the automation system. The next parameters have been identified to affect comfort and energy performance: indoor temperature and humidity, natural lighting, user activity level and power consumed by electrical devices. Environmental parameters (temperature, humidity and natural lighting) present a direct influence on energy and comfort conditions but, in addition to them, thermal conditions also depend on the user activity level and the number of users in the same space. Depending on the indoor space (such as a corridor or a dining room), the comfort conditions are different and, therefore, the energy needed too. Moreover, the heat dissipated by electrical devices also affects thermal conditions.

One of the most relevant inputs to the energy efficiency subsystem is the human activity level, which is provided by our indoor localization mechanism integrated in PLATERO and presented in [35]. This is based on an RFID/IR data fusion mechanism. Hence, location and presence information allows the system to esti-mate the human activity level and control the HVAC and lighting equipment accordingly.

4.6.2 Building Intelligence

Traditional approaches rely on predictive models too complex and based on static perceptions of the environment. Controlling such systems requires facing the inaccuracy of sensors, the lack of adequate models for many processes and the non-deterministic aspects of human behavior. There is an important research area that proposes artificial intelligence as an alternative to solving the drawbacks of non-deterministic models. Among others, Neural Networks (NNs), Fuzzy Logic Systems (FLSs) and Genetic Algorithms (GAs) could be applicable. NNs have been widely used for energy prediction and modeling in buildings [36, 37]. However, NNs are regarded as black boxes that are not very compressible for humans and they usually have a certain complexity in their definition. This brings a complex problem to the scope of building automation, since it is focused on the interaction with the user. Therefore, being able to make the intelligent system understandable to the end user is an issue. Due to this, the intelligent system integrated in PLATERO is based on fuzzy logic, since it offers a framework for representing imprecise and uncertain knowledge in a similar way in which people make their decisions. Fuzzy systems provide mechanisms to approximate rea-soning, which allows dealing with vague and incomplete information. In addition, fuzzy controllers exhibit robustness with regard to noise and variations of system parameters. However, these systems have a well known problem related to the determination of their parameters, therefore they need a learning mechanism to gather them. As an extension, machine learning algorithms have been proposed as a solution for learning the parameters of fuzzy systems [38], which adapt the system to the dynamic conditions of the environment and users.

4.6.3 A Fuzzy Framework for Intelligent Energy Control in Buildings

Our purpose is to provide an energy management building based on using the surplus energy generated to satisfy extra energy requirements for more comfort-able temperatures, lighting and ventilation. This way a trade-off between energy saving and user comfort is ensured, taking into account the daily operation mode in the building (i.e. net-negative/zero/positive).

Firstly, we need to model comfort conditions that help us to predict accurately the dynamic response of the system. The smart control system proposed to cope with the decision making process has been developed in the form of an intelligent fuzzy system which takes into account all the requirements of minimum energy consumption while a specific level of comfort is ensured. Figure 3 illustrates a

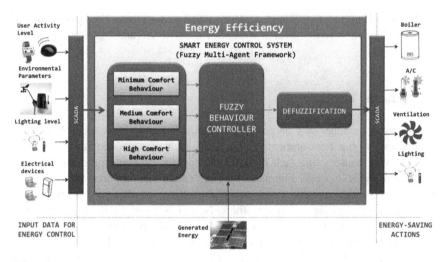

Fig. 3 Energy efficiency subsystem

schema of the energy efficiency subsystem. This is split into three stages. The first one is based on fuzzy models that represent the three kinds of comfort conditions: minimum, medium and high comfort. These models provide different estimations of the energy consumed according to the input data and the comfort conditions. The second phase is the responsible for deciding which behavior to choose (minimum, medium or high level of comfort) depending on the estimated consumption calculated in the first stage and the energy generated by the building. And, finally, the phase of defuzzification is in charge of translating the fuzzy clusters into numerical values for the HVAC and the lighting settings to be passed to the SCADA.

The process to extract the fuzzy models from a set of data $Z = z_1, ..., z_n$ is called fuzzy modeling. The success on applying fuzzy models to system identification has been widely demonstrated in the literature [39, 40]. Generally, the fuzzy modeling process comprises three main phases. The first task is to decide the number of rules of the fuzzy model. Afterwards, in the rule generation phase, the goal is to obtain a first approximation of the model in the form of fuzzy clusters. Each cluster gives an initial fuzzy rule. Then, the next phase is the identification of parameters, which is in charge of the fine adjustment of the parameters of the fuzzy sets in order to obtain a model with good approximation capabilities. In order to implement these models, we have used the Adaptive Neuro Fuzzy Inference System (ANFIS) proposed by Jang in [41], which combines the back propagation learning and the least squares error methods to tune the antecedents and the consequents, respectively.

5 Prototyping the System

The PLATERO components described in the previous section have been developed to validate the architecture. All the prototypes presented here are part of a real deployment of the system in a testbed building targeted to improve energy efficiency, although some hardware components are showed out of their installation point to make easier the explanation. Regarding application screenshots, they comprise real scenarios of usage.

The reference building chosen has been the Technology Transfer Center at University of Murcia,[1] which was designed as a smart environment since the early stages of design. The roof of the building is covered by a set of solar panels, and the interior has been automated to make the most of the power used. Each floor has a HAM to control all common areas, whereas each working zone has another one to monitor the water and power consumption, adapt the lighting according to natural light, detect fires and floods, or automatically switch on/off several devices. A control panel has been installed in each working zone to manage these capabilities using a friendly HMI. This can be seen in Fig. 6a, together with the adapted electric panel and the optional manual control of the air conditioning. All HAMs in the building provide control access capabilities, by using smart cards, and a local monitoring/management point has been set-up at the reception position, accessing the SCADA-Web.

The three main advantages of the wide automation system of the reference building, powered by the PLATERO platform, are:

- Energy balance, combining solar energy, common external energy and an auxiliary diesel generator.
- Tele-maintenance, tele-monitoring and automatic control of devices and subsystems of the building (air conditioning, lifts, pumps, alarms, etc).
- Access control through personal identification based on RFID (Radio Frequency Identification)/NFC (Near Field Communication) technologies.

Then next subsections explain in more details the different prototypes developed for this study case, which instantiate the functionalities widely presented in the previous section.

5.1 HAM Devices

The home automation module is based on the SIROCO 3.0 (System for Integral contROl and COmmunications) hardware architecture, designed at the University of Murcia for automation purposes. The first generation of the SIROCO platform was presented in [29]. The different modules that comprise the unit can be seen in

---

[1] http://www.um.es/otri/?opc=cttfuentealamo

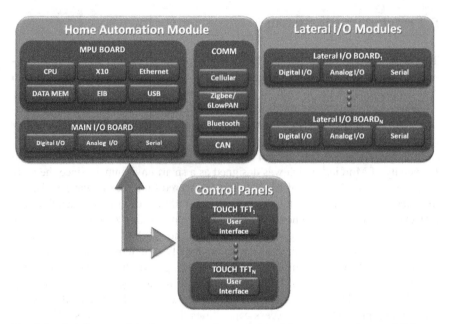

Fig. 4 Logical diagram of the home automation module and its communication capabilities

Fig. 4. SIROCO is a modular system highly adaptable that gives a self-sufficient platform to perform management and monitoring tasks. It offers the option of installing a low-cost solution or a complex one, extending the base system with the required modules.

The third generation of SIROCO hardware MPU series is based on a 32-bit microcontroller. The MPU (Main Processor Unit) Board is equipped with a set of I/O channels:

- 16 basic I/O ports (analog and digital).
- Communication ports: Ethernet, USB, CAN 2.0B, RS-485 and three RS-232.
- Possibility of adding extra memory through microSD card or USB flash drive.

The HAM is additionally provided with extended networking capabilities. Specific domotics communications are provided by an X10 module and an EIB controller, connected both through a serial interface. Furthermore the MPU board can be extended with additional communication boards (if needed) through the serial and USB ports. The CAN bus support offers an alternative to EIB when a more flexible communication channel with wired sensors is needed.

The main I/O board provides extra wired interfaces with appliances, sensors and actuators adding up to 16 lateral I/O boards connected to the main I/O board. With this configuration, complex control schemes can be tackled.

The hardware unit developed following the previous design can be seen in Fig. 5. The main I/O ports are visible (CAN/serial, Ethernet and USB), and it can

Fig. 5 Home automation module developed

be observed the compacted case chosen, as compared with the former prototypes described in [29].

5.2 Control Panels

The second generation of control panels are based on the MPU board of the previous SIROCO architecture and have been upgraded to include a TFT 7" touch screen, as compared with the first units presented in [29]. They guarantee a familiar HMI limited to automated devices in the surroundings (connected to the same HAM). Users can define configuration profiles, which contain a set of device states and actions to be performed under certain conditions. Moreover, the house alarm can be armed/disarmed by a defined control panel. Any panel, however, can be used to activate panic, security or fire alarms at any time. Additionally, when an alarm is activated by the HAM (due to sensor measurements) or manually, control panels warn users via acoustic and visual messages.

Figure 6a shows the control panel developed for the prototype building, while Fig. 6b shows a screenshot of the HMI integrated, where the user is reviewing the configuration of blinds and awnings of a laboratory in the reference building. Two blinds are used, the first has been set-up to close and open automatically, depending on lighting conditions, whereas the behavior of the second one has been programmed at different times of day. Currently, they are opened at 50 % and 60 %, respectively.

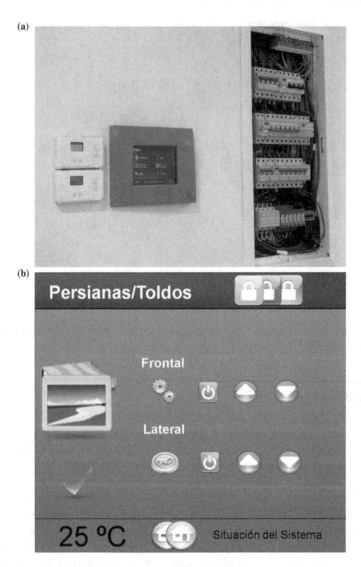

Fig. 6 Control panel prototype. **a** Control panel installed in the reference building. **b** HMI of the integrated software

5.3 SCADA-Web Management Application

Control panels provide a local management of certain spaces in a building, but remote access to the whole system is also possible through the SCADA-Web application. This software can be used to monitor and control indoor spaces, but also to store incidents, manage machinery services, access control and system status reporting.

Users, administrators and technical personnel, by using this JNLP application, obtain an personalized 3D view of the building depending on their access type, and can manage the automation systems as if he/she were physically there. Figure 7a shows a screenshot of the application when a technician is accessing the main view of the building. Here the five different administrative domains can be accessed by pressing on the desired floor. Moreover, important events are listed on the left part of the window and the user can click on them to directly access the device emitting the alert.

In Fig. 7b it can be seen the view for both a common user (e.g. the caretaker of the building) or a technician. The application is showing the status of the bathroom available in the first floor of the building (a map view is depicted on the bottom right part of the window). The status of the lighting system and the information provided by the presence and flooding sensors are showed here. Apart from the monitoring features, the user can press on the different subsystems (e.g. lighting) for changing the current state; thus, this view also serves for managing automated devices.

Security staff receive fire alarms from the building through the SCADA-Web application, and they can monitor in real time the fire sensor deployment along the building. In case of fire, an effective and timely response is possible thanks to the precise information about the incident given by the platform.

5.4 PLATERO Platform Setting-Up

When accessing the SCADA, if the user authenticates as an administrator, the JNLP application downloaded provides additional functionalities for allowing specialists to access the building configuration (system set-up in Fig. 1). Figure 8 shows a screenshot of the application while the administrator is establishing the keys to be used in the communication with one of the HAMs installed in the reference building. The software also monitor X10, EIB and UDP communications with the HAM.

The software enables the installer to configure the different partitions and zones of the building domain managed by the HAM, set the devices connected to the system, and define the remote accesses allowed from outside. All this information is stored in the HAM database, and then replicated in the SCADA logic. The HMI allows the installation of initial profiles and actions to be performed under certain events detected by sensors. All settings can also be saved for application to other HAMs.

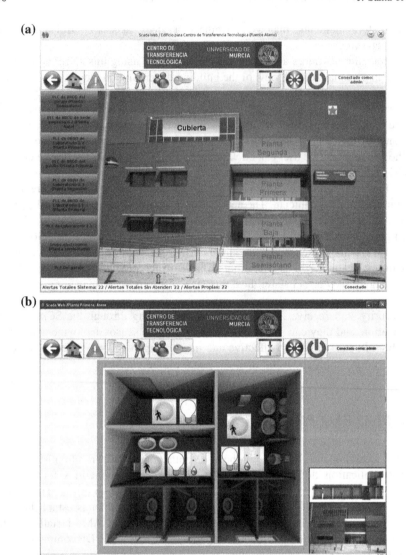

Fig. 7 JNLP application with 3D HMI for local/remote management through the SCADA-web access. **a** Overall building view. **b** Monitoring the bathroom of the 1st floor

5.5 Saving Energy in the Reference Building

The proposed intelligent system is able to save energy according to the user behavior, environmental parameters and the desired comfort level. The energy efficiency module has been developed using the reference building where the PLATERO platform is being exploited. It has been implemented as an independent

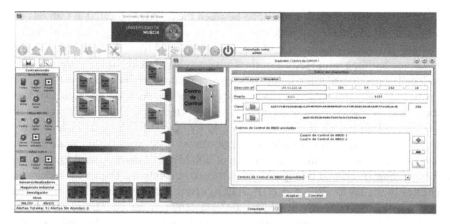

Fig. 8 Screenshot of the HAM set-up software

software that gathers information from the JNLP application about the energy production and consumption of all devices.

The consumption parameters considered are the power and water, although the last one is not directly used by the system for the moment. The power consumed in the different spaces of the building are collected by digital power meters connected with the installed HAM. The feedback about the power generated by the building is obtained from the solar panels, since the inverter is also automated and connected with one of the HAMs.

All spaces of the building (laboratories) have been automated to offer save energy chances. Furthermore, temperature, humidity, presence and lighting sensors have been distributed among all rooms and corridors. Since each laboratory includes a HAM, this does not only let the ambient intelligence part of the system to monitor the area, but also to send actuation commands to HVAC devices (A/C, boiling and lighting, for the moment) with the aim of saving on power consumption.

5.6 Remarks about the System Operation

The reference building considered as testbed at University of Murcia is open since March, 2010. The whole sensor and actuator infrastructure connected with HAMs was present since the first day, and the building has been monitored and managed using PLATERO since that date (the former name of the system was DOMOSEC). During these three years of operation the SCADA platform has been updated to offer more functionalities and being the entry point to the building for both gather sensor data or manage specific devices. The hardware operation has been correct and in-situ maintenance has only been needed to substitute some of the batteries used by HAMs.

About 66 MB are diary exchanged between the SCADA, located at Campus de Espinardo in Murcia (main university campus), and the building, placed in Fuente Álamo. This means an average of 6,400 bps. An optic fibre network connects the reference building with the rest of the University of Murcia but the system is able to locally operate under the failure of the external network by using control panels as the interface and HAMs as local processing units. As soon as the network is recovered, all the changes in the system are reported to the SCADA. Moreover, it is worth noting that all the devices managed by PLATERO have a database entry in the SCADA distributed logic, although almost a 20 % of the database size is used by the base platform. Hence, a total of 500 MB are currently used to maintain the building status and configuration in the main database. The set of replicated entities in the back-office part of PLATERO further assures the reliability of the system, which is currently hosted by the data center of the University of Murcia.

The current users of the system installed in this reference building are the researchers working in the laboratories, who interface with the system through the control panel installed in each laboratory; the caretaker, accessing a SCADA view mainly oriented to monitoring; and our research group, accessing the SCADA for both tele-assistance and system set-up. Additionally, an external security company is connected to the system to receive fire alarms and unauthorized accesses to the building. The acceptability of the system by the users has been really good and most of them use the system to monitor their power and water consumption, enable the automatic management of the lighting, and enable/disable certain sockets. This is attributed to the profile of the users, most of them researchers and engineers used to information technologies. However, the most noticeable advantages are found in the monitoring capabilities, the remote management of building subsystems (water supply, electrical panels, air conditioning, etc.) and the capability to dynamically adjust the comfort conditions according to the power consumption, the user requirements and the energy generated by the building.

6 Conclusion

After a review of the needs of a proper platform for enabling energy saves in the building automation domain, a home automation architecture has been presented, which gives an integral solution for most domotics and security necessities. The system integrates innovative modules which can be combined to cope with a whole building automation solution ready to offer multitude of pervasive services. Thus, a potential set of home automation functionalities, along with essential prototypes, have been developed in a testbed building that proves the potential of the system. The energy efficiency capabilities of the PLATERO platform provides an intelligent management of the power consumed, according to the desired level of comfort, the environmental conditions and the power generated by the building.

PLATERO involves a completely modular and flexible platform composed of a building infrastructure and a set of optional remote software elements providing added-value services in the indoor domain. The in-building network is centralised in the embedded solution provided by the home automation module. It is able to work with most common domotic specifications, in order to connect with sensors, actuators and the rest of devices installed in the building. Moreover, a UDP-based protocol is used over an IP network to connect the networked elements of the building (i.e. HAM and control panels) with the SCADA. This base system can be adapted to several applications out of pure building automation, such as greenhouses, elderly attention, eHealth or energy efficiency.

In conclusion, the platform presents a cost-effective, practical and novel indoor automation architecture ready to deploy pervasive services.

A current line of work is focused on the expansion of the system to networks of buildings and other smart city infrastructures as street lighting, where saving energy is a key aspect to consider. Nowadays, several buildings are being controlled at the University of Murcia using the same PLATERO/SCADA platform, and some other private companies are considering its installation in the short-term.

Acknowledgments This work has been sponsored by the European Seventh Framework Program, through the IoT6 project (contract 288445); the Seneca Foundation, by means of the GERM program (04552/GERM/06) and the FPI program (grant 15493/FPI/10); and the Spanish Ministry of Education and Science, thanks to the the FPU program (grant AP2009-3981).

References

1. D. Clements-Croome, D.J. Croome, *Intelligent Buildings: Design, Management and Operation* (Thomas Telford, London, 2004), pp. 221–228
2. K. Sangani, Home automation—it's no place like home. IET Eng. Technol. **1**, 46–48 (2006)
3. S.S. Intille, Designing a home of the future. IEEE Pervasive Comput. **1**, 76–82 (2002)
4. K. Wacks, Home systems standards: achievements and challenges. IEEE Commun. Mag. **40**, 152–159 (2002)
5. J. Walko, Home control. IET Comput. Control Eng. J. **17**, 16–19 (2006)
6. International Energy Agency, *Key World Energy Statistics* (IEA Report, Paris, 2010)
7. European Commission, DIRECTIVE 2010/31/EU OF THE EUROPEAN PARLIAMENT AND OF THE COUNCIL of 19 May 2010 on the energy performance of buildings (recast). Official J. Eur. Union **53**(153), 13–34 (2010)
8. L. Pérez-Lombard, J. Ortiz, C. Pout, A review on buildings energy consumption information. Energy Build. **40**, 394–398 (2008)
9. D.H.A.L. van Dijk, E.E. Khalil, Energy efficiency in buildings. ISO Focus 22–26 (2009). September 2009
10. M. Hazas, A. Friday, J. Scott, Look back before leaping forward: four decades of domestic energy inquiry. IEEE Pervasive Comput. **10**, 13–19 (2011)
11. A.I. Dounis, C. Caraiscos, Advanced control systems engineering for energy and comfort management in a building environment-a review. Renew. Sustain. Energy Rev. **13**, 1246–1261 (2009)
12. A.R. Al-Ali, M. Al-Rousan, Java-based home automation system. IEEE Tran. Consum. Electron. **50**, 498–504 (2004)
13. A.Z. Alkar, U. Buhur, An internet based wireless home automation system for multifunctional devices. IEEE Tran. Consum. Electron. **51**, 1169–1174 (2005)

14. R.J. Caleira, in *A Web-Based Approach to the Specification and Programming of Home Automation Systems*. 12th Mediterranean Electrotechnical Conference (IEEE Press, New York, 2004) pp. 693–696
15. J. Nehmer, M. Becker, A. Karshmer, R. Lamm, in *Living Assistance Systems—An Ambient Intelligence Approach*. ACM International Conference Software Engineering (ACM Press, New York, 2006), pp. 43–50.
16. A.J. Jara, M.A. Zamora, A.G. Skarmeta, in *A Wearable System for Tele-Monitoring and Tele-Assistance of Patients with Integration of Solutions from Chronobiology for Prediction of Illness*. Ambient Intelligence Perspectives (IOS Press, Lansdale, 2008), pp. 221–228.
17. S. Helal, W. Mann, H. El-Zabadani, J. King, Y. Kaddoura, E. Jansen, The gator tech smart house: a programmable pervasive space. Computer **38**, 50–60 (2005)
18. A. Meliones, D. Economou, I. Grammatikakis, A. Kameas, C. Goumopoulos, in *A Context Aware Connected Home Platform for Pervasive Applications*. Second IEEE International Conference Self-Adaptive and Self-Organizing Systems Workshops (IEEE Press, New York, 2008) pp. 120–125.
19. D.T.J. O'Sullivan, M.M. Keane, D. Kelliher, R.J. Hitchcock, Improving building operation by tracking performance metrics throughout the building lifecycle (BLC). Energy Build. **36**, 1075–1090 (2004)
20. Z. Chen, D. Clements-Croome, J. Hong, H. Li, Q. Xu, A multicriteria lifespan energy efficiency approach to intelligent building assessment. Energy Build. **38**, 393–409 (2006)
21. C. Diakaki, E. Grigoroudis, D. Kolokotsa, Towards a multi-objective optimization approach for improving energy efficiency in buildings. Energy Build. **40**, 1747–1754 (2008)
22. D. Kolokotsa, D. Rovas, E. Kosmatopoulos, K. Kalaitzakis, A roadmap towards intelligent net zero- and positive-energy buildings. Solar Energy **85**, 3067–3084 (2011)
23. J. Figueiredo, J. Martins, Energy production system management—renewable energy power supply integration with building automation system. Energy Convers. Manag. **51**, 1120–1126 (2010)
24. G. Escrivá-Escrivá, C. Álvarez-Bel, E. Peñalvo-López, New indices to assess building energy efficiency at the use stage. Energy Build. **43**, 476–484 (2011)
25. V. Sundramoorthy, G. Cooper, N. Linge, Q. Liu, Domesticating energy-monitoring systems: challenges and design concerns. IEEE Pervasive Comput. **10**, 20–27 (2011)
26. D. Bonino, F. Corno, F. Razzak, Enabling machine understandable exchange of energy consumption information in intelligent domotic environments. Energy Build. **43**, 1392–1402 (2011)
27. D. Han, J. Lim, Design and implementation of smart home energy management systems based on ZigBee. IEEE Trans. Consum. Electron. **56**, 1417–1425 (2010)
28. P. Oksa, M. Soini, L. Sydänheimo, M. Kivikoski, Kilavi platform for wireless building automation. Energy Build. **40**, 1721–1730 (2008)
29. M.A. Zamora-Izquierdo, J. Santa, A.F. Gomez-Skarmeta, An integral and networked home automation solution for indoor ambient intelligence. IEEE Pervasive Comput. **9**, 66–77 (2010)
30. P. Pellegrino, D. Bonino, F. Corno, in *Domotic House Gateway*. ACM Symposium Applied Computing (ACM Press, New York, 2006), pp. 1915–1920
31. F.T.H. den Hartog, M. Balm, C.M. de Jong, J.J.B. Kwaaitaai, Convergence of residential gateway technology. IEEE Commun. Mag. **42**, 138–143 (2004)
32. K. Myoung, J. Heo, W.H. Kwon, D. Kim, in *Design and Implementation of Home Network Control Protocol on OSGi for Home Automation System*. 7th International Conference Advanced Communication Technology (IEEE Press, New York, 2005), pp. 1163–1168
33. P. Bergstrom, K. Driscoll, J. Kimball, Making home automation communications secure. Computer **34**, 50–56 (2001)
34. Centre Europeen de Normalisation, Indoor environmental input parameters for design and assesment of energy performance of buildings—addressing indoor air quality, thermal environment, lighting and acoustics. EN 15251 (2006)

35. M.V. Moreno, M.A. Zamora, J. Santa, in *An Indoor Localization Mechanism Based on RFID and IR Data in Ambient Intelligent Environments*. The Sixth International Conference Innovative Mobile and Internet Services in Ubiquitous Computing (IEEE Press, New York, 2012), pp. 805–810
36. S. Kalogirou, Applications of artificial neural-networks for energy systems. Appl. Energy **67**, 17–35 (2000)
37. D. Cohen, M. Krarti, in *A Neural Network Modeling Approach Applied to Energy Conservation Retrofits*. International Conference Building Simulation, pp. 423–430 (1995)
38. B. Egilegor, J. Uribe, G. Arregi, E. Pradilla, L. Susperregi, A fuzzy control adapted by a neural network to maintain a dwelling within thermal comfort. Build. Simul. **2**, 87–94 (1997)
39. Y.W. Teng, W. Wang, Constructing a user-friendly GA-based fuzzy system directly from numerical data. IEEE Tran. Syst. Man Cybern. Part B **34**, 2060–2070 (2004)
40. R. Babuska, *Fuzzy Modeling and Identification. International Series in Intelligent Technologies* (Kluwer Academic Publishers, Dordrecht, 1998)
41. J. Jang, Anfis: adaptive-network-based fuzzy inference system. IEEE Tran. Syst. Man Cybern. Part B **23**, 665–685 (1993)

Collective Reasoning Over Shared Concepts for the Linguistic Atlas of Sicily

Dario Pirrone, Giuseppe Russo, Antonio Gentile and Roberto Pirrone

Abstract In this chapter, collective intelligence principles are applied in the context of the Linguistic Atlas of Sicily (ALS - Atlante Linguistico Siciliano), an interdisciplinary research focusing on the study of the Italian language as it is spoken in Sicily, and its correlation with the Sicilian dialect and other regional varieties spoken in Sicily. The project has been developed over the past two decades and includes a complex information system supporting linguistic research; recently it has grown to allow research scientists to cooperate in an integrated environment to produce significant scientific advances in the field of ethnologic and sociolinguistic research. An interoperable infrastructure has been implemented and organized to allow exchange of information and knowledge between researchers providing tools and methodologies to allow collective reasoning over shared concepts. The project uses different types of data (structured, unstructured, multimedia) that require tight integration and interoperability. Additionally, the framework allows for data aggregation into shared concepts that can be exchanged between researchers and constitute a common knowledge base for the entire research community of the ALS.

D. Pirrone · G. Russo · A. Gentile (✉) · R. Pirrone
Dipartimento di Ingegneria Chimica Gestionale Informatica Meccanica,
Universita' degli Studi Palermo, Viale delle Scienze 90128 Palermo, Italy
e-mail: antonio.gentile@unipa.it

D. Pirrone
e-mail: dario.pirrone@unipa.it

G. Russo
e-mail: giuseppe.russo@unipa.it

R. Pirrone
e-mail: roberto.pirrone@unipa.it

F. Xhafa and N. Bessis (eds.), *Inter-cooperative Collective Intelligence:*   403
Techniques and Applications, Studies in Computational Intelligence 495,
DOI: 10.1007/978-3-642-35016-0_15, © Springer-Verlag Berlin Heidelberg 2014

1 Introduction

In this chapter, collective intelligence principles are applied in the context of the Linguistic Atlas of Sicily, an interdisciplinary research focusing on the study of the Italian language as it is spoken in Sicily, and its correlation with the Sicilian dialect and other regional varieties spoken in Sicily. The project has been developed over the past two decades and includes a complex information system supporting linguistic research; recently it has grown to allow research scientists to cooperate in an integrated environment to produce significant scientific advances in the field of ethnologic and sociolinguistic research. The definition of an integrated methodology able to perform a comprehensive analysis in the different fields of sociolinguistic is the ultimate goal of the Linguistic Atlas of Sicily project. This is one of the first sociolinguistic projects across Italy and Europe. It was started in the early nineties from an intuition of professor Ruffino [1]. Sociolinguistics is the the study and the explanation of possible connections between society behaviors and language evidences. Sociolinguistics study the language as a dynamic and complex field changing over space and time. Sociolinguistics incorporate knowledge and information from many other disciplines, such as anthropology, sociology, psychology and lately computer science and information processing. The project is currently a joint effort with research units in Palermo, Catania and Messina leaded by the Dipartimento di Scienze Filologiche e Linguistiche of the University of Palermo. As stated, the project goal from an operative point of view is to exploit different type of phenomena related to phonetic, lexical, morpho-syntactic and textual aspects of language. The focus is on the relations between the evolution over space and time of the usage of regional Italian and Sicilian dialect. A relevant aspect is the dissemination of the produced results as a way to keep track of the social evolution through language. The dissemination is performed through specific publications and reports. The on-going collaboration process is a perfect example a domain hybridizing process, enabling the training on-the-field of a joint group of researchers who, coming from the peculiarly different scientific and cultural domains pertaining to the project, participate to the constitution of the core of a local humanities computing community. The project is extensively interdisciplinary and the progress made from a technology point of view are mostly in the direction of building an inter-operable infrastructure. The infrastructure is organized to allow exchange of information and knowledge between users providing tools and methodologies to allow collective reasoning over shared concepts. The project uses different types of data (structured, unstructured, multimedia) so a first aspect is related to data integration and interoperability. Another important aspect is the exchange of information resulting in the process of data aggregation. Information visualization is the last step of the overall process.

The sociolinguistics field has been investigated over the years in different nations especially where the preservation of a national identity is an urgency. Numerous projects have been realized and some are still working. Usually the dimension of the project is related to the purposes of the project itself. A common goal is to produce atlas representing particular linguistic occurrences. A good example is the Linguistic Atlas of Dolomitic Ladinian and neighbouring Dialects (https://www.sbg.ac.at/rom/people/proj/ald/ald_home.htm). This project is related to preservation and study of ladinian dialect which is an ancient language spoken in some european regions. Another relevant example is the Linguistic (and ethnographic) Atlas of Castille - La Mancha (http://www.linguas.net/alecman/ http://www.linguas.net/). In this atlas of the central region of the Iberian Peninsula, phonetic, lexical and grammatical information is offered. The data has been collected in situ, over the course of several years, through interviews carried out by questionnaires and through recorded conversations. A transnational project is the PRESEEA project. PRESEEA is the "Project for the Sociolinguistic Study of Spanish from Spain and America". The goal is to coordinate sociolinguistic researchers from Spain and the Hispanic America in order to make possible comparisons between different studies and materials, as well as a basic information exchange. The main aim is to create a spoken language Corpus with sufficient guarantees and rich in terms of linguistic information. A more complete project has been carried out over the year from the Copenaghen University. The project is called Lanchart (http://lanchart.hum.ku.dk/aboutlanchart/ http://lanchart.hum.ku.dk/aboutlanchart/) and is working for over three decades with the definition of a multidisciplinary centre to keep track of linguistic evolution in danish language.

The rest of the chapter is organized as follows: next section is deputed to present related works in the various technological fields involved. The third section presents the information organization in the ALS framework. Next section is used to present the main components of the framework. A case study is detailed in the fifth section while the last section presents conclusions and future works

2 Related Works

The Linguistic Atlas of Sicily project has a technological infrastructure able to perform different tasks according to users' needs and specifications. Due to this reason different research fields have been investigated too. The main involved aspects are related to data integration from different sources and collective intelligence building.

2.1 Collective Intelligence Building

Collective intelligence can be defined as the surplus of intelligence created by a group or a community emerging by the contributors' interactions using a system and ultimately the Internet . To produce collective intelligence a key component is a *Collective Knowledge System* (CKS) as a system where "machines enable the collection and harvesting of large amounts of human-generated knowledge" [2]. A CKS incorporates some important elements that are intelligent users able to produce knowledge, an environment allowing exchange of relevant information and social interactions between users and a service of knowledge retrieval such as a search engine to find relevant information. One of the most important features of such service is the capability to produce emergent knowledge which is the knowledge that is not directly found by humans in their investigations. The definition of such systems involves the ability to keep track of interactions between users in the network which is known as social network analysis. This lead to a new way to consider networks of people producing, modifying and exchanging information. The digital environment become in such way a Digital Ecosystem [3]. Numerous attempts to produce and realize systems according to the definition of collective intelligence has been carried on in recent years. The main applications involve definition of social networks explicitly modeled according to the Collective Intelligence principles [4] or the application of collective intelligent principles to build systems able to perform specific tasks [5–7].

2.2 Data Integration

Data integration is a preliminary task to perform to allow information exchange and knowledge discovery. A common way to allow data integration is to use a neutral format. To this purpose a wide range of XML-related technologies has been investigated. XML [8] is becoming de-facto the universal language for data exchange over the web. Some standard interfaces like SAX [9] and DOM [10] have been defined to directly access the content of an XML document. The XML structure naturally discriminates between data and data structures. At present XML has a series of related technologies and formalisms useful to perform particular operations. Some examples are the DTD [11], and the XML Schema [12], used to check and to validate the format of a particular XML document. To obtain a XML document from another one XSLT [13] transformations are used. XPath [14] and XQuery [15] are used to query XML databases. The family of related technologies is growing and is able to perform more and more complex operations. The utilization of XML to perform annotation process starts from the past decade [16]. The widely accepted approach is intended to define an abstract level of annotation that is independent from the physical organization of the process. The major problem is the multiple overlapping hierarchies, a peculiar characteristic of the nature of linguistic information. In the annotation process a single element can be analyzed in many ways and this leads to overlapping layers of annotations. An approach able to deal with the multidimensional nature in the annotation process is required. Many systems to perform annotation like GATE [17], ATLAS [18], or AWA [19] have been developed. To solve the problem of the standoff annotation a system able to switch between the different annotation schemas is required. The other peculiar problem of the presented framework is related to the integration of different sources. An XML based approach [20, 21] to integrate different data sources is to prefer where some data are in an XML format or can be easily produced in an equivalent way. The problem of finding new and useful patterns in data is known in literature as Knowledge Discovery in Database (KDD) [22]. Due to the nature of the treated data the problem of finding useful data starts with a preliminary condition: data are stored in different modalities so they can be strictly structured (database record), semi-structured (XML files and repositories), or totally unstructured like multimedia data. An approach able to combine different types of data is necessary. A natural way to approach this type of problems is to relay raw data to a form of controlled representation of knowledge such as an ontology [23]. In recent years many works utilize this approach in different areas with good results [24–27]. One of the main issues in this approach is that the knowledge base of the system is static so an iterative and incremental approach able to perform a growth in the knowledge base is required.

3 Information Organization in the ALS framework

The ALS (Linguistic Sicilian Atlas) framework has been developed as a virtual laboratory and deployed as a web application. Each component of the framework is deputed to the definition of a set of functionalities that support the researchers in their jobs. In Fig. 1 the main flow of the process is depicted in order to define the pipeline to acquire, transform and use data related to the project.

3.1 Data Acquisition and Organization

Usually, the first step is data acquisition. Data have been acquired over the years trough a set of digitally recorded interviews that are derived from a multi-part questionnaire. The usual methodology developed in the ALS project allows to compare different results over time. The interviewed population is organized for families: a fixed number (usually five) of related people is chosen from different generations (usually three). The same number of families for a particular geographic point in Sicily is selected in order to have compatible clusters. In a standard questionnaire there are three main parts: a biographic section, a metalinguistic section, and a linguistic section. The first section is composed of questions focused about personal information, statistics, level of instruction (personal and familiar), cultural consumes and other similar information useful to have a social characterization. The metalinguistic section is composed of questions investigating the perception and the self-perception about the two different linguistic codes (Italian language and Sicilian dialect) and the ability to move within the two codes. The third section regards specifically language and translation skills

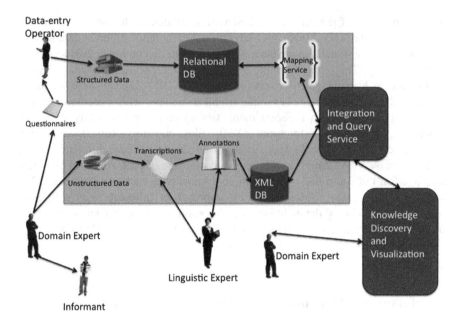

Fig. 1 The main interactions of the overall project and the main interactions between users and system components

between the two codes or other related exercises. After data acquisition, the information from the first and second section is stored into a relational database, while the third section is transcribed and than annotated directly by domain experts. In practice, a set of attributes is related to the transcript, to mark the linguistic phenomena. The result is a valid XML document responding to a XML schema, reflecting the level of annotation. There are two main components in a schema: a common part used to apply general information that is relevant for the retrieval part of the process, and a specific part that is defined by the domain expert. Different annotation schemas have been realized. Schemas for phonetic annotation work at a single word level, lexical annotation works at a phrase level and textual annotation works on a bigger portion of text. The last one makes the expert able to keep track of the interaction progress as well as people behaviors during the interview. Future definitions of new schemes are not precluded. New schemas can allow researchers to perform text analysis at a different level of abstraction, still not defined. The XML documents resulting fro annotation, are stored in an XML-based DBMS. The last part of the process is the retrieval of useful information. This process is defined in two main steps:

1. Defining an integration process in conjunction with a query service, which is able to retrieve information both from structured and semi-structured sources.
2. Knowledge discovery and visualization service, which is able to return useful information for users, to aggregate in a convenient way and to show

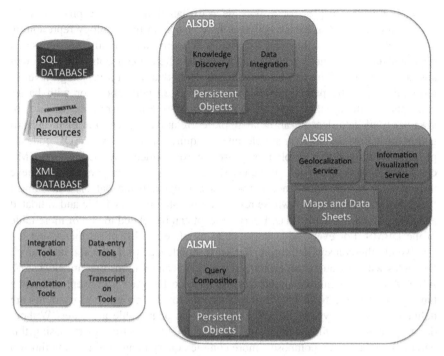

Fig. 2 The three main components of the ALS project, the complementary tools and the data sources of the framework

information on a map for displaying geographic correlation between information items.

The entire framework is arranged into different sub-systems (Fig. 2). The main components are: the ALS Data Base (ALSDB) that manages relational data, the ALS Meta Language (ALSML) that manages semi-structured data, and the ALSGIS (ALS Geographic Information Systems) for visualization. More external components have been realized to allow data entry or annotation, and are complementary to the ALS framework.

3.2 Information Definition Process

The process is an iterative one, and it allows cross validation between different users. An important feature is the possibility to share and prove hypothesis based on empiric evidences found in the interrogation phase. The organization of the framework allows this exchange between different researchers. The ALSDB is a central part in the process of knowledge discovery. A full automatic procedure able to retrieve the information, according to the user needs, has been realized. The reader is referred to [28] for a more deep explanation. The core process is the definition of a correspondence between the concepts in the ontology representing high-level hypothesis in the various domains of interest and the different views of a relational database. The user can be interested in different concepts with a different abstraction degree. S/he can reason with concepts that are at a very high level of abstraction (i.e. the perception of the user, his self-perception) or with basics concepts (i.e. the speakers age, his degree of study). For this reason, the level of granularity of data doesnot allow a simple correlation between the main investigation variables, and even a simple query requires many lines of SQL code. A possible solution to this problem is to use a correspondence XML file. The XML correspondences and the data of the DBMS are the inputs of the system to retrieve information. A correspondence between a concept and the database is defined through a mapping process, whose results are writ-ten in a XML file and validated according an ad-hoc built DTD. Every concept can be related to one or more table in the DBMS. For every table is possible to define the list of attributes that are involved in the concept definition. The correspondence is set also at a table level: the tables with attributes can be related with Boolean relations that are NOT, OR, AND. After the definition of the correspondence between a concept and some tables it is possible to define some constraints for the mapping. Constraints are defined at a class level in the DTD with the list of possible attributes. Without defining constraints the mapping results in a default selection query. Using this paradigm, the user can compose a more complex query in an intuitive way throw a simple web interface, without writing lines of SQL code. The web interface allows users to refine results and save all the needed information. The result of a query is a set of tuples matching the selection criteria. The results can be saved as a new concept and be used as a starting point for further investigations. In this way the system produces an incremental knowledge base that is controlled by the domain experts. This knowledge base is accessible to other researchers to be validate and used in his/her domain of interest. The ALSDB component works on a persistence layer data, composed by a set of XML files. For each concept, one persistent XML file exists that contains all the entries matching with the criteria defined inside it. This has several advantages. From the software engineering point of view, if objects are persistent, they can be easily defined as entities in the project (the domain classes). Another advantage is a better portability of the application: the presence of an abstraction layer from data gives the possibility to change easily their access policies, the interfaces or the data drivers. The described persistence level has been implemented using the iBatis persistence framework [29].

4 The Main Framework Components

4.1 The ALSDB Component

The web interface for query composition is shown in Fig. 3. In the top left column the user can select the possible answers to a single question, defined into the

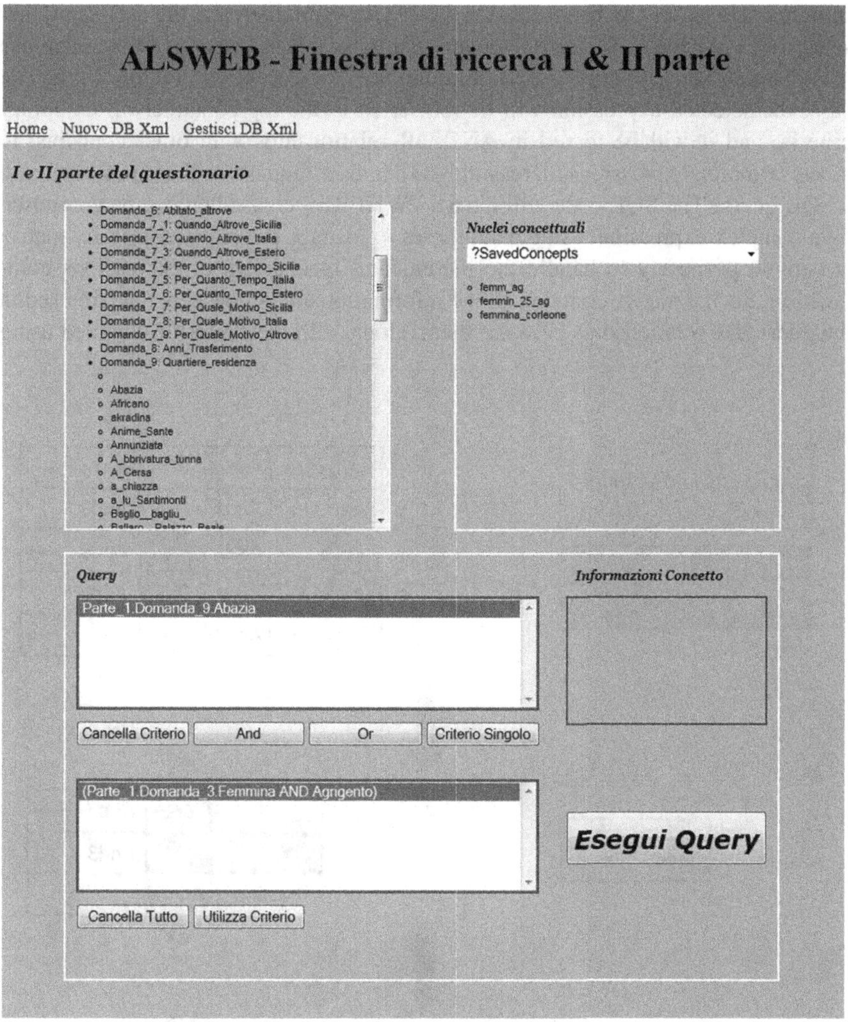

Fig. 3 The web interface for the query composition of relational data

questionnaire. In the top right column the user can choose an item in a list of concepts. Many types of concepts are available: some of them have been saved by other users or they are related to towns, scholar levels, family typologies, and so on. The selected concepts are added to the query panel to perform actual investigations. This GUI allows to define complex research criteria by mixing concepts through boolean relationships. The user selects some criteria that are inserted in the container of the temporary criteria to be aggregated with other ones. When selecting more than one concept, the user can put them in logical relationship (AND, OR), and insert them in the bottom container of the composed criteria.

Infinite expressions can be inserted in the criteria container, and the user can either select one criterion to perform the query using the "Run Query" button, or re-use it in conjunction with other criteria through the "Use criterion" button. In the last case, the selected expression will be reinserted in the top container of temporary criteria, and it will be mixed in AND-OR relationships with other concepts, to create complex expressions increasingly. Then user requests will be translated into a SQL query. The SQL code will have a "WHERE" clause that can be composed by a complex expression, using parentheses to manage priorities. A priority matrix is defined purposely to handle the incremental mechanism of expressions composition. Such a matrix contains all the information about concepts that are used in the query, the relationships between them, along with their priorities defined using

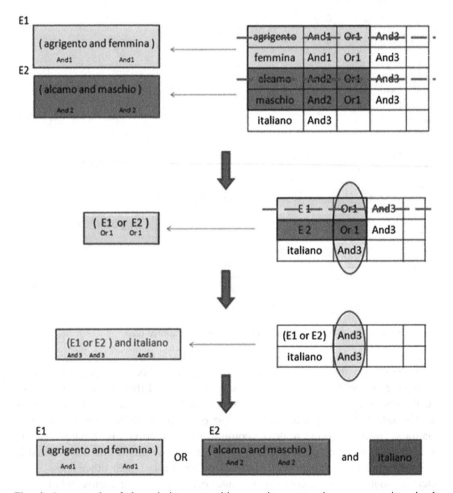

Fig. 4 An example of the priority composition matrix construction process using simple concepts

parentheses. This matrix is created incrementally while the user combines the concepts using the GUI. The key point of the query generation service is precisely the priority matrix whose structure and contents are needed to generate the XML correspondence file and to build dynamic query (Fig. 4).

4.2 The Transcription and Annotation Tools

The annotation tools are used to transform portions of the interview (in third part of questionnaires) into semi-structured documents, which contain information (annotation labels) about the different levels of research investigation. A tool is used to produce transcriptions directly from listening audio. Such a tool has been realized starting from the Wavesurfer software [30], an open source tool for sound visualization and manipulation. WaveSurfer is highly customizable and adaptable to any type of analysis and audio processing the researcher wants to perform. The original version has many different configurations, each of which changes the interface by showing the most frequently used tools for any particular field of interest.

To customize WaveSurfer to ALS transcription needs, a new configuration called "ALS Transcription" has been created (Fig. 5), which consists of many panels. The first one displays the spectrogram of the audio file to be annotated. A spectrogram is a graphical representation of the sound intensity as a function of time and frequency. The second panel shows the waveform of the signal that is a graphical representation of the sound levels, and can be used as a reference for the timing labels positioning. Another panel displays a time scale that acts as a reference for all other panels, which are aligned to it. The last two panels are

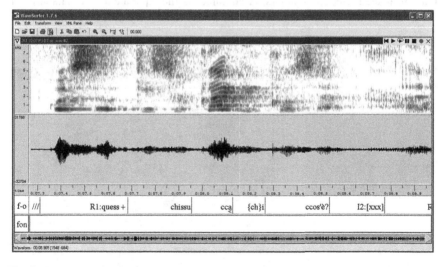

Fig. 5 The implemented ALS transcription panels on WaveSurfer

specifically designed for creating and managing phono-orthographic and phonetic transcriptions. Using this customized version of WaveSurfer, the linguistic researchers can make a careful transcription of interviews. They mark a temporal interval, specifying timing labels inside one of last two panels described above, and type the related text. To save transcription, the software creates a single text file in which the start time, the final instant and the text for each label are stored. To annotate linguistic phenomena, the domain expert works directly on the XML files that are generated from the WaveSurfer transcription, using a wrapper. This tool transforms timing labels and transcript text into valid XML files. It also allows to add to the document some information about the interviewee. The software shows a small form window with a set of input fields (Fig. 6). The most part of the fields are automatically filled by the wrapper, taking the information stored in both the relational database and some configuration files.

Now the researcher can perform the annotation. S/he can use one of the defined annotation levels, simply associating the correct schema to XML documents produced by wrapper. So s/he can label the text using the tags set and related attributes, defined into the chosen XML schema. To help this operation, researchers use a XML Editor software, which makes it easier to insert annotation tags with auto-completion of code. The document (Fig. 7)is then saved by the user as a file that will be uploaded successively into the ALSML Component.

4.3 The ALSML Component

The ALSML component is deputed to manage the XML annotated documents. The system interfaces with the Oracle Berkeley DB XML [31] to store XML files of transcriptions. In general, a native XML database organizes files into folders that are indeed called containers. A container is the equivalent of a database for relational DBMS. To manage these containers, the XQuery [15] language is used.

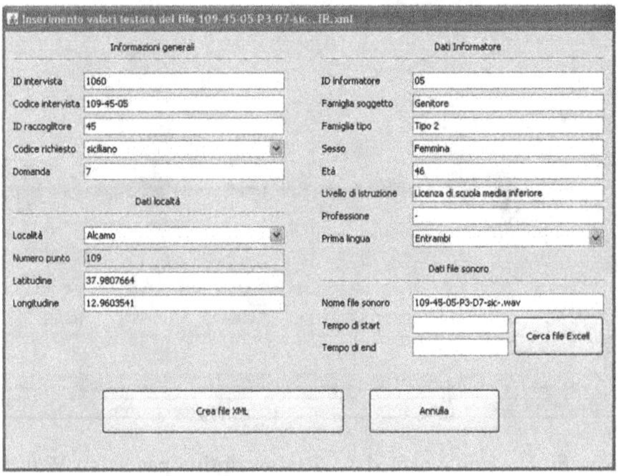

Fig. 6 The Wrapper Interface

Fig. 7 A portion of a XML file compliant to a proper schema, which reflects a specific annotation

This is a programming language specified by W3C and intended to query XML documents and databases. XQuery uses the syntax of XPath expressions for selecting specific portions of XML documents, with the addition of the so-called For Let Where OrderBy Return (FLWOR) expressions to make complex queries. In particular, the FLWOR constructs played a crucial role in building the complex queries that are necessary to the investigation of linguistic phenomena. In addition to basic functions for managing containers (i.e. add, remove, and show documents) the ALSML component provides the users with a graphical web interface to define the investigation on data. The end user can define a very complex query without writing a single line of XQuery code, but simply using the graphical interface (Fig. 8). The interface is built dynamically in relation to the current annotation level that has been chosen by the linguistic researcher. All fields about the annotation tags, contained within the first two panels, are populated with the information encoded in the XML schema associated with current container. The remaining fields are populated with data contained in the headers of XML documents (see Transcription and annotation paragraph). By selecting the values of fields presented in the GUI, the user can define an interrogation, which will become a query to be submitted to the system. To this purpose, the first operation for the researcher is determining the so-called "unitary tag of research" among all possible linguistic tags. This tag represents the entity the user is searching for, as

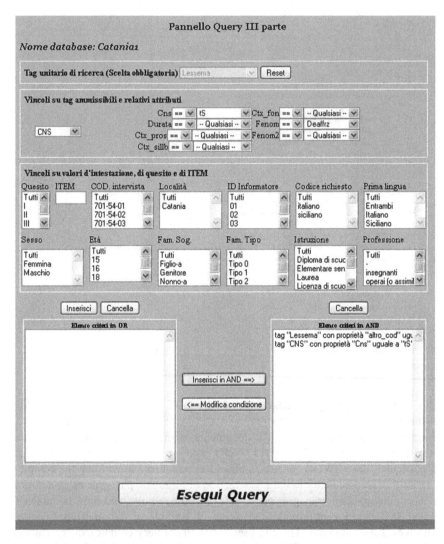

Fig. 8 The graphical interface of the ALSML component of the framework

the smallest piece of information to be detected. Only after this choice, the user can define constraints that characterize her research. These constraints can refer to tags (and their attributes) that have a direct lineage with the "unitary tag of research" including the unitary tag itself; otherwise the result of the query would be null. At a technical level, the choice of this tag represents a crucial phase in the composition of XQuery. As stated previously, Xquery consists of different FLWOR nested constructs (Fig. 9). The degree of relationship, between the "unitary tag of research" and another tag belongs to a constraint, and determines to which level of the FLWOR expression the same constraint must be locked.

```
● Senza titolo1.xquery ×                                                         ◁ ▷ ▦
  1 for $doc in collection("file_DB/Giulia")
  2 let $annotazione := $doc//annotazione
  3 let $INTESTAZIONE := $annotazione/INTESTAZIONE
  4 return
  5     for $frase in $annotazione//ITEM
  6     return
  7         for $tagUnitario in $frase//Lessema
  8         where ($tagUnitario[@altro_cod="ibrido"])
  9             and ($tagUnitario//CNS[@Cns="tS" and @Fenom="Deaffrz"])
 10         return <radice>
 11                 <id nome_file="" id_intervista="{data($INTESTAZIONE/ID_intervista)}"
 12                 cod_intervista="{data($INTESTAZIONE/COD_intervista)}"
 13                 localita="{data    ($INTESTAZIONE/Localita/Nome)}"
 14                 fam_sog="{data($INTESTAZIONE/fam_sog)}" fam_tipo="{data($INTESTAZIONE/fam_tipo)}"
 15                 sesso="{data($INTESTAZIONE/sesso)}" eta="{data($INTESTAZIONE/eta)}"
 16                 livello_istruzione="{data($INTESTAZIONE/livello_istruzione)}"
 17                 professione="{data($INTESTAZIONE/professione)}"
 18                 audio_nome="{data($INTESTAZIONE/File_sonoro/nome)}"
 19                 audio_tsstart="{data($INTESTAZIONE/File_sonoro/tsstart)}"
 20                 audio_tsend="{data($INTESTAZIONE/File_sonoro/tsend)}"
 21                 Lessema="{distinct-values($frase//Lessema)[1]}"
 22                 CNS="{distinct-values($frase//CNS)[1]}"
 23                 ITEM_tdstart="{data($tagUnitario/ancestor-or-self::*[@tdstart][last()]/@tdstart)}"
 24                 ITEM_tdend="{data($tagUnitario/ancestor-or-self::*[@tdstart][last()]/@tdend)}" />
 25                 {$tagUnitario}
 26         </radice>
```

Fig. 9 An example of an automatically built FLWOR nested construct

The "unitary tag of research" is also the atomic entity returned by the system as result of an investigation. In fact, as in the Fig. 10, the results page simply shows a list of tags.

After selecting this important tag, the user can compose complex search criteria using boolean logic. To this aim, the panel has two panes: the OR pane and the AND pane. Such panes are used to enter the search parameters to compose the query. The user can add multiple criteria to the OR panel. Such OR sets may be connected in a AND relationship with other ones that are in the AND pane by clicking on the central insert button. Furthermore, the user can edit a criterion that is in the AND pane moving it to OR pane through the "edit condition" button. All of these features (creation of the criteria and generation of the corresponding XQuery) are handled at client side, using JavaScript, to relieve the web server of computational burden produced by such a work. The interface detailed above gives the user great versatility in query construction; the system can generate very complex FLWOR structures. All the process is transparent to the user who does not know the complexity of XQuery generated by the system.

After defining the criteria, the system queries the Berkeley DB and displays results in a new window. As specified above, such a window (Fig. 10) contains the list of tags that match the search criteria specified by user. This set of tags is divided into groups; each of them is accompanied by a small header summarizing the information about the informant which heads the tags group. Another peculiarity of this page is the presence of an audio player to listen the audio registration of an interview or a small portion of this. The result window contains also statistic information about the research, useful to linguistic researchers for their investigations. The ALSML Component allows to increment the knowledge base of the system because it provides the possibility to package the query results into a

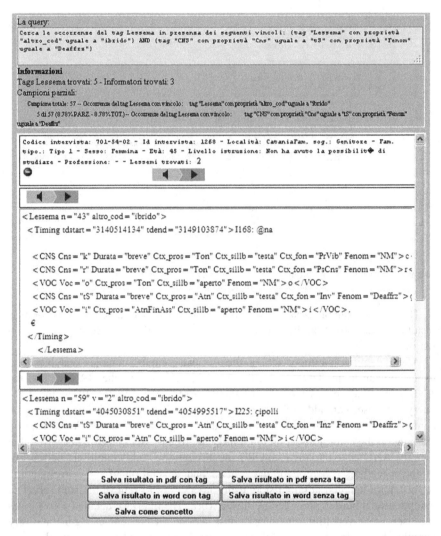

Fig. 10 The results page as a list of tags

concept that can be utilized to query the relational database. After saving the concept, its possible to find it in the query interface of ALSDB component (Fig. 11) where it can be mixed with other concepts to generate a new one.

This feature allows to correlate the investigation variables belonging to first, second and third section of the ALS questionnaire. In this way, the system allows information exchanging about the saved concepts in order to build a common knowledge base regardless the creation process used by the researchers. Moreover, reasoning at different levels of abstraction is enabled. The process is intentionally managed at the user level to maintain a requisite of the project: the researchers

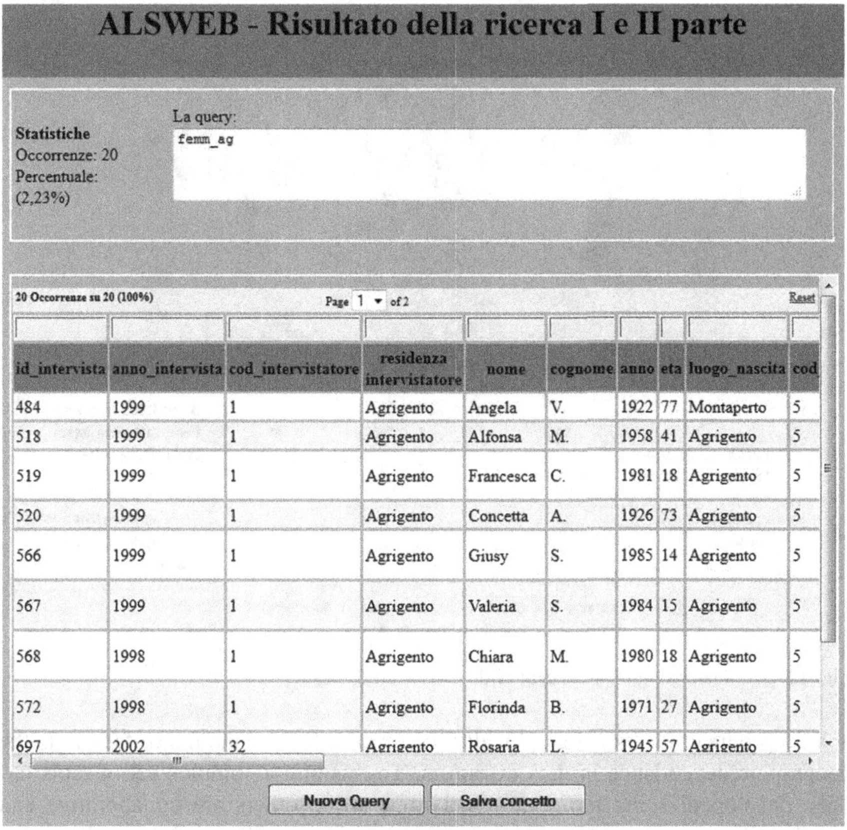

ALSWEB - Risultato della ricerca I e II parte

Statistiche Occorrenze: 20 Percentuale: (2,23%)	La query: `femm_ag`

20 Occorrenze su 20 (100%)   Page 1 ▼ of 2   Reset

id_intervista	anno_intervista	cod_intervistatore	residenza intervistatore	nome	cognome	anno	eta	luogo_nascita	cod
484	1999	1	Agrigento	Angela	V.	1922	77	Montaperto	5
518	1999	1	Agrigento	Alfonsa	M.	1958	41	Agrigento	5
519	1999	1	Agrigento	Francesca	C.	1981	18	Agrigento	5
520	1999	1	Agrigento	Concetta	A.	1926	73	Agrigento	5
566	1999	1	Agrigento	Giusy	S.	1985	14	Agrigento	5
567	1999	1	Agrigento	Valeria	S.	1984	15	Agrigento	5
568	1998	1	Agrigento	Chiara	M.	1980	18	Agrigento	5
572	1998	1	Agrigento	Florinda	B.	1971	27	Agrigento	5
697	2002	32	Agrigento	Rosaria	L.	1945	57	Agrigento	5

Nuova Query   Salva concetto

Fig. 11 The results of an interrogation process and the option to save a new named concept

want to have direct control and access to data, and want to be guided by their analysis and aggregation processes. Stored Concepts can be used freely if they are compliant to new findings.

4.4 The ALSGIS Component

The visualization tools are a relevant aspect of the project to support decisions and disseminate results. Information visualization is a very important key to achieve better research results. As stated in [32] the visualization of information can be seen as a method of computing because it enables researchers to see the unseen and enrich the process of scientific discovery. Nowadays Geographic Information Systems (GISs) are increasingly being used for effective accessibility to spatial data [33]. The ALSGIS is the part of the project related to tools and methodology

420   D. Pirrone et al.

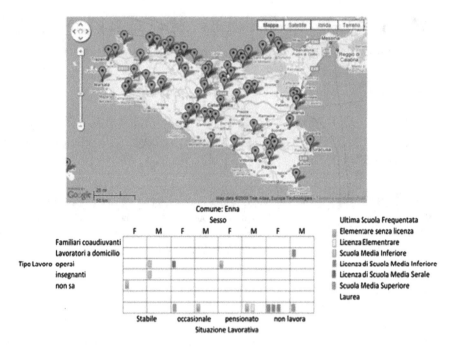

Fig. 12 A visualization of a concept with several attributes

able to produce results geographically referenced. One of the major factors is the geographic dependence in data evolution. To this aim a complex set of tools has been developed: some tools are stand-alone while others are collaborative and web-based. As previously stated, data exploration can be performed following several directions of investigation and different modalities. This has lead to a methodology where the interaction with users is the main driver of the process. There are two main types of outputs for the process: maps and data sheets. The researchers have to build their own data visualization that is strictly related to the type of research to perform. A general paradigm of data visualization through visual tools defines different levels of exploration: usually users are focused on a general overview of the over-all information, then they process the data with proper filters and zooming functions and at last they ask for more details. This organization is also known as the Shneiderman's Visual Information Seeking Mantra : "overview first, zoom and filter, details-on-demand" [34]. This approach is particularly suited to obtain useful data organization and visualization. To obtain a set of geographically referenced visualization many tools that are freely available have been personalized. A first experience has been performed with the combination of Google Maps [35] and Google Charts [36] (Fig. 12).

The visualization of raw data in data sheets has to face the problem of data aggregation at run-time. The framework provides different tools to perform fast and reliable data visualization like the possibility to incorporate visualizations such as the ones defined in the Many Eyes tool [37] (Fig. 13).

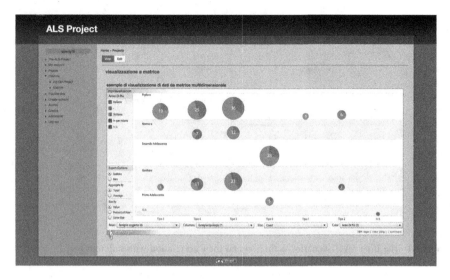

Fig. 13 An example of data visualization using a data sheet in ManyEyes

5 Case Study

This section reports a case study where all the features in the framework are involved. It's used to better explain how the framework allows collaborative work between users. A typical scenario involves different actors, and different instances of the same actor. The interaction between different users is managed by the framework that exposes the new concepts defined from a single researcher. One of the main advantages of the process is that shared concepts defined by a single researcher are shared with the empirical evidences related to her investigation (that is the underlying data). In this way, the story of producing a concept is presented immediately to the other researchers. The typical user is an expert in the socio-linguistic field or has a particular expertise in one of the annotation levels. The figure (Fig. 14) illustrates the main scenarios as well as the relations between different users. The researchers usually perform query operations and validate results either in a direct way by means of observation of the results or in an indirect way via the definition of new concepts in the framework. The definition of new concepts leads to new knowledge in the framework that other users have to validate in order to be employed. This validation process has been intentionally left to single users. In this way, reasoning about new concepts takes into account two main variables: the user, which creates a new concept, and the level of granularity of the concept itself. Concept definition is related to the family of saved concept via a direct descent relation. Each new concept is an offspring of another one that has been defined previously. As a consequence, it is always possible to determine how a new concept has been built. This case study reports the creation of a concept related to the interviewee self-perception defined according to the way he felt his

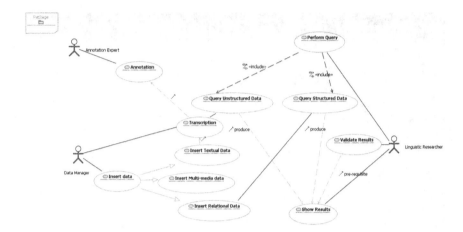

Fig. 14 The high level scenarios definition and their correlation with different users

spoken language. Such a concept represents a particular subclass of a linguistic variable known as *diatopia*, which is the correlation between linguistic variations and the geographic area of origin. The presented question is intended to specify how the user perceives himself as different in relation to language traits in neighbor towns such as the intonation of spoken words, the phonetic aspects of pronunciation, and other relevant aspects. The process of data acquisition is performed through the ALSML component of the infrastructure with a new level of annotation that is strictly related to the researcher, which is able to categorize the answers of users according to the main classes that have been defined. In the presented scenario, the researcher was able to define seven different main classes with six/seven subclasses for each class. The first step of the process is the observation of the raw results like the number of occurrences of each class according to the defined classification. After the first analysis, which is useful to evaluate the classification ontology empirically, the researcher starts to search new possible correlations with other classifications defined from different users. This process is performed using the ALSDB component of the system which incorporates the quantitative results of previous observations. The researcher can use either all the classification defined from other researchers or only one aspect of the classification. This is an important side effect of the classification process, which builds close classes in the population for any given classification. As an example, in the presentation scenario the researcher can relate his classification not only to usual built-in variables like gender or level of education, but also with entirely different domains like the domains that have been defined earlier that regard the degree of formalism in the spoken language through the definition of public and private language differences used for other researches. This process can be iterated over different qualitative and/or quantitative variables. The result is a different classification of the same population that can be visualized in the cartography component of the framework. In this way, the process of concepts definition can be performed in a collaborative environment with an high degree of interaction with the users and the framework. The overall process has been proved to be useful by the involved research community as a non invasive way to add and formalize knowledge to allow the growth of the collective intelligence in a common environment.

6 Conclusions and Future Works

In this chapter the Linguistic Atlas of Sicily framework has been presented. The definition of an integrated methodology able to perform a comprehensive analysis in the different fields of sociolinguistic is the ultimate goal of the Linguistic Atlas of Sicily project. The definition of the subsystems, the model of data integration and visualization have been explained . The system is intended to allow exchange of information and interoperability between users in the field of sociolinguistics. The framework is defined according to collective intelligence principles to allow the emergence of new knowledge provided by the interactions between the users and the system. One of the main advantages for users is the transparency of the processes related to information retrieval and fusion starting from heterogeneous data. The framework is able to produce and process structured, semi-structured and unstructured data according to users' needs. An important feature is the control of data source that the users have. In this way the researchers involved in the processes can always perform a check and a verification of the data underlying the concepts definition. Another important feature is the possibility to use different concepts derived from different users: this allows cross-validation and creation of new knowledge. Future works are in the direction of a more explicit definition of the process of knowledge sharing with the introduction of a search engine service. This service will be used from researchers to address particular findings that are relevant to other users. The preliminary effort is the definition of a fair trade-off between the increasing of computational load for the users who has to annotate somehow the created concepts and the effective benefits of new and automatic concepts highlighted by the search engine.

Acknowledgments This work has been partially funded under the Programma di Rilevante Interesse Nazionale (PRIN) 2009, entitled: "Atlante Linguistico della Sicilia: variazione linguistica e etnodialettologia fra nuovi modelli e nuove aree di investigazione" Authors would like to thank the ALS development team lead by prof. Giovanni Ruffino, and are especially indebted with Prof. Maria D'Agostino, Prof. Marina Castiglione, Dr. Luisa Amenta, Dr. Vito Matranga, Dr. Roberto Sottile, Dr. Giuseppe Paternostro and the many Ph.D. students at ALS for the extenuating and thoughtful discussions on the definition of the framework.

References

1. G. Ruffino, M. D'Agostino, I Rilevamenti Sociovariazionali. Linee Progettuali, Centro studi filologici e linguistici siciliani (2005)
2. T. Gruber, Collective knowledge systems: where the social web meets the semantic web, Web Semantics: Science. Services and Agents on the, World Wide Web (2008)
3. P. Di Maio, Digital ecosystems, collective intelligence, ontology and the 2nd law of thermodynamics. Digital Ecosystems and Technologies. DEST 2008, in 2nd IEEE International Conference (2008), pp. 144–147
4. H.H. Lek, D.C.C. Poo, N.K. Agarwal, Knowledge community (K-Comm): towards a digital ecosystem with collective intelligence, Digital Ecosystems and Technologies, DEST '09, in 3rd IEEE International Conference (2009), pp. 211–216
5. V. Lertnattee, S. Chomya, T. Theeramunkong, V. Sornlertlamvanich, Applying collective intelligence for search improvement on Thai herbal information, computer and information technology, CIT '09. in Ninth IEEE International Conference on 2, 178–183 (2009)
6. H. Mizuyama, Y. Maeda, A prediction market system using SIPS and generalized LMSR for collective-knowledge-based demand forecasting, Computers and Industrial Engineering (CIE), in 2010 40th International Conference. pp. 1–6
7. P.E. Maher, J.L. Kourik, A knowledge management system for disseminating semi-structured information in a worldwide university, Management of Engineering and Technology, PICMET 2008. in Portland International Conference (2008), pp. 1936–1942
8. T. Bray, J. Paoli, C. Sperberg-McQueen, E. Maler, Extensible markup language (xml), 1.0 2nd ed. W3C Recommendation (2006), Available: http://www.w3.org/TR/2006/REC-xml-20060816/
9. The sax project home page. Available: http://www.saxproject.org/
10. L. Wood, A. L. Hors, V. Apparao, S. Byrne, M. Champion, S. Isaacs, I. Jacobs, G. Nicol, J. Robie, R. Sutor, C. Wilson, Document object model (dom) level 1 specification (second edition) version 1.0 (2000), Available http://www.w3.org/TR/2000/WD-DOM-Level-1-20000929/
11. J. Bosak, T. Bray, D. Connolly, E. Maler, G. Nicol, C. M. Sperberg-McQueen, L. Wood, and J. Clark, Dtd (document type definition) (2000), Available: http://www.w3.org/XML/1998/06/xmlspec-report.htm
12. D. C. Fallside and P. Walmsley, XML schema part 0: primer second edition (2004), http://www.w3.org/TR/2004/REC-xmlschema-0-20041028/
13. M. Kay, Xsl transformations (xslt) version 2.0 (2007), Available http://www.w3.org/TR/2007/REC-xslt20-20070123/
14. S. DeRose and J. Clark, XML path language (XPath) version 1.0 (1999), http://www.w3.org/TR/1999/REC-xpath-19991116
15. J. Siméon, D. Chamberlin, D. Florescu, S. Boag, M. F. Fernández, and J. Robie, XQuery 1.0: an XML query language 2nd ed. (2010), Available http://www.w3.org/TR/xquery/
16. J. Carletta, D. McKelvie, and A. Isard, Supporting linguistic annotation using xml and stylesheets, in READINGS IN CORPUS LINGUISTIC (2002)
17. H. Cunningham, Y. Wilks, R. J. Gaizauskas, Gate: a general architecture for text engineering, in Proceedings of 16th Conference on Computational Linguistic: association of computaional linguistic (1996)
18. S. Bird, D. Day, J. Garofolo, J. Henderson, C. Laprun, M. Liberman, Atlas: a flexible and extensible architecture for linguistic annotation, in Proceedings of the Second International Conference on Language Resources and Evaluation, pp. 1699–1706, Paris: European Language Resources Association (2000)
19. X. Artola, A. D. de Ilarraza, A. Soroa, A. Sologaistoa, Dealing with complex linguistic annotations within a language processing framework, Audio, Speech, and Language Processing, IEEE Transactions, vol 17, no 5, (2009) pp. 904–915

20. T. Feng, H. Xiao-bing, and W. Feng-bo, The heterogeneous data integration based on xml in coal enterprise, vol 1 (2008), pp. 438–441
21. D. Draper, A. Y. HaLevy, and D. S. Weld, The nimble xml data integration system, in In ICDE (2001), pp. 155–160
22. U. Fayyad, Knowledge discovery in databases: an overview, Relational Data Mining (2000), pp. 28–45
23. T. R. Gruber, A translation approach to portable ontology specifications, Knowledge Acquisition, vol 5(2) (1993), pp. 199–221. Available http://tomgruber.org/writing/ontolingua-kaj-1993.pdf
24. R.T. Alves, M.R. Delgado, A.A. Freitas, Knowledge discovery with artificial immune systems for hierarchical multi-label classification of protein functions, in Fuzzy Systems (FUZZ), 2010 IEEE International Conference (2010), pp. 1–8
25. Z. Peng, B. Yang, W. Qu, One knowledge discovery approach for product conceptual design, in Artificial Intelligence, International Joint Conference on. Artificial Intelligence (2009), pp. 109–112
26. A. Barb and C.-R. Shyu, Ontology driven content mining and semantic queries for satellite image databases, in Geoscience and Remote Sensing Symposium, 2008. IGARSS 2008. IEEE International, vol 3, (2008), pp. III-503–III-506
27. C. Diamantini, D. Potena, Semantic annotation and services for kdd tools sharing and reuse, in International Conference on Data Mining: Workshops 2008. ICDMW '08 (2008), pp. 761–770
28. G. Russo, A. Gentile, R. Pirrone, V. Cannella, XML-based knowledge discovery for the Linguistic Atlas of Sicily (ALS) project, in International Conference on Complex, Intelligent, and Software Intensive Systems: CISIS (2009), pp. 98–104
29. C. Begin, B. Goodin, L. Meadors, Ibatis in Action, (Manning Publications Co., Greenwich, CT, USA, 2007)
30. K. Sjlander, J. Beskow, Wavesurfer: an open source speech tool (2000), Available http://www.speech.kth.se/wavesurfer/index.html
31. D. Brian, The Definitive Guide to Berkeley DB XML, (Apress, Berkely, CA, USA, 2006)
32. B.H. McCormick, T.A. DeFanti, M.D. Brown (eds.) Visualization in scientific computing and computer graphics, in ACM Special Interest Group on GRAPHics and Interactive Techniques Conferenze, vol 6(21) (1987)
33. T. Jing, X. Juan, W. Li, Open source software approach for internet Gis and its application, in Intelligent Information Technology Applications, 2007 Workshop, vol 3 (2008), pp. 264–268
34. B. Shneiderman, Designing the User Interface, 3rd edn. (Addison Wesley Longman, 1998)
35. Google Maps API Available https://developers.google.com/maps/?hl=it
36. Google Charts API Available https://developers.google.com/chart/
37. Many Eyes, Data visualisation tools from IBM. http://www-958.ibm.com/software/data/cognos/manyeyes/
38. U. Fayyad, G. Piatetsky-shapiro, P. Smyth, Knowledge discovery and data mining: towards a unifying framework, in Proceedings of the Second International Conference on Knowledge Discovery and Data Mining, (1996) pp. 82–88

Glossary

6LowPAN IPv6 over Low Powered Wireless Personal Area Networks

ACT Any-contact time

A/C Air Conditioning

AES Advanced Encryption Standard

AI Artificial Intelligence

ANFIS Adaptive Neuro Fuzzy Inference System

AP Access point

API Application Programming Interface

ASR Asynchronous Service Request.

BSSID Basic service set identifier.

CA Cloud Agency.

CAN Cloud Agency Notification.

CAN Controller Area Network

CDMI Cloud Data Management Interface.

CI Comfort Index

C-fields Fields that are copied to the transformed packet without modification.

CGL Community-Guided Learning

Channel Table Lists the transformations of *Channels* established in a Tunnel.

Channel An established transformation associated with a unique S-field list.

F. Xhafa and N. Bessis (eds.), *Inter-cooperative Collective Intelligence:*
Techniques and Applications, Studies in Computational Intelligence 495,
DOI: 10.1007/978-3-642-35016-0, © Springer-Verlag Berlin Heidelberg 2014

Cloud Computing A market-oriented distributed computing paradigm consisting of a collection of inter-connected and virtualized computers that are dynamically provisioned and presented as one or more unified computing resources based on service-level agreements established through negotiation between the service provider and consumers.

Cloud A large pool of easily usable and accessible virtualized resources (such as hardware, development platforms and/or services). These resources can be dynamically reconfigured to optimum resource utilization. This pool of resources is typically exploited by a pay-per-user model in which guarantees are offered by the Infrastructure Provider by means of customized SLA's.

CPU Central Processing Unit.

CT Contact time.

D2R Database To RDF (utility).

D2RQ Database To RDF Query (utility).

DB Database

DBA Database Administrator.

DBIA Database Interface Agent.

DBIE Database Interface Environment.

DBS Database Subsystem.

DCP Data Collection Point

DEO Data Entry Operator.

Direct Channel (*DC*), a *Channel* in a Direct Tunnel.

Direct Tunnels (*DT*) a tunnel that is directed towards branch networks; optimal for "many-to-few" communications.

DOMOSEC Domotics and Security

e-Business Electronic Business.

e-Governance Electronic Governance.

e-Health Electronic Health.

EIB EEuropean Installation Bus

Entrance A tunnel entrance entity (device or software) that transforms a frame from the *Transported* network into a packet usable in the *Transport* network.

Exit A tunnel exit entity that transforms a packet used in the *Transport* network back into the frame used in the *Transported* network.

FCFS First come first served.

FLS Fuzzy Logic System

FP7-ICT Seventh Framework Program.

GA Genetic Algorithm.

GCEI Generation Consumption Effectiveness Index.

G-D logic Good-Dominant Logic.

Grid Computing Basically a paradigm that aims at enabling access to high performance distributed resources in a simple and standard way.

GSMA Global System for Mobile Communications Association.

GUI Graphical User Interface.

HAM Home Automation Module

Hederata The encapsulation headers of IP packet besides native Ethernet header may include VLAN tags MPLS headers etc. All fields preceding an IP packet included as the last part of the Ethernet frame are referred to collectively as *hederata* (see Fig. 2.)

HMI Human Machine Interface

HTTP Hypertex Transfer Protocol

HVAC Heating, Ventilation and Air Conditioning

Hybrid Cloud A Cloud Computing environment in which an organization provides and manages some resources in-house and has others provided externally.

I/O Input/Output

IACT Inter-any-contact time.

ICT Inter-contact time.

IHIP Intangibility-Heterogeneity-Inseparability-Perishability.

Indirect Channel (*IC*), a *Channel* in an Indirect Tunnel.

Indirect Tunnels (*IT*) a tunnel that is directed towards central and global networks; optimal for "few-to-many" communications.

Infrastructure-as-a-Service (IaaS) A provision model in which an organization outsources the equipment used to support operations, including storage, hardware, servers and networking components, where the service provider owns the equipment and is responsible for housing, running and maintaining it and the client typically pays on a per-use basis.

IoIT Internet of Intelligent Things.

IoT Internet of Things.

IPv4 Internet Protocol version 4.

IPv6 Internet Protocol version 6.

ISO International Organization for Standardization

IT Information Systems.

JNLP Java Networking Launching Protocol

M2M Machine-to-Machine Communications.

MAS Multiagent System.

m-Business Mobile Business.

mOSAIC Open source API and Platform for multiple Clouds FP7-ICT project.

MPU Main Processor Unit

MTBF Mean Time Between Failures.

NICE A frame that fulfills all criteria set in Definition 1 NICE frames are eligible for zero-overhead tunneling.

NFC Near Field Communication

NN Neural Network

OCCI Open Cloud Computing Infrastructure

OGSA Open Grid Services Architecture

ON Opportunistic network.

OOSE Object-Oriented Software Engineering

OWL Web Ontology Language

Platform-as-a-Service (PaaS) A Cloud service model which offers an environment on which developers create and deploy applications and do not necessarily need to know how many processors or how much memory that applications will be using.

Private Cloud A marketing term for a proprietary computing architecture that provides hosted services to a limited number of people behind a firewall.

Pseudowire A logical networking channel that emulates properties of a physical wire connecting two nodes disregarding the actual physical topology.

Public Cloud One based on the standard Cloud Computing model, in which a service provider makes resources, such as applications and storage, available to the general public over the Internet. Public Cloud services may be free or offered on a pay-per-usage model.

QoS Quality of Service

RaaS Robotics as a Service

RDB Relational database

RDF Resource Description Framework

REST Representational Transfer State

RF Radio Frequency

RFID Radio Frequency Identification

ROS Robot Operating System

SaaS Software as a service

SCADA Supervisory Control and Data Acquisition.

SCI Social and Community Intelligence.

S-D Logic Service-Dominant Logic.

S-fields also "Saved fields" Layer 2-3 information in the tunneled frame that needs to be preserved for accurate recreation on tunnel *Exit*.

SGCC Study Group on Cloud Computing.

SHA Study Group on Cloud Computing.

SHAP Superior Home Automation Protocol

SIROCO System for Integral Control and Communications.

SIoT Social Internet Of Things

SLA Service Level Agreement.

SNIA Storage Networking Industry Association

SNL Simplified Natural Language.

SOA Service Oriented Architecture

SOCCI Service-Oriented Cloud-Computing Infrastructure

Software-as-a-Service (SaaS) A software distribution model in which applications are hosted by a vendor or service provider and made available to customers over a network, typically the Internet.

SPARQL Protocol and RDF Query Language.

SQAS Semantic Query Access System.

SSID Service set identifier

SSL Secure Socket Layer

SSR Synchronous Service Request

SW Semantic Web

Tfi Time of failure *i*.

Tc Communication Period

T-fields packet fields that must incorporate values necessary for tunneling the packet over transport network towards the *Exit*.

Transport network. The network carrying the tunneled packets. Internet is an example of *Transport* network.

Transported network The network from which network traffic is to be transported into a tunnel. An example would be a company network.

Tunnel A simplex logical connection from one *Entrance* host to one *Exit* host that is used for ZERO protocol operation.

UGLY A frame that does not fulfill all criteria set in Definition 1 UGLY frames are not eligible for zero-overhead tunneling.

UPnP Universal Plug and Play.

UIA User Interface Agent.

UIE User Interface Environment.

US User Subsystem.

VISION VISION Cloud FP7-ICT project.

VL Virtual location.

VM Virtual Machine.

WSN Wireless Sensor Networks.

X-fields Also "Index fields", often unused fields in the packet's Layer 3-4 PDUs.

YARP Yet Another Robotic Platform.

ZERO ENC IP packet with a UDP datagram that encapsulates the transported frame.

ZERO IP IP packet into which a NICE frame is transformed for zero-overhead forwarding in the *Transported* network.

Zero server a ZERO protocol prototype implementation in a userland process.

ZERO SYN like ZERO ENC encapsulates the transported frame incapsulated in a UDP datagram ZERO SYN has one extra field for *index* value that allows to syncronise the S-fields of encapsulated frame to the tunnel *Exit*.

Author Index

F. Xhafa and N. Bessis (eds.), *Inter-cooperative Collective Intelligence:*
Techniques and Applications, Studies in Computational Intelligence 495,
DOI: 10.1007/978-3-642-35016-0, © Springer-Verlag Berlin Heidelberg 2014

433

Index

F. Xhafa and N. Bessis (eds.), *Inter-cooperative Collective Intelligence:
Techniques and Applications*, Studies in Computational Intelligence 495,
DOI: 10.1007/978-3-642-35016-0, © Springer-Verlag Berlin Heidelberg 2014

Printed in the United States
By Bookmasters